international®
AIR POWER
REVIEW

AIRtime Publishing
United States of America • United Kingdom

international® AIR POWER REVIEW

Published quarterly by AIRtime Publishing Inc.
191 Post Road West, Westport, CT 06880
Tel (203) 454-4773 • Fax (203) 226-5967

© 2005 AIRtime Publishing Inc.
Su-22 and Valiant cutaways © Mike Badrocke
Su-20/22 profiles © Tom Cooper

Photos and other illustrations are the copyright of their respective owners

ISBN: 1-880588-97-8 (softbound)
    1-880588-98-6 (casebound)
Softbound Edition ISSN 1473-9917

Publisher
    Mel Williams

Editor
    David Donald   e-mail: dkdonald@hotmail.co.uk

US Desk
    Tom Kaminski

Russia/CIS Desk
    Piotr Butowski, Zaur Eylanbekov

Europe and Rest of World Desk
    John Fricker, Jon Lake, Marnix Sap

Correspondents
    Argentina: Santiago Rivas and Juan Carlos Cicalesi
    Australia: Nigel Pittaway
    Belgium: Dirk Lamarque
    Brazil: Claudio Lucchesi
    Bulgaria: Alexander Mladenov
    Canada: Jeff Rankin-Lowe
    France: Henri-Pierre Grolleau
    Germany: Frank Vetter, Bernd Vetter
    India: Pushpindar Singh
    Israel: Shlomo Aloni
    Italy: Luigino Caliaro
    Japan: Yoshitomo Aoki
    Netherlands: Tieme Festner
    Romania: Danut Vlad
    Spain: Salvador Mafé Huertas
    USA: Rick Burgess, Brad Elward, Mark Farmer (North Pacific
    region), Peter Mersky, Bill Sweetman

Artists
    Mike Badrocke, Piotr Butowski, Tom Cooper, Chris Davey

Controller
    Linda DeAngelis

Sales Director
    Jill Brooks

Origination by Chroma Graphics, Singapore
Printed in Singapore by KHL Printing

International Air Power Review is published quarterly in two editions (Softbound and Deluxe Casebound) and is available by subscription or as single volumes. Please see details opposite.

Acknowledgments

Thanks to Tom Cooper, Heinz Berger and Frank Rozendaal for help with the 'Fitter' article

The authors of the Royal Jordanian Air Force article would like to thank all the people involved for their help and assistance during the preparation of this feature, especially our guide Harar Mismar, Commander of the RJAF General Hussein Al-Bis, and all pilots and squadron commanders that made our visit unforgettable. Shukran to you all!

The author of the HT-18 feature would like to thank CDRs Gerald Briggs, John Fleming, Pam Kunze, LT COL Joe Richards, MAJs Michael Marko and Karl Stoetzer, LCDR Eric Sturgill, CAPTs Sean Dickman and Andreas Lavato (both USMC), LT Matt Knowles, LTJG Shaun Robertson, and the many others that helped.

Thanks to Lt Col Valerie Trefts, SSgt Ryan Hansen, Col Donald Watrous, Lt Col Dawn Dunlop, Mr Tom Fuller, Mr Bo Sills and Mr Joe Stout for their assistance with the 'Desert Testers' feature

Contact and Ordering Information
(hours: 10 am - 4.30 pm EST, Mon-Fri)
addresses, telephone and fax numbers
International Air Power Review, P.O. Box 5074, Westport, CT 06881, USA
    Tel (203) 454-4773 • Fax (203) 226-5967
    Toll free within USA and Canada: 1 800 359-3003
    Toll free from the United Kingdom, Belgium, Denmark, France, Germany,
        Holland, Ireland, Italy, Luxembourg, Norway, Portugal, Sweden and
        Switzerland (5-6 hours ahead): 00 800 7573-7573
    Toll free from Finland (6 hours ahead): 990 800 7573-7573
    Toll free from Australia (13-15 hours ahead): 0011 800 7573-7573
    Toll free from New Zealand (17 hours ahead): 00 800 7573-7573
    Toll free from Japan (14 hours ahead): 001 800 7573-7573

website
    www.airtimepublishing.com
e-mails
    airpower@airtimepublishing.com
    inquiries@airtimepublishing.com

Subscription & Back Volume Rates
**One-year subscription (4 quarterly volumes),
inclusive of ship. & hdlg./post. & pack.:**
    Softbound Edition
    USA $59.95, UK £48, Europe EUR 88, Canada Cdn $99,
    Rest of World US $119 (air)
    Deluxe Casebound Edition
    USA $79.95, UK £68, Europe EUR 120, Canada Cdn $132,
    Rest of World US $155 (air)

**Two-year subscription (8 quarterly volumes),
inclusive of ship. & hdlg./post. & pack.:**
    Softbound Edition
    USA $112, UK £92, Europe EUR 169, Canada Cdn $187,
    Rest of World US $219 (air)
    Deluxe Casebound Edition
    USA $149, UK £130, Europe EUR 230, Canada Cdn $246,
    Rest of World US $295 (air)

Single/back volumes by mail (each):
    Softbound Edition
    USA $19.95, UK £14.95, Europe EUR 22
    (incl. ship. & hdlg./post. & pack.)
    Deluxe Casebound Edition
    US $24.95, UK £18.95, Europe EUR 30
    (incl. ship. & hdlg./post. & pack.)

Prices are subject to change without notice. Canadian residents please add GST. Connecticut residents please add sales tax.

Volume Eighteen

CONTENTS

MAJOR FEATURES PLANNED FOR VOLUME NINETEEN

Focus Aircraft: Bell-Boeing V-22 Osprey, **Warplane Classic:** Junkers Ju 88, **Variant Briefing:** Supermarine Swift, **Air Power Analysis:** Germany, **Air Combat:** B-52 close air support, **Special feature:** 1st Fighter Wing F/A-22s

PROGRAMME UPDATE

Bell-Boeing V-22 Osprey

On 28 September the Pentagon's Defense Acquisition Board (DAB) approved a 'so-called' Milestone III decision which allows the Bell-Boeing V-22 Osprey to enter full-rate production (FRP). The decision allows the Osprey Team to increase from the current low-rate initial production (LRIP) of 11 aircraft per year up to 48 by 2012.

In preparation for delivery of the first Block B MV-22B, tiltrotor test squadron VMX-22 transferred the first of its Block A Ospreys to tiltrotor training squadron VMMT-204 at MCAS New River, North Carolina, on 28 September. The Osprey fleet readiness squadron (FRS) subsequently launched its first MV-22B sortie in mid-October. The mission ended an operational pause that has been in place since December 2000 when the fleet was grounded. VMX-22 will receive the first Block B Osprey in December and the first fleet squadrons will also be equipped with the later version which incorporates various improvements over the Block A training aircraft. Among the improvements in the Block B aircraft are a retractable refuelling probe, updated avionics, navigation and communications equipment, a ramp-mounted gun and a rescue hoist. Marine tiltrotor squadron VMM-263 will reach initial operational capability (IOC) in 2007. MCAS New River will receive 9-13 new MV-22Bs annually over the next three years, and deliveries will subsequently ramp up to 24 and 36 per year. The USMC will field two new Osprey squadrons annually, beginning with VMM-263, VMM-162 and VMM-266.

The CV-22B is scheduled to achieve IOC in 2009 following an operational evaluation (OPEVAL) that will focus on those systems that are unique to the USAF variant. USAF crews will initially undergo training with VMMT-204, before transferring to the 71st Special Operations Squadron at Kirtland AFB, New Mexico, for instruction in systems and tactics unique to the CV-22. The USAF plans to conduct an operational utility evaluation in summer 2006 that will validate a preliminary training flight envelope. Once this has been completed, USAF pilots that have transitioned to the Osprey with VMMT-204 will begin training on the CV-22.

Also, the V-22 Integrated Test Team concluded a 'high-hot' testing detachment in Santa Fe, New Mexico, which validated the MV-22B's perfor-

The USAF took delivery of its first production CV-22B at Bell Helicopter's Amarillo, Texas, facility during September. It was later flown to NAS Patuxent River, Maryland, for electromagnetic testing.

PROJECT DEVELOPMENT

Argentina

Pampa programme resumed

Having been suspended in 2001 for economic reasons, Lockheed Martin Aircraft Argentina's IA 63 Pampa intermediate trainer and light-attack aircraft programme was recently revived by the Buenos Aires government. LMAA president Alberto Buthet said that of some 20 original IA 63s delivered to the Argentine air force (FAA) from 1987, 11 would be upgraded to current Phase II AT-63 light attack standards. They would be supplemented by November 2007 by the first six of 12 more new-build Phase III AT-63s from resumed production. The last six will be offered to Argentinian or foreign services.

AT-63 multi-role capabilities are achieved from installation of Elbit and Honeywell 1553B integrated digital avionics, with mission computer, glass cockpit and head-up displays, plus GPS/INS, HOTAS, data-transfer and stores management systems. Optional replacements include uprating the original 15.57-kN (3,500-lb) Honeywell TFE731-2C-2N turbofan by the 16.68-kN (4,200-lb) TFE731-40, for improved performance at increased take-off weights.

The first FAA AT-63 was rolled-out last December, and started flying on 22 June. Production is continuing at two per month. Lockheed Martin APG-67 fire-control radar, plus five hardpoints for additional air-to-air and air-to-surface roles, may also be installed. Other AT-6 features include servo-assisted flight-controls, and quick-connection power-plant attachments allowing a complete engine change within an hour.

The Argentine navy is also interested in an initial eight IA 63s, to start replacing its 16 Aermacchi MB.326GBs. Claiming turboprop-class operating costs with jet performance, LMAA hopes for at least 15 per cent of a potential total world light jet-trainer market. This would represent about 180 export aircraft, probably initially from Latin America, where Colombia has already expressed interest.

Australia

Wedgetail news

Boeing recently completed the aircraft performance and flight handling test programme with the first Australian Project Wedgetail 737 airborne early warning and control (AEW&C) aircraft. During the tests, the Wedgetail logged more than 245 flights and more than 500 flight hours. Testing included take-off performance, flight handling and simulated air-to-air refuelling sorties that were conducted at Edwards AFB, California. Flight testing of the airborne early warning and control mission system, including the Multi-role Electronically Scanned Array (MESA) radar, began in late August. Australia has

The 27th Fighter Squadron deployed six F/A-22As to Hill AFB, Utah, from Langley AFB, Virginia, on 15 October, demonstrating the unit's ability to deploy from its home station to unfamiliar territory and effectively fly missions. The squadron, which is a component of the 1st Fighter Wing, is the first operational unit equipped with the F/A-22A. The deployment is part of the USAF's effort to certify the Raptor's Initial Operational Capability (IOC) by December. The squadron subsequently conducted the Raptor's first operational weapons employment sorties on the Utah Test and Training Range on 18 October, when five F/A-22As delivered nine inert 454-kg (1,000-lb) GPS-guided GBU-32 joint direct attack munitions (JDAM).

The APG-81 AESA radar for the F-35 has flown in Northrop Grumman's BAC 1-11 test aircraft (above). The array of transmit/receive modules is fixed and angled upwards (right) so that any radar returns are deflected away from the illuminating radar. The first prototype F-35A was essentially complete by the summer of 2005, but will not fly for another year following extensive ground tests.

mance envelope at high altitudes and air temperatures, and high gross takeoff weights. An additional test detachment will deploy to Nellis AFB, Nevada in February 2006.

Lockheed Martin F-35

Northrop Grumman recently conducted the initial flight tests of the advanced fire-control radar under development for the Lockheed Martin F-35 Joint Strike Fighter (JSF) aircraft. Designated the AN/APG-81, the active electronically scanned array radar (AESA) is being tested aboard the contractor's BAC 1-11 test aircraft. The search, air-track and synthetic-aperture radar mode capabilities of the radar were successfully evaluated against airborne and ground-based targets in the vicinity of the contractor's facility, adjacent to the Baltimore Washington International Airport in Maryland. The first APG-81 radar system will be delivered to Lockheed Martin's Fort Worth, Texas, facility in November. It will subsequently be installed into the JSF Mission Systems Integration Lab and tested with other mission avionics systems.

The first of nine prototype F-35 Joint Strike Fighters will make its initial flight in late 2006. The conventional take-off and landing (CTOL) F-35A variant, known as 'A-1', will be followed by the first of four short take-off and vertical landing (STOVL) F-35Bs in the third quarter of 2007. The second and third CTOL models, known as 'A-2' and 'A-6', will join the test programme in the first quarter of 2008. These so-called 'optimised' aircraft will differ from 'A-1' in that they will be equipped with some of the structural modifications incorporated in the F-35B and the F-35C carrier version (CV). Two F-35Cs will be the last to fly, joining the flight test programme in the first quarter of 2009. In addition to the nine flying aircraft, the programme includes seven non-flying ground test airframes. Lockheed Martin will also operate a Boeing 737-300 airliner modified as a JSF Cooperative Avionics Test Bed (CATB) that will support inflight evaluations of fighter's systems, will begin flying in mid 2006.

purchased six aircraft for its Wedgetail fleet and delivery of the first two, which are being modified in Seattle, is scheduled for 2006. The remaining four will follow by 2008.

China

MiG-21 development projects
The People's Liberation Army Air Force (PLAAF) is expecting deliveries from early 2006 from the Guizhou Aviation Industry Group of the first five JL-9 Mountain Eagle two-seat ground-attack and combat-trainer aircraft, for operational trials. Developed from the FT-7 version of the MiG-21U, with a J-7E cranked delta-wing behind lateral intakes for its 63.25-kN (14,220-lb) WP13F afterburning turbojet, the JL-9 began flight development on 13 December 2003.

Its equipment includes radar, full digital avionics, and provision for air refuelling. According to state media reports, the PL-9 is scheduled for both PLAAF and Chinese Naval Aviation operation, from a five-year production plan, which includes provision for export sales to developing countries.

Development by Chengdu Aircraft (CAC), in co-operation with the MiG OKB and Pakistan, of the Klimov RD-93-powered FC-1 or JF-17 Thunder variant of the MiG-21, which also first-flew in 2003, has reportedly been delayed for required modifications to the fourth prototype. These include revised flank intakes, wing strake extensions, and increased vertical tail area.

CAC plans to build two more FC-1 development aircraft by 2007, prior to joint production with Pakistan's Kamra Aeronautical Complex of an initial batch of 16 for the PAF, from total requirements for 134.

India

AL-55s to power HJT-36s
Initial production of the Hindustan Aeronautics Ltd (HAL) HJT-36 Sitara intermediate jet trainer will now start in 2007, about three years later than originally planned. This follows a late 2004 decision to replace the original 14.12-kN (3,175-lb) SNECMA Larzac 04-H20 turbofans in the two prototype HJT-36s with recently-developed 21.58-kN (4,851-lb) Ufa Machinery-building Production Organisation (UMPO)/ Saturn AL-55 powerplants.

Inter-government ratification of this agreement for licence transfer and technical assistance for Indian AL-55I production occurred in August at Moscow's MAKS-05 international air show. UMPO general director, Valery Lesunov, said that the contract, signed between Russia's Rosoboronexport arms sales agency, and the Indian government in May, would cover initial sales of 180-250 AL-55I engines, costing some $200m or more, for IAF HJT-36s.

The AL-55 is still under development by LPO Saturn, but the IAF plans early acquisition of 187 HJT-36s, equipped with full digital avionics and tandem Russian Zvezda K-36CT lightweight zero-zero ejection-seats, to replace ageing HAL HJT-16 Kiran basic trainers. The Indian navy wants another 24. HJT-36 flight-development started on 7 March 2003, soon after HAL received its initial IAF production contract. UMPO is also developing an afterburning 34.34-kN (7,720-lb) AL-55F version, which India is considering installing in upgraded IAF HAL-built MiG-27ML ground-attack fighters.

South Korea

Production T-50 roll-out
On 30 August Korea Aerospace Industries (KAI) and Lockheed Martin achieved a major milestone in their joint T-50 Golden Eagle advanced jet trainer programme, with roll-out of the first production aircraft at Sachon. The T-50 is being developed and marketed world-wide with Lockheed Martin aid, and is claimed to enhance pilot training at lower overall costs from reduced flight-times for operational fighter qualifications. The RoKAF has established a T-50 aviation training battalion for 2007 initial deployment.

From the prototype's first flight on 1 August 2002, 1,119 of 1,146 scheduled development sorties had been completed by last July. Six aircraft were then in final assembly, for initial RoKAF deliveries from October, two months ahead of contract. Current RoKAF orders total 50 T-50 advanced trainers and 44 A-50 ground attack aircraft, while an additional 200 A-50s could be ordered in the 2007-12 timeframe to replace that many ageing Northrop F-5A/B/E/Fs. KAI and LM estimate potential sales of at least another 600 A/T-50s elsewhere.

A pair of Block II F/A-18F Super Hornets were recently rolled unfinished from the Boeing assembly line into a separate hangar at the St Louis, Missouri, facility where they are undergoing conversion into EA-18G developmental test aircraft. As part of the conversion the aircraft will be fitted with the ALQ-218(v)2 tactical receiver, communications countermeasures set, interference cancellation, ALQ-99 tactical jamming system pods and the multi-mission advanced tactical terminal (MATT). EA-1, which began life as the 134th F/A-18F, will fly for the first time in September 2006. EA-2 will follow in November 2006.

The US Army has selected the General Atomics Aeronautical Systems Warrior UAV as the winner of its Extended Range Multi Purpose (ERMP) air vehicle contest. The contractor has received a $214.4 million contract associated with the project's System Development and Demonstration (SDD) phase. The Army plans to acquire 11 Warrior systems, which will each comprise 12 air vehicles, five ground control stations and associated support equipment. Based on the USAF's RQ-1 Predator, the long-range Warrior is powered by a diesel engine. The ERMP system will achieve initial operational capability in 2009 and will initially replace the army's RQ/MQ-5B Hunters.

Northrop Grumman recently conducted a series of tests involving the RQ-8A Fire Scout Vertical Takeoff Unmanned Aerial Vehicle (VTUAV) at the US Army's Yuma Proving Grounds in Arizona. As part of the tests the RQ-8A successfully launched two 70-mm (2.75-in) Mk 66 unguided rockets. Northrop Grumman is developing a more advanced multi-function model, designated the MQ-8B, that will support the US Navy's planned Littoral Combat Ship (LCS) beginning in 2009. The multi-mission Fire Scout will be equipped with electro-optical/infrared sensors and a laser designator that will allow the VTUAV to locate, track and designate targets.

Poland

Rolls-Royce to power SW-4

In September, Rolls-Royce signed a letter of intent (LOI) with Polish helicopter manufacturer PZL-Swidnik at the 2005 UK Helitech exhibition at Duxford, to acquire 10 313-kW (420-shp) 250-C20R turboshaft engines. These will be delivered during 2006, for installation in PZL-Swidnik's SW-4 light single-engined helicopters, for which the Polish company has already received national air force (PWLOP) orders for 30 for training use, and from several export customers.

PZL-Swidnik is now investigating potential development of a growth version of the SW-4, and has also signed an MoU with Rolls-Royce to investigate powering the new variant with the 485-kW (650-shp) Model 250-C30

Poland

Typhoon tropical trials

High-temperature operational trials were started this summer by EADS CASA Military Aircraft (ECMA) of a Eurofighter Typhoon at the Spanish air force (EdA) base of Morón (near Seville), in southern Spain. Following Cold Environmental Trials completed last winter near the Arctic Circle in Sweden, the instrumented production IPA4 Typhoon and a mobile telemetry station were transferred from ECMA's HQ site at Getafe, near Madrid, to

Morón, for an equally ambitious test programme.

Eurofighter gives the user air forces an integrated Network Enabled Capability (NEC), indispensible for their future interoperability. The EdA actively supported the Typhoon trials by locating its MIDS (Multiple Information Distribution System) ground station at Morón, and flying one or two operational Eurofighters in joint missions with IPA 4. ECMA placed its MIDS station at Talavera Air Base, near Portugal, to create the required network for favourable data-link tests.

With unusually high summer ground temperatures of up to 43°C, Morón provided optimum conditions for successful programme completion in environmental extremes. The adjacent Atlantic Ocean also offered air corridors for low-level supersonic flights required in the planned trials. ECMA continued the high-temperatures test programme begun in the summer of 2004, but in the new Block 2B software aircraft configuration, incorporating the FLIR sensor, and defensive aids sub-systems (DASS).

Further trials were also made of the Typhoon's new automatic Manoeuvre Limitation System (MLS), which prevents the pilot exceeding flight parameter limits, for carefree-handling. Other test items in the 41 flights and numerous ground-checks included transonic aircraft handling, and aeroelastic behaviour data-acquisition at speeds up to 1390 km/h (750 kt). Further flight-

trials in the Block 2B type acceptance programme, completed in September 2005, included air-refuelling from an EdA KC-130H Hercules.

Block 2B is the third of several software stages in Tranche 1 Typhoon production. Block 1 covers only basic air defence roles with MBDA ASRAAMs and Raytheon AIM-120C AMRAAMs, while Block 2 adds Link 16 data-transfer and initial DASS capabilities. Blocks 3 and 4 add further upgrades, while Block 5 will confer limited air-to-surface capabilities, including integrating Raytheon Paveway II laser/GPS-guided bombs.

United Kingdom

Sentinel R.Mk 1 progress

Having installed, as prime contractor, the first dual-mode active-array radar for the RAF's Airborne Stand-Off Radar (ASTOR) programme in a modified twin-turbofan Bombardier Global Express Sentinel R.Mk 1, Raytheon Systems Ltd (RSL) started its flight development from Hawarden airfield at Broughton in North Wales, on 25 July. RSL is undertaking most of the major modification and integration work on the ASTOR system at its Broughton facility, where the aircraft (ZJ691) was the second Sentinel R.Mk 1 of five planned to fly, for this UK ground-surveillance and reconnaissance programme.

The first, modified and flown at Greenville in Texas, was completing final integration work there before flying to the UK. Conversion of the remaining three ASTOR aircraft is well advanced at Broughton.

As key element in the system, the ASTOR dual-mode radar can generate synthetic aperture (SAR) imagery of various resolutions, with advanced moving target indicator (MTI) modes. The first Sentinel R.Mk 1 is being used by RSL to complete all aero validation, ground environmental, non-radar mission equipment installation, and icing trial objectives, for initial service with No 5(AC) Squadron in 2007.

RAF Typhoon update

Official figures released in July indicated that the 14 Eurofighter Typhoons then delivered to the RAF had accumulated 1,529 flying hours in 1,270 sorties. As the RAF's Typhoon Operational Conversion Unit (OCU), No. 29(R) Squadron, was receiving an average of one new aircraft every five weeks at

Coningsby, in Lincolnshire, to where it was transferred following completion of BAE System's Case White initial training contract on 1 July. It joined No. 17 Squadron, the RAF Typhoon Operational Test and Evaluation unit, also based at Warton for the previous 15 months, until its 1 April 2005 departure to Coningsby.

Following initial deliveries of two-seat Typhoon T.Mk 1s, the first single-seat Batch 2 Typhoon F.Mk 2s in Tranche 1 were delivered to the four participating nations early this year.

These signed the Tranche 2 production contract in December 2004 for another 236 Typhoons, including Germany 68, Italy 46, Spain 33, and UK 89, for deliveries between 2008-15. If RAF procurement reaches the planned 232 aircraft, the Typhoon will equip seven front-line squadrons, the Operational Conversion Unit (OCU) and Operational Evaluation Unit (OEU). These units will operate 137 Typhoons in four air defence, two swing-role, and one offensive support squadrons, with the remaining 95 held in reserve as attrition and rotational replacements.

Nimrod MRA.Mk 4 progress

BAE Systems reported steady progress in July with its Nimrod MRA.Mk 4 flight-trials programme, in less than a year after its autumn 2004 inaugural flight. PA01, the lead development aircraft, was joined in mid-December 2004 by PA02, equipped for missions-systems testing. A third and final development aircraft (PA03) joined the MRA.Mk 4 development fleet later this year, having first flown on 29 August. That afternoon, it flew from BAE Systems' Woodford site, near Manchester, to the company's Warton base, near Preston, after a 75 minute sortie.

In mid-July, BAE Systems submitted a final and comprehensive bid for production of a 12-aircraft Nimrod MRA.Mk 4 fleet to the UK MoD, for approval expected around the year-end. The bid includes conversion to full production standards of the three MRA.Mk 4 development aircraft at the 2007 conclusion of their flight trials programme. MRA.Mk 4 deliveries are then scheduled to RAF Kinloss between 2010-2012.

In early August, PA02 left the UK for the first time, flying to Sicily for preliminary hot-weather trials. All preparations for Nimrod MRA.Mk 4 manufacture are now concentrated at BAE Systems

Lockheed Martin has been pole-testing this full-scale model of the Northrop Grumman X-47B at its Helendale radar cross-section testing facility in southern California. Two X-47B demonstrators have been ordered in support of the J-UCAS (Joint Unmannd Combat Aircraft System) programme, along with three Boeing X-45Cs. Note the Skunk Works badge under the wing.

Woodford site, where scheduled groundings of PA01 and PA02 were awaited for installation of design modifications. Extensive strip and survey work was also being undertaken at Woodford on four additional RAF Nimrod MR.Mk 2 fuselage shells, for their major remanufacture as MRA.Mk 4s. Jigs have also been installed at Woodford for assembly of MRA.Mk 4 wings and other major components.

United States

Raptor progress
Manufacturing Development (EMD) testing of the F/A-22A concluded at Edwards AFB, California, recently, and the programme surpassed 2,592 flight hours. The F/A-22A subsequently began Follow-on Operational Test and Evaluation (FOT&E), which includes the final milestones that must be met before the fighter achieves initial operational capability (IOC), in late 2005. Conducted by the Air Force Operational Test and Evaluation Center (AFOTEC), the FOT&E will test the Raptor in several event-based operational battle space scenarios that include evaluations of its air-to-ground capabilities and its suitability for deployment by C-17s. Seven operationally representative Raptors are supporting the FOT&E efforts, which are being conducted at Nellis AFB, Nevada, White Sands Missile Range, New Mexico, and the Utah Test and Training Range.

As of mid-September 2005 Lockheed

Martin had completed final assembly of 62 of the 83 F/A-22A fighters on contract, and had delivered 47 Raptors to the USAF.

USAF tanker team formed
Northrop Grumman Corporation has announced plans to compete as the prime contractor for the USAF's next-generation air refuelling tanker. As the prime contractor, Northrop Grumman will include EADS as a principal subcontractor and teammate. Based on the Airbus A330 airliner, the contractor's KC-30 tanker will be tailored to meet the USAF's requirements. The US content of the programme will reportedly exceed 50 percent. Northrop Grumman subsequently announced plans for a production site adjacent to the EADS site at Brookley Downtown Airport in Mobile, Alabama.

USAF helicopter request
The USAF's Aeronautical Systems Center (ASC) has released a request for proposals (RFP) associated with its $8 billion CSAR-X project. Previously referred to as the personnel recovery vehicle (PRV) programme, the CSAR-X includes the purchase of 141 aircraft which will replace the service's current fleet of 104 HH-60Gs. Contenders for the project include Boeing with a development of the MH-47G, Lockheed Martin/AgustaWestland with the US101 and Sikorsky/Boeing with the H-92. The Bell-Boeing V-22 Team has dropped plans to respond to the RFP, citing the

The mission design series (MDS) designation VH-71A has been assigned to the Lockheed Martin US101 helicopter, which had previously been known by the project name VXX. The as-yet unnamed next-generation Presidential transport, which is based on the AgustaWestland EH101 helicopter, will replace VH-3D and VH-60N helicopters currently used by Marine Helicopter Squadron HMX-1 beginning in 2009. The US Navy plans to purchase 26 examples, including three test and 23 production aircraft. Here the first US101 test article arrives at Lockheed Martin's Owego, New York, facility. It will support testing locally until it transfers to NAS Patuxent River, Maryland, in 2006.

tiltrotor's excess capabilities for the mission. The CSAR-X acquisition will be a spiral development that includes two separate aircraft configurations known as the Block 0 and Block 10. Development will begin 2006 with low-rate initial production (LRIP) starting in FY08. The first operational unit will achieve initial operational capability (IOC) with the Block 0 version in 2011. The Block 10 will follow in 2014 and the production Block 0 aircraft would be upgraded to the advanced configuration by 2018. The initial phase includes the purchase of five Block 0 production-representative test vehicles (PRTV) that will be delivered in 2007/08. The first LRIP purchases will follow beginning in 2009. Two Block 10 PRTVs will be ordered in 2008 with LRIP following in 2012.

CH-47F construction underway
Boeing has begun construction of the first new CH-47F cargo helicopter in support of the Army's ongoing modernisation programme. In addition to new avionics and engines, the new version features structural improvements that include air transportability modifications, intended to reduce the time required to prepare the aircraft for cargo transport, and advanced corrosion protection. The CH-47F is powered by a pair of 3630-kW (4,868-shp) Honeywell T55-GA-714A engines, has a top speed of speed more than 282 km/h (175 mph) and can carry payloads greater than 9525 kg (21,000 lb). The aircraft's internal auxiliary fuel tanks provide a mission radius that exceeds 400 miles (644 km).

UPGRADES AND MODIFICATIONS

Australia

F-111 missile tests
The Royal Australian Air Force recently conducted the first test firings of the AGM-142E missile from an F-111C, modified by Boeing Australia Limited, over the Woomera Test Range in South Australia. Equipped with an imaging infrared seeker, the AGM-142E stand-off missile is capable of engaging non- and semi-hardened targets at ranges greater than 48 km (30 miles). The missile will provide the RAAF with a long-range strike capability until the service can field upgraded F/A-18s that will be equipped with new long-range weapons. The RAAF intends to retain the F-111C/G in service though at least 2010. Weighing approximately 1302 kg (2,870 lb), the missile is powered by a single-stage solid rocket motor and can be armed with either a general-purpose blast fragmentation or a penetrating warhead. It will be introduced into service in 2006.

Belarus

Russian fighters upgraded
Some 48 Belarusian air force (BVVS) MiG-29S and eight two-seat MiG-29UB combat trainers are being equipped with new digital mission computers, software and associated Russkaya Avionika avionics. These include multi-function colour cockpit displays, Phazotron N019 radar upgrades, and a fixed air-refuelling probe, for additional precision-guided

air-to-surface weapons capabilities.

The redesignated MiG-29BMs are also being joined by 18/4 BVVS Sukhoi Su-27/UBs with similar upgrades, initially flown in a modified two-seat Su-27UBM1 flight-development aircraft. They are similar to the upgraded single-seat Su-27SKM exhibited at last summer's Paris Air Show, as a further development of the Russian VVS Su-27SM, delivered from KnAAPO from December 2003, and export Su-27SKs.

Undertaken jointly with MiG RAC, Russkaya Avionika, and the Sukhoi OKB, most of the BVVS aircraft upgrade work is being undertaken by the Belarus national 558 Aircraft Repair Plant, at Baranovichi. This has also upgraded BVVS Mil Mi-8MTs to Russian Mi-8MTKO standards, with similarly enhanced digital avionics, including a chin-mounted sensor turret.

Morocco

Mirage upgrades planned
A visit by French Prime Minister Dominique de Villepin to Morocco in September, to meet King Mohammed VI was aimed at providing "a new dynamic" in France's relationship with its former North African territory. Its outcome included a preliminary agreement for the Thales group to upgrade about 20 of the Royal Moroccan air force's (QJMM) 35 Mirage F1CH/EH combat aircraft. French officials said that Morocco opted to modernise its

Mirages, for which a contract was expected soon, rather than buy F-16s.

New Zealand

Kiwi 'Combis'
The Royal New Zealand Air Force has selected Singapore Technologies Aerospace's (ST Aero) to convert its two B757-220 series airliners into 'combi' aircraft capable of carrying troops or cargo, or a combination of both. The aircraft will undergo conversion at ST Aero's facility in Mobile, Alabama, beginning in March 2006. The B757s are capable of transporting up to 200 troops, and once modified they will be able to carry up to 11 pallets or 4,990 kg (11,000 lb) of cargo.

First P-3K upgrades completed
In August, L-3 Communications announced that its Integrated Systems

(L-3 IS) subsidiary had delivered the first upgraded RNZAF P-3K Orion maritime patrol aircraft to the New Zealand Ministry of Defence, two months before scheduled in late October 2005. L-3 IS received an October 2004 contract for the RNZAF's P-3K Systems Upgrade Project, for enhancements to the Orions' mission, communications, and navigation systems.

The first three of the six aircraft upgrade programme have early installations of L-3 Communications' Wescam MX-20 imaging turret system with both video and infra-red sensors, as an interim benefit prior to the full aircraft upgrade. Work continues on the additional aircraft, and includes replacement of the data-management, sensor, communications and navigation systems, plus provision of associated ground equipment. P-3K upgrade contract completion is due in 2010.

Seen at Sliac Air Base on 28 August 2005 are these two newly updated Slovak Aero L-39s. This update has been executed at Trencin during recent months, installing two new MFDs in the cockpit, instrumentation calibrated in feet and knots rather than metric, and new antennas under the fuselage. Serial 1730 is an L-39ZAM and 5301 is an L-39CM. Five more L-39CMs will have been produced by the end of 2005.

MAKS 2005

Held in August, the MAKS 2005 show at Zhukovskiy provided the Russian aerospace industry with a showcase for its products. The Sukhoi Su-47 Berkut made a welcome return, flying with a pair of 'Flankers' (above). The major debutante was Kazan's Ansat-2RC light attack helicopter (left), which made its first flight on 29 July and flew during the show. Based on the civil Ansat, the 2RC has a Pratt & Whitney Canada PW207K engine and has a TOES-521 FLIR/TV/laser turret. Armament comprises a single 12.7-mm (0.5-in) machine-gun on the starboard side, and stub wings for the carriage of rocket pods or up to four light missiles.

Undoubted star of the flying display was the MiG-29OVT (left), which put on a staggering demonstration of thrust-vectoring manoeuvring. RSK MiG announced that future production versions would incorporate this feature, and would be known as MiG-35s. The company also displayed its MiG-29M2 two-seat multi-role demonstrator (above).

Nigeria

G.222 refurbishment plans

Finmeccanica's Alenia Aeronautica received a $74.5 million two-year Nigerian air force contract in July for refurbishment, assistance and logistic support of the NAF's Aeritalia G.222 twin-turboprop tactical transports. The NAF received five G.222s from 1984, but like many of its aircraft, these achieved low utilisations in recent years through funding and spares problems.

Apart from building new infrastructures at the military base of Ilorin, 185 miles (300 km) north of Lagos, Alenia Aeronautica will supply full-scale maintenance equipment for the NAF's G.222s. The Italian company will also train Nigerian pilots, technicians, engineers, loadmasters and logistic support staff, and supply an additional ex-Italian air force G.222.

Thailand

L-39s returned to service

On 29 September Aero Vodochody redelivered the last of 25 L-39ZA armed jet trainers after major overhauls to the RTAF. The Czech company originally delivered 43 L-39ZA/ARTs to Thailand between 1994-96, and in September 2002 received the overhaul contract, now completed on schedule from a Thai assembly line.

Around 6,500 of some 12,000 components were returned to parent companies for specialised repairs or overhauls. The RTAF L-39 fleet is regularly operated in temperatures around 40°C and around 95 per cent humidity, and had flown over 60,000 hours in this environment.

United Arab Emirates

Ex-Libyan Chinooks upgraded

AgustaWestland has started work in Italy on refurbishment and upgrades of the first two of 12 Elicotteri-Meridionali CH-47C Chinooks acquired by the UAE from Libya in 2003. These were originally licence-built in Italy for a Libyan army contract for 14 from 1976, but have been grounded for many years from former US arms embargos.

A follow-on contract has now been reported for a second batch of four conversions by AgustaWestland of the UAE's ex-Libyan Chinooks to CH-47D+ standard, with new avionics, EW systems, airframe overhaul, composite rotor-blades, and 28 uprated 2795-kW (3,750-shp) T55-712E turboshafts. Boeing, Honeywell, Piaggio Aero Industries, and South Africa's Avitronics are also involved in this programme.

United States

Orion delivered

Lockheed Martin delivered the first Update II.5 series P-3C modified under the Anti-Surface Warfare Improvement Program (AIP) to the US Navy on 12 July. The aircraft was initially delivered to Air Test and Evaluation Squadron VX-20 at NAS Patuxent River, Maryland, and will support upcoming trials.

C-5M progress

Lockheed Martin technicians recently installed the first engine pylon associated with the C-5 reliability enhancement and re-engining program (RERP) on a test aircraft. The RERP is the second portion of a two-phased C-5 modernisation effort that will provide the airlifter with more than 70 improvements including new engines and avionics. The modernization programme will be carried out on 112 Galaxies including both C-5A and C-5B models and will allow the C-5M to remain in service through 2040. Designed and built by Goodrich Aerospace, the pylons will support new General Electric CF6-80C2 engines that are each capable of delivering more than 222.4 kN (50,000 lb). The engines will provide the C-5M with 22 per cent more thrust and a 30 per cent reduction in takeoff distances, allowing the aircraft to carry more than 122470 kg (270,000 lb) and to take off and land in as little as 1524 m (5,000 ft). RERP follows the AMP avionics modernisation programme. The first fully modified C-5M is scheduled to fly in spring 2006.

HC-130J modifications

The US Coast Guard has awarded Integrated Coast Guard Systems (ICGS) a $117.5m contract to provide six HC-130J long-range search (LRS) aircraft with mission equipment. As part of the modifications the aircraft, which are currently restricted to transport duties, will be equipped with an electro optical/infrared (EO/IR) turret, a UHF/VHF direction-finder system, and an airborne automatic identification system (AIS).

Additionally, the aircraft will receive a maritime surface search radar. Current plans call for the first HC-130J to enter modification in January 2007 and return to service nine months later. Delivery of the final aircraft is scheduled to occur in July 2008.

Navy tankers

The US Navy is examining the possibility of equipping its 20 C-130Ts with refuelling pods. It recently fitted an aircraft, assigned to VR-64 at NAS Willow Grove, Pennsylvania, with the pods, which subsequently refuelled a pair of F/A-18s in flight. Although the C-130Ts were delivered with the necessary plumbing required to install the pods, the underwing hardpoints needed to carry the equipment were not fitted. Unlike the USMC KC-130Ts, the VR-64 aircraft was not equipped with internal auxiliary fuel tanks that increase offload fuel capacity.

PROCUREMENT AND DELIVERIES

Bahrain

Hawk leads new training system

BAE Systems test-pilot Pete Wilson made a successful initial 73-minute flight at Warton on 26 August, some nine months ahead of schedule, of the first of six Hawk Mk 129s (BT001) ordered by the Royal Bahraini air force (RBAF). Derived from Australia's updated Hawk 127, the Mk 129 is the first productionised Hawk with BAE Systems' operational flight programme integrated with the new FADEC-equipped Rolls Royce/Turboméca Adour Mk 951.

The first two Mk 129s will initially remain with the Technical Academy at Warton, for use in a six-month training course for Bahraini aircraft technicians, before acceptance. All six will be delivered by December 2006, as part of BAE's equipment and training package, to establish an RBAF Air Training Wing (ATW). BAE also aims to continue assisting and supporting the ATW throughout its lifetime.

The ATW will take in raw recruits for pilot training, before progressing to RBAF F-16s, or helicopter conversion. Unusually, the RBAF is considering adopting a two-aircraft (Slingsby Firefly/Hawk) approach, using flight-simulators to bridge the gap, rather than the customary turboprop advanced-training element. Only Finland and New Zealand have previously implemented this two-type trainer approach.

This is based on a four-phase programme, working closely with the customer to achieve an indigenous aviation training capability. Phase 1 will be delivery of pilot aptitude-testing equipment. Phase 2, the supply of three Slingsby Firefly T67M-260 aircraft for primary and basic training, with three instructor pilots, one field service representative and a support package. Phase 3, provision of synthetics (full dome simulator plus avionics part-task trainer) to provide the bridge from Firefly to advanced jet-training. Phase 4 will be delivery of all six Hawks for this role, plus a comprehensive two-year field-service and technical support package.

Bangladesh

Fighters purchased

Bangladesh has purchased 12 single seat F-7 and four two-seat FT-7 fighters from China at a cost $118 million. The F-7 is an export version of the Chengdu J-7, which is based on the MiG-21.

Brazil

Surplus AdA Mirages acquired

Agreement was signed by Brazilian Foreign Affairs minister Celso Amorim and French Defence Minister Michelle Alliot-Marie on 18 July for FAB purchase of 12 surplus AdA Dassault Mirage 2000B/C air-defence fighters. Equipped with RDI pulse-Doppler radar, the Mirages are being acquired from a $102.5m contract, which includes ground support equipment and personnel training.

Originally approved in the spring, the contract was reportedly delayed by high costs sought by the French government for Brazilian M53 engine test-bed requirements. The Mirage 2000s are expected to re-equip the FAB's Mirage IIIE air defence squadron, following recent deferment on cost grounds of Brazil's F-X requirement for 12-24 new combat aircraft.

Colombia

More Black Hawks ordered

A contract worth up to $100m was being finalised in July for Colombian air force acquisition of eight Sikorsky UH-60L Black Hawk transport helicopters and associated support equipment. These are effectively attrition replacements for nine of 59 UH-60A/Ls delivered to the FAC from 1987 for general duties, related mainly to drug interdiction missions.

Czech Republic

New VIP transports planned

Funding totalling Euro132.7m ($161m) was authorised earlier this year to re-equip the military-operated government transport fleet. Three Tupolev Tu-154B-2s, two Yakovlev Yak-40s, a Bombardier Challenger 601 and a few Russian helicopters, will be replaced by two large twin-turbofan and two medium-sized transports, plus two utility helicopters.

This follows a $184 million contract signed in September 2004 by the Czech state-owned Letecke Opravny Malesice (LOM) with Russia's Rosoboronexport for additional military helicopters, with funding from Russian trade debts. These covered earlier deliveries to the Czech Republic air force (VSACR) of seven upgraded Mil Mi-24V attack helicopters, followed in May by the first three of 10 new Mi-35 export versions on order. The VSACR is expecting further deliveries of 16 Mil Mi-171Sh transport helicopters in the coming year.

Czech Republic

More F-16s ordered

As forecast, following the Greek government's April deferment of possible procurement of 60 Eurofighter Typhoons, it announced plans in July to purchase 40 F-16 Lockheed Martin F-16C/D Block 52+ advanced combat aircraft, with F100-PW-229 engines, APG-68(V)9 radars, and associated arms and equipment. On 25 October US Congress was notified of a proposed FMS contract for this programme, worth up to $3.1bn with all options. These will supplement previous Hellenic air force (HAF) receipts of 80 earlier F-16C/Ds and 60 Block 52 versions.

Thirty HAF F-16C/Ds plus plus 10 options were expected to cost some Euro1.1bn ($1.326bn), with the balance of funding covering advanced arms and equipment, including helmet cueing and NVG systems, 10 LANTIRN targeting and 11 reconnaissance pods; four AGM-154C Joint Stand off Weapons (JSOWs); and six Joint Direct Attack Munitions (JDAMs) with three BLU-109 and three Mk 84 bomb bodies.

The Greek government emphasised that the new F-16 order was to replace HAF attrition losses, and meet interim requirements for third-generation fighters. "An even more important HAF need" remained for (a reduced total of) 40 new fourth-generation combat aircraft, to be discussed later this year. Johann Heitzmann, EADS Military Aircraft president, said the Greek F-16 order "did not pre-empt anything in terms of HAF Eurofighter procurement".

First army NH90 flies

The first of 20 NH90 tactical transport helicopters, equipped with Rolls-Royce Turboméca RTM322-01/9 engines, and ordered in August 2003 for Hellenic Army Aviation (EAS) from a Euro657m ($792.3m) contract, began flight-tests at Eurocopter's Marignane facility on 13 July. This was the sixth production NH90 to fly, following earlier examples for Germany, Finland, Italy and Sweden.

The EAS will receive its first NH90s by late 2005, with programme completion, including options, planned by 2010. The Greek contract includes offset agreements involving Hellenic Aerospace Industry (HAI) participation.

More than 40 NH90s are currently being built on assembly lines in France, Germany, Italy and Finland, for a total of 357 firm orders, 86 options and more than 45 selections, by 13 countries and 17 armed forces.

Hungary

Gripen weapons selected

Multi-role capabilities of recently delivered Hungarian Air Defence and Aviation Command (LERP) JAS 39C/D Gripen combat aircraft are being extended from integration of new targeting and precision-guided weapons systems. Clearance was nearing completion in August to use Rafael Litening III laser-designation pods (LDP) with Raytheon GBU-10, GBU-12 and GBU-16 Paveway II laser-guided bombs (LGBs), to achieve a full night-attack/all-weather precision-guided munition (PGM) capability.

The Gripen development team then began integration in May of the ramjet-

The first 'new production' MH-60R helicopter was ferried to Lockheed Martin's facility in Owego in early October. Built by Sikorsky Aircraft in Stratford, Connecticut, the aircraft, which was accepted by the Navy in August, was flown to Owego as a bare airframe. Lockheed Martin is currently installing the helicopter's multimode radar, dipping sonar, acoustics equipment and other systems. It will deliver the aircraft to the Navy in about five months.

Lockheed Martin delivered the first F-16D to the Royal Air Force of Oman (RAFO) at its Fort Worth, Texas, facility after a first flight on 8 July. The fighter is the first of 12 advanced Block 50 fighters ordered by the Sultanate of Oman's Ministry of Defence in May 2002. The Peace A'sama A'safiya (Clear Skies) FMS programme includes eight single-seat F-16Cs and four two-seat F-16Ds, powered by General Electric F110-GE-129 engines. The initial aircraft was followed by the first F-16C, and both fighters were ferried to Oman in mid-October.

powered MBDA Meteor advanced beyond-visual-range air-to-air missile (BVRAAM), to supplement medium-range Raytheon AIM-120 AMRAAMs. Gripen International will undertake all the initial live-firing trials of the Meteor, from late 2005. The Hungarian MoD also plans to acquire 40 AIM-120C-5 AMRAAMs, for which little integration will be needed, since the Gripen is already cleared for the AIM-120B.

Hungary is also inviting tenders for more air-to-air and air-to-ground weapons. These are expected to include the Raytheon AGM-65 Maverick ASM, and a new short-range AAM, for which the Gripen is already cleared. Tests are also starting in the Gripen of the Cobra helmet-mounted display (HMD), to extend the aircraft's agile air-to-air and precision-guided weapons capabilities.

India

US aircraft and arms interest

In late summer, Defence Minister Shri Pranab Mukherjee confirmed that Boeing and Lockheed Martin had been approved to respond with offers of the F/A-18 and F-16 to India's Request for Proposals (RFPs) for 126 IAF medium multi-role combat aircraft. No date had then been given, however, for RFP issue.

India was expected to be offered both surplus and new F-16s and F-18s, the latter including the latest F/A-18E/Fs being produced only for the US Navy. With either Raytheon APG-73 or actively electronically-scanned array (AESA) APG-79 radar, the F/A-18E/Fs could cost the IAF up to $53m per aircraft.

Other contestants included the Dassault Mirage 2000-5 and Rafale, MiG-29M/M2, and Sweden's JAS 39 Gripen. The MiG-29M has been redesignated the MiG-35, and is being offered for Indian licenced production, with multi-axis thrust-vectoring, now undergoing flight-development on the MiG-29MOVT.

More Il-38s considered

The Indian government is still considering acquisition of two Ilyushin Il-38 four-turboprop maritime patrol and surveillance aircraft from Russia, according to Defence Minister Mukherjee. These are required to replace two similar Indian Navy aircraft lost in a fly-past collision in October 2002, and supplement three remaining IN Il-38s. They are currently being upgraded and refurbished in Russia to extend their operating lives by another 15 years.

Mukherjee also said that, although the IAF will acquire three Perm PS-90A-76 turbofan-engined Ilyushin Il-76s equipped with Israeli Phalcon airborne early-warning and control radar and mission systems from a 2004 order, India's DRDO sanctioned a design and development programme for indigenous AEW&C aircraft in October 2004. Preliminary designs of various AEW&C sub-systems had already been completed, he added, and completion of development activities to start user trials was planned within 78 months.

Legacy Jets accepted

Embraer recently delivered three specially configured Legacy Executive aircraft to the Indian government. The aircraft, which comprised two for the air force and one that will be used by India's Border Security Force, are part of an order for five aircraft. The remaining Legacy Executives, which are based on the ERJ 135, will be delivered in October. India holds options on two further examples.

Boeing Business Jets for VIPs

The Indian Government has announced plans to purchase three Boeing 737-700 business executive jets. The transports, which will be delivered by January 2009, will feature missile defence equipment. They will replace the two Boeing 737-200s purchased in 1983 and will primarily be used for international trips by the president, the vice-president and the prime minister.

Tanker/transport plans

The Indian Air Force has announced plans to acquire six additional Il-78 air refuelling aircraft. The air force, which previously bought six tankers from Uzbekistan, cites the need to strengthen its ability to act as an 'expeditionary force' while responding to natural disasters and emergencies in the region, as the reason for the purchase. The IAF also plans to upgrade its ageing An-32 medium transport fleet and is also considering the purchase of other transport aircraft.

Italy

SF-260EA deliveries begin

Aermacchi delivered the first two of 30 new SF-260EA piston-engined primary trainers to the Italian air force (AMI) in August, from the Euro33m ($40m) 2004 contract. One SF-260EA has been used to qualify four 70th Wing Pilot Instructors at Latina Air Force Base.

Having first flown on 20 October 2004, with a new avionics suite and enlarged bubble canopy, the prototype SF-260EA obtained National Armaments Directorate type certification on 10 May. The AMI will receive 19 SF-260EAs from 2006, for pilot screening and primary training.

Jordan

Il-76s ordered

Agreement was reached on 17 August during a meeting between Rosoboronexport director-general S.V. Chemezov, and Jordanian armed forces C-in-C King Abdullah II, to supply two new Ilyushin Il-76MF stretched military transports fitted with 142.34-kN (32,000-lb) thrust Aviadvigatel Perm PS-90A-76 turbofans, to the RJAF. This is the first export contract for the Il-76MF, flown in prototype form since August 1995. Funding has only recently been approved for the first two of an initial Russian Federation air force requirement for 10 by 2010, with up to 90 more later.

In early 2003 the Ilyushin Aviation Complex, which includes the Design Bureau (OKB) in Russia; VASO (Voronezh Aircraft Manufacturing Enterprise); and TAPOiCh (Tashkent Chkalov Aircraft Manufacturing Plant) in Uzbekistan, had completed prototype Il-76MF flight-development, after some 1,500 test sorties. Installation of PS-90A turbofans and new digital avionics in existing Il-76MDs allows them to meet ICAO Stage 4 noise and air traffic management standards for international flights, with improved performance and combat readiness.

South Korea

First F-15Ks arrive

Flight deliveries started in early October of the first two of 40 F-15K Eagle multi-role combat aircraft from Boeing's St Louis production line, from the $5.5bn 2002 contract. After they landed at Hickam AFB, Hawaii, Boeing KC-135 Stratotankers from the Hawaii Air National Guard air-refuelled the two fighters twice on their way to Andersen AFB, Guam, before they continued to RoKAF's Seongnam AFB.

Boeing test pilots and RoKAF pilots took the first two F-15s through various performance tests during their 20-hour flight-time, covering 15962 km (8,618.8 nm) on route to Hawaii. The two F-15Ks made their Korean debut at the Seoul Air Show in October, before formal RoKAF acceptance in November. Two more were expected by the year-end; eight by 2006; 16 by 2007; and 12 by 2008. The F-15Ks will become operational from January 2007, and replace all the RoKAF's obsolete MDC F-4D/E Phantoms by 2011. The RoKAF plans to introduce 40 more F-15K-level aircraft after 2008.

Kuwait

More Hornets planned

After extensive KAF evaluations of several contenders, the US Defense Security Co-operation Agency notified Congress in August of a proposed Foreign Military Sale (FMS) to Kuwait of a Boeing F/A-18 combat aircraft package. With technical services, plus associated armament, equipment and services through 2011, this could be worth up to $295m.

In August the KLu took delivery of its first Pilatus PC-7 after corrosion inspection and repainting in a black scheme by RUAG at Oberpfaffenhofen. The scheme was adopted to improve conspicuity in the air, with a yellow stripe added to make the aircraft more visible on the runway for air traffic controllers.

Seoul air show 2005

The Seoul air show in October provided an interesting glance at some of South Korea's new aircraft, which have been procured from a variety of sources. From Russia comes the Ilyushin Il-103 light trainer (left, local designation T-103) and Kamov Ka-32 (above). The US is supplying 40 F-15Ks for a heavy multi-role fighter competition, while the local aviation industry has produced the KO-1 light attack/weapons trainer (below). The F-15K (below left) was one of the first two to be flown to Korea: the first two aircraft from the line remain in the US for tests.

Further contractual details were then still awaited, although Kuwait has had a long-term requirement for at least 20 F/A-18E/F Super Hornets. These will supplement an initial batch of 32/8 F-18C/D Hornets, powered by F404-GE-402 enhanced performance engines, and delivered to Al Jabar Air Base, near Kuwait City in 1992-93, from an August 1988 letter of offer and acceptance (LOA).

Mexico

Navy receives Panthers
Two Turboméca Arriel 2C-powered Eurocopter AS565SB Panther utility helicopters were recently delivered to Mexican Naval Aviation (AAM) from a 2003 contract which included a further eight options. As well as featuring a starboard-door SAR hoist, they are equipped for coastguard, drug interdiction, surveillance, and troop transport.

Netherlands

Dutch Hercs
The Royal Netherlands Air Force has reportedly purchased three EC-130Q aircraft at a cost of $65 million. The aircraft were previously operated as strategic communications aircraft by the US Navy. The air force reportedly plans to make two of the aircraft airworthy while retaining the third as a source of spare parts. The three aircraft are currently registered to Derco Aerospace in Milwaukee, Wisconsin, but remain in storage in Tucson, Arizona.

New Zealand

New helicopters sought
Earlier NZ government plans to spend $NZ400-561m ($278.74-391m) on new military helicopters were followed in August by more detailed proposals for RNZAF Bell UH-1B Iroquois and Bell 47G-B-3 Sioux replacements. These will include 8-12 NHIndustries NH90s to replace the RNZAF's 14 UH-1Bs for its medium utility helicopter (MUH) requirement, and a new light-utility/trainer helicopter to supplant five aged piston-engined Bell 47G-B3s.

Welcoming these plans, RNZAF C-in-C Air Vice-Marshal John Hamilton said that negotiations remained to be finalised with NH Industries to finalise costs, aircraft totals, and delivery dates.

Nigeria

Fighters ordered
The China National Aero-Technology Import and Export Corporation (CATIC) will deliver 15 Chengdu F/FT-7NI multi-role combat aircraft to Nigeria beginning in June 2006. The first to arrive in Nigeria will be three two-seat FT-7NIs, which will be followed in October 2006 by the first of 12 single-seat F-7NI. As part of the $251.3 million deal 12 Nigerian pilots, including four instructors, will undergo conversion training in China next year.

Pakistan

Orion delivery nears
Pakistan will take delivery of its initial pair of refurbished P-3Cs from the US Navy in December 2005. In total, eight Orions will be delivered at a cost of $970 million. The initial group of Pakistani aviators began training with VP-30 at NAS Jacksonville, Florida, in September. Technicians are also updating two aircraft already in service with 28 Squadron at Karachi. Pakistan took ownership of the eight aircraft during August.

Two F-16s ordered
While continuing discussions with the US since last April, for possible FMS procurement of 55 Block 50/52 Lockheed Martin F-16s, plus 20 options, the Pakistan government placed a mid-2005 order for only two refurbished F-16A/Bs. These were due for delivery in October, to supplement some 20 Block 15 F-16As and 10 F-16Bs received in 1983.

Partial funding for additional F-16s was to be from a five-year US military aid package promised to Pakistan for its anti-terrorism support. Additional funds are required from Pakistan's defence budget, although unlikely to be sufficient for all 55 aircraft.

Raytheon Systems received a $29,443,350 modification to a previous FMS contract in October for 300 AIM-9M tactical and 10 captive air-training missiles for PAF F-16s, for delivery by August 2007.

Airborne surveillance system
Saab AB announced signature of a SEK8.3bn ($1.055bn) contract in October, in a two-third/one-third partnership with Ericsson Microwave Systems, to supply an airborne surveillance system for Pakistan. This will comprise a twin-turboprop Saab 2000 carrying Ericsson's Erieye planar radar system, for continuous surveillance of national air, territory and sea areas. With existing ground-based radars, it will provide a more detailed picture to counter air and ground threats, and conduct SAR operations.

Singapore

US details winning F-15 bid
Selection of the Boeing F-15T in preference to the similarly short-listed Dassault Rafale for the Republic of Singapore air force's next-generation fighter replacement programme (NFRP) requirement was presaged in August, when the US Defense Security Co-operation Agency notified Congress of a possible FMS package to the Singapore government. Worth up to $741m, this included weapons, logistics, and training support, plus associated equipments.

No indication was then given of aircraft totals, believed to involve 20-24, although perhaps initially only 12. Contract negotiations with Boeing had then still to be finalised, but funding was not expected to be a problem, since Singapore's $S9.26bn ($5.8bn) defence budget represented 31.2 per cent of national spending, among the highest in Asia.

The US listed Major Defence Equipment (MDE) items as including 200 Raytheon AIM-120C AMRAAMs; 200 Raytheon AIM-9X Sidewinders; 50 Mk82 GBU-38 Boeing JDAMs with BLU-111 warheads; 44 AN/AVS-9(V) night-vision goggles; 24 Link 16 multi-functional information distribution system/low-volume terminals (MIDS/LVTs); 30 Raytheon AGM-154A-1 JSOWs with BLU-111s; and 30 AGM-154C JSOWs.

The proposed US F-15 package also includes 300,000 20-mm (0.787-in) practice rounds; 100 KMU-556 GBU-31 JDAM tail-kit assemblies; four Mk 82 and Mk 84 bomb practice trainers; plus testing, integration, and support equipment.

South Africa

A109s delivered
The South African Air Force took delivery of the first four of 30 new AgustaWestland A109 light utility helicopters (LUH) at Bloemspruit AB on 19 October. Delivery of the A109s will allow the SAAF to begin replacing its elderly Alouette III fleet. The first eight pilots will start their training on the new helicopter with 87 Helicopter Flying School in February 2006.

Thailand

More F-16s planned
Further procurement is planned of Lockheed Martin F-16C/Ds for the RTAF from a $1.13bn programme,

Archangel 2005

In September the Hellenic Air Force organised its first large-scale international air show at the base of Tanagra (home to 114 PM and Mirage 2000s, plus Hellenic Aerospace Industries facilities). On display were most of the aircraft in Greek military and government service, including the latest types, and a high turn-out from international guests. Among the special schemes on display were a A-7E Corsair II from 335 Mira with tiger stripes blending into the Greek flag on the rear fuselage (below right), and a spectacular F-4E (SRA) Phantom II from 337 Mira at Larissa (below). Those F-4Es that have not been upgraded under the Peace Icarus

Above: Greece's new CFT-equipped F-16 Block 52+s were on display at Tanagra, including this aircraft surrounded by weaponry. The latter included EADS/LFK AFDS stand-off glide bomb, AGM-88 HARM, AIM-120 AMRAAM and GBU-12 LGB.

according to its C-in-C, Air Chief Marshal Kongsak Wanthana, in a June statement. F-16 totals then remained to be finalised in the programme, which would include weapons and support items, on a barter basis for Thai agriculture products. The F-16s are required to replace recently decommissioned Rockwell OV-10C Broncos and some 12/30 Northrop F-5A/Bs and F-5E/Fs now being withdrawn.

The RTAF had previously acquired 36 new and 15/1 refurbished Block 15OCU/ADF F-16A/Bs in four Peace Naresuan programmes between 1988-2003, the last with Falcon Up structural upgrades and other modifications, plus two Block 10OCU F-16As for spares reclamation. Last January, the RTAF also received Singapore's remaining three Block 15OCU F-16As and four F-16Bs, donated in return for RSAF continuation of training facilities at Thailand's Udon Thani air base.

Turkey

Common avionics for F-16s
Lockheed Martin at Fort Worth, Texas, was awarded an $83.539m firm fixed-price FMS contract in July, to begin developing the Turkish government's Peace Onyx III F-16 mission systems modernisation programme. Some $67.73m was initially obligated for this programme, worth about $800m in all, for mid-2006 ratification, when prelimi-

nary work will be completed. The upgrade programme will create a common avionics configuration for the Turkish air force's (THK) 102/15 F-16C/D Block 40 and 60/20 Block 50 aircraft, plus more modest modifications to its 34/9 F-16C/D Block 30s.

Turkish upgrades include Northrop Grumman APG-68(V)9 multi-mode radar, currently installed in new Advanced Block 50/52 F-16s; colour cockpit displays and recorders; new core avionics processors; VSI's Joint Helmet-Mounted Cueing System; Link 16 data-link; advanced IFF interrogator/transponder; integrated precision navigation; enhanced Self-Protection Electronic Warfare System (SPEWS II); and compatibility with several new weapons and targeting systems. As programme prime contractor, Lockheed Martin is providing all required upgrade facilities, logistic support, and training, for 37 THK Block 30, 76 Block 50, and four Block 40 kits, plus options for another 100 Block 40 kits. These will be installed in Turkey by TUSAS Aerospace Industries (TAI), with Lockheed Martin's help. Initial THK operational capability with the new configurations is scheduled for 2011.

More Seahawks planned
In June, Sikorsky Aircraft announced signature of an MoU with the Turkish government for 12 new S-70B Seahawk ASW/ASuW helicopters, plus options

for five more, for delivery from 2008. These will supplement eight S-70Bs received by the Turkish navy from 2002, while the Land Forces also operate more than 100 S-70A Black Hawk transport and utility helicopters.

Other international Seahawk customers include Australia, Greece, Japan, Spain and Thailand. Sikorsky is also under contract to deliver six S-70B Seahawk helicopters to Singapore from 2008.

Attack helicopter revived
Turkey's ambitious but long-delayed mid-1990s programme for planned procurement of an initial 30 attack helicopters was further extended in September, when the Ankara government postponed a critical bid deadline to 8 November. Some contenders, mainly in the US, had reportedly asked for a deadline extension, to discuss disputed clauses in the contract specifications.

Agreement had been reached in 2000 for joint Turkish production and assembly of 50 upgraded AH-1Zs attack helicopters, but four years of contract negotiations ended in 2004, following major price and licensing disputes. The SSM then opened a fresh international competition to meet increasingly urgent Turkish army requirements to reinforce its shrinking force of only seven Bell AH-1W Super Cobras, and about 20 earlier Cobra versions, for operations against resumed PKK terrorist attacks.

The SSM issued requests for proposals (RFPs) for the $385m programme for up to 50 helicopters in February, and revised their terms in May, to attract more potential contenders. These included the Bell Helicopter Textron AH-1Z, Boeing AH-64D, and an armed Sikorsky S-70 version, plus the Eurocopter Tiger, Italy's AgustaWestland A-129 Mangusta, Moscow Helicopter Plant's Mi-28, and South Africa's Denel Rooivalk.

JSOW procurement plans
In September, the US Defense Security Co-operation Agency notified Congress of a possible $35m Foreign Military Sale to Turkey of Raytheon AGM-154A/C

JSOW glide-bombs, plus associated equipment and services. The initial THK order will comprise 50 AGM-154A-1s with 500-lb BLU-111 Mk 82 blast fragmentation bomb warheads, and 54 AGM-154C Unitary JSOWs. Also included are four dummy air-training AGM-154A-1s; three captive flight-vehicles; and three simulation units.

United States

Airlift project accelerated
The US Army has accelerated development of its Future Cargo Aircraft (FCA) and now expects to make a decision on its $1.3 billion programme by June 2006. The FCA will ultimately replace the Army's fleet of 43 C-23B/B+ Sherpa airlifters in service with the Army National Guard. The service initially plans to purchase 22 aircraft and take options on a further 11. Proposals are expected from L-3 Communications/Alenia and Raytheon/EADS North America for the C-27J and C-295.

USAF acquires transports
The 319th Special Operations Squadron (SOS) was activated as part of the 16th Special Operations Wing (SOW) at Hurlburt Field, Florida on 1 October 2005. The squadron, which is assigned to the 16th Operations Group, will be equipped with six single-engined U-28A utility aircraft and approximately 45 personnel will be assigned.

The U-28A is a militarised version of the Swiss-built Pilatus PC-12/45 and is capable of operating from short, unimproved runways. The six former civil aircraft are currently registered to Sierra Nevada Corporation in Sparks, Nevada, which is likely equipping them with specialised equipment before delivering them to the USAF. The system integration contractor owns Aviation Resources DE Inc. (ARDI) in Hagerstown, Maryland, which specialises in aircraft repair and modification. The single-engined PC-12 is powered by a 895-kW (1,200-shp) Pratt & Whitney Canada PT6A-67B and can carry a payload of 1465 kg (3,229 lb). It is capable of being flown by one or two pilots.

The USAF ended operations at Rhein Main AB, Germany, when it held a ceremony that marked the closure of its airlift hub on 10 October. The base, which shared its runways with Frankfurt International Airport, had been the USAF's primary European hub since 1945. Rhein Main's operational career came to a close on 1 October when the airlift support mission transitioned to Ramstein AB and Spangdahlem AB. The base's final military and commercial missions departed earlier, on September 26 and 30 respectively. To mark the closure, C-17A 98-0049 operated by the 62nd Airlift Wing was named the Spirit of Rhein Main, and was unveiled during the deactivation ceremony.

Flying just before the show after a major overhaul, this is one of two Dakotas left in Greek service – the last serving C-47s in Europe. It is intended that this aircraft will continue to operate with a historic flight.

Highlight of the show from an international aspect was the NAMC (Nanchang) K-8E trainer from the Egyptian Air Force, which was supported by a C-130. It wears the colours of the national aerobatic display team. China is offering the K-8 to the Greeks as a Buckeye replacement.

The first Alenia C-27J Spartan for Greece was on display, as was a CL-415 configured for SAR duties.

Typhoons came from Spain (above) and Germany. A Rafale and Saab Gripens also attended.

Wearing civil test registration, this is the first EMBRAER EMB-145 AEW for Greece.

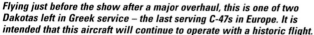

AIR ARM REVIEW

Australia

Army helicopter changes

The army will soon release a request for tender associated with a plan to commercialise its rotary-wing flying training programme. The service hopes to select a contractor in mid-2006 and commence training at the Army Aviation Centre, at Oakey Army Air Field, Queensland in 2007. The programme will continue to use the service's Bell 206B-1s, which will remain in service until 2015. Additional Kiowas will become available as the aircraft assigned to 161 and 162 Squadrons are replaced by the Tiger ARH in 2007/8. The winning contractor will also conduct maintenance training and loadmaster and search and rescue instruction. The School of Army Aviation will continue to conduct type transition, armament and instruction training.

The Army's 162 Reconnaissance Squadron made its last flight at Laverack Barracks on 13 October prior to relocating to its new home at Darwin. The squadron had been based in Townsville since 1971.

The Army recently announced plans to relocate 200 personnel and 12 S-70A Black Hawk helicopters from Townsville, Queensland, to Holsworthy Barracks near Sydney. The move, which will allow Australia's Special Forces to better respond to terrorist threats on the country's east coast, will take place in late 2006.

Canada

Transport Plan Formulated

Air Command is developing plans for the purchase of new transport and search & rescue (SAR) aircraft and heavy-lift helicopters at a cost of $5.1 billion. The proposal includes 15-20 Lockheed Martin C-130Js and 15 new fixed-wing search and rescue aircraft, which would replace the CAF's fleet of CC-130E/H airlifters and CC-115 Buffalo SAR aircraft. The plan also

includes the purchase of 20 CH-47 Chinook helicopters. Under the proposal the acquisition would be spread over the next five to 10 years. Canada already has formulated plans to replace the Buffalos and is believed to favour the Alenia C-27J for this mission.

SAR shuffle

1 Canadian Air Division has temporarily transferred two CH-146 Griffon helicopters to 8 Wing/CFB Trenton where they will stand in for the wing's CH-149 Cormorants. The Griffons are expected to remain at Trenton until spring 2006 but could remain there longer. The CH-149s have suffered poor availability as the result of cracks in the tail rotor assembly and shortages of spares main gearboxes and main rotor heads.

Czech Republic

MiG-21s retired

Following recent initial deliveries of 14 leased JAS 39 Gripens, the VSACR retired the last of its MiG-21MFs in a formal tour by nine of these Russian fighters over Namest nad Oslavou, Pardubice and Pilsen air bases on 12 July. In all, 467 MiG-21s were delivered to the former Czechoslovakia from 1962, at least 100 being lost in accidents, together with 42 pilots. Over 20 ex-VSACR MiG-21s, including two-seat MiG-21UMs, have already been sold to unspecified customers, and offers invited for another 11.

The arrival at Caslav air base on 31 August of two two-seat JAS 39D Gripens from Sweden followed six JAS 39Cs in April, and six more in August, to complete VSACR deliveries. On lease until at least 2015, these had

then flown for more than 300 hours, and Czech Gripen pilot and technician training was continuing with F7 Såtenäs Wing in Sweden.

France

2006 defence budget details

Military budget proposals for 2006 presented in September by Defence Minister Michèle Alliot-Marie rose by 3.3 per cent over the previous year, to Euro47bn ($56.54bn), and represent over 2 per cent of GDP. Euro16bn was allocated for equipment procurement, having increased from Euro14.9bn in 2004, and Euro250m for external operations, compared with only Euro100m in the previous year. R&D funding was also increased from Euro1.38bn to Euro1.47bn in 2006.

The French navy gets priority from 2006 equipment funding, notably Euro926m towards initial development of the new PA2 conventionally-powered aircraft-carrier programme. This is the first official go-ahead for this programme, for which joint design studies were launched in mid-2004 by DCN and Thales Navale, with so-far unspecified co-operation to achieve 80-90 per cent commonality with Britain's CVF project. Additional specific 2006 budget procurement includes another 14 Dassault Rafale combat aircraft, with new weapons systems for these and AdA Mirage 2000s; 160 ship-launched MBDA SCALP-EG cruise missiles; 272 MBDA MICA, plus another 130 planned, medium-range AAMs; an initial batch of 16 SAGEM AASM precision-guided attack weapons; new-generation submarine-launched M51 ballistic

missiles; and nuclear air-strike ASMPAs.

New Rafale funding increases total French orders to 134, and production through 2011. Formation of the first operational AdA Rafale unit, Escadrille de Chasse 1/7 was scheduled in late 2005 at St Dizier, to include initial deliveries of enhanced multi-role single-seat Rafale C F2s, following their operational evaluation at Mont-de-Marsan.

AdA DC-8s retired

Replacement was announced in September of two veteran AdA McDonnell Douglas DC-8-72CFs operated since 1982 for long-range military personnel and light cargo transport, by two four CFM56 turbofan-powered Airbus A340-200s. These are reportedly being acquired from Austrian Airlines on an initial five-year lease, with options for another four years, through TAP Portugal, which will also undertake their maintenance and technical support.

India

Training squadron activated

The navy activated Indian Naval Air Squadron (INAS) 552 to train pilots for its Sea Harrier fleet recently. Its Sea Harrier pilots previously converted to the type in the UK, and then undertook operational training with the Sea Harrier Operational Flying Training Unit (SHOFTU) in India. The latter unit is part of INAS 551.

Italy

737AEW&C considered

The Ministry of Defence has begun discussions with Boeing regarding the

This Croatian AF MiG-21UMD received these patriotic markings to lead the flypast at the celebrations marking the 10th anniversary of Operation Oluja (Storm), in which Croatian forces liberated the Serb-held village of Knin and the surrounding area.

Fresh from overhaul and carrying R-73 missiles, this Su-25UB wears the markings of the Georgian air force. 'Frogfoot' work is undertaken at the TAM factory in Tbilisi that manufactured Su-25s and produced the Su-25KM Scorpion upgrade in conjunction with Elbit of Israel.

purchase of the B737 airborne early warning and control (AEW&C) system. The project includes three aircraft, with an option for a fourth and related support systems. The sale will likely be structured similarly to the Italy's KC-767A tanker purchase that included 100 per cent offsets for local industry. Italy has already shown interest in joining the US Navy's P-8 multi mission maritime aircraft (MMA) programme and could purchase six to eight aircraft. The P-8A and 737 AEW&C share the same B737-700 airframe, which would offer the air arm considerable commonality.

Russia

Budget boost for military forces
Substantial increases of 22 per cent in 2006 defence spending were announced in August, to help modernise Russia's armed forces. Defence Minister Sergei Ivanov confirmed draft military budget allocations of Rbls668.3bn ($23.7bn) in 2005, or about 2.75 per cent of projected GNP. Some Rbls225bn ($7.9bn) of this total will be spent on weapons and equipment, or about $1.3bn more than in 2004. The proposed budget represents a third consecutive year of significant real-terms increase in Russian defence spending.

However, following the 1998 merger of Russia's Air Defence Forces and Air Forces into a single Russian Federation air force (RFAF), its current chief of staff, Colonel-General Boris Cheltsov, recently complained of "tangible reductions in air defence and combat capabilities, as a result of systematic underfunding". He said that in 2004, combat pilots could fly only 20-25 per cent of their

required annual totals. Increases in combat training flight- hours were not expected in 2006, following further reductions in RFAF fuel allocations.

Nevertheless, Cheltsov said that the RFAF now possessed sufficient potential to repel aggression on a local or regional scale, with both conventional and nuclear weapons by strategic aviation and general-purpose forces. RFAF budget benefits will include funding for another 17 Sukhoi Su-27SM upgrades, compared with 11 in 2005.

New smart weapons now being developed include a conventional modification of the nuclear Raduga Kh-55 or AS-15 'Kent' Tomahawk-type cruise-missile, designated Kh-555, to arm strategic attack aircraft such as the Tupolev Tu-95 or Tu-160. Its range is over 2500 km (1,349 nm), and it uses first-generation cruise-missile targeting, which includes a radar-based digital terrain-matching system, independent from a satellite-based navigation system.

Funding problems have also delayed the start of the RFAF's PAK FA next-generation combat aircraft development until about 2008, according to Mikhail Pogosyan, CEO of Sukhoi Aviation Holding Company. Speaking at the Moscow MAKS air show at Zhukovsky in August, Pogosyan said that financial approval was still awaited for the PAK FA programme, which was already unlikely to start production before 2015.

NPO Saturn and Ufa, with about 10 per cent state assistance, have reportedly invested about Rbls1bn ($35.43m) in developing five new prototype AL-31-derived engines to power the PAK FA, starting with flight trials in a Sukhoi Su-30, and an advanced radar is being proposed by Phazotron-NIIR.

One of Taiwan's F-16A Block 20s is seen landing. It is fully armed with wingtip AIM-120s and live AIM-9s on the outer underwing pylons.

Singapore

Hueys stand down
The Republic of Singapore Air Force's 120 Squadron conducted its last UH-1H flight at Sembawang AB on 11 July. Formed in 1969, 120 Squadron, which is the RSAF's oldest operational unit, continues to operate five Hueys from Brunei International Airport as part of a permanent detachment, however these will be withdrawn later this year.

United States

EW Hornet stationing plan
The US Navy has formally announced plans to station the electronic warfare variant of the Boeing Super Hornet at NAS Whidbey Island, Washington. Developed from the two-seat F/A-18F, the EA-18G will be a direct replacement for the navy's carrier-based EA-6B Prowler. The EA-18G, unofficially called the 'Growler', will begin replacing the Prowler when the first of 10 squadrons at Whidbey transitions in 2008. The Navy's current plans include the purchase of 57 EA-18Gs and its proposed budget for 2006 includes $310 million for the first four examples.

Aries II leaves Europe
Naval Station Rota, Spain marked the departure of the last EP-3E operated by VQ-2 during ceremonies held on 19 September 2005. The squadron's final aircraft departed for its new home at NAS Whidbey Island, Washington, on 21 September.

Viking changes
Sea control squadron VS-31 was transferred from CVW-7 to CVW-17 where it replaced VS-30, which will be deactivated in December 2005. Both squadrons are stationed at NAS Jacksonville, Florida. The move follows the assignment of two Super Hornet squadrons to CVW-7.

On 30 September Commander, Sea Control Wing Atlantic at NAS Jacksonville, assumed control of VS-33

and VS-41 at NAS North Island, California. The move was the result of the disestablishment of Commander, Sea Control Wing Pacific, at North Island, which had been responsible for all of the Pacific Fleet S-3B squadrons. The wing was originally established on 23 April 1993 when its predecessor Antisubmarine Warfare Wing Pacific was disestablished.

In related news Naval Air Depot North Island completed its last S-3B overhaul on 30 September when it delivered the last Viking to undergo the integrated maintenance concept (IMC) programme. Implemented in 2000, the IMC replaced the earlier scheduled depot level maintenance (SDLM) programme. Between 1994, when North Island assumed the responsibility for the Viking from NADEP Alameda, and 2005 the depot completed 97 S-3Bs through SDLMs and 112 aircraft underwent the IMC.

UDP plans
Although the US Navy indicates that VMFA-115 will work up and deploy with CVW-3, USMC sources indicate that the squadron will be detached and will deploy to Japan in support of the unit deployment program (UDP). In related news, VFA-97 will make one further UDP deployment to Japan before being deactivated as the second of three navy squadrons that will stand down as a result of the USN/USMC TACAIR Integration programme. VFA-94 is currently scheduled to join the UDP deployment schedule.

ICAP III fielded
The 'Cougars' of electronic attack squadron VAQ-139 have received the last of four updated EA-6Bs and the squadron is preparing to make its first deployment with the Improved Capability III (ICAP III) version of the Prowler aboard the USS *John C. Stennis* (CVN 74) in January 2006. In related news, the 'Rooks' of VAQ-137 received the first of four ICAP III Prowlers on 1 September. Northrop Grumman is currently under contract to update 10 EA-6Bs to ICAP III configuration, and the final two examples will be assigned to the 'Vikings' of VAQ-129 in support of training requirements. With the exception of a single Naval Reserve squadron all of the navy's Prowlers are home ported at NAS Whidbey Island, Washington.

VIP unit deactivates
An era came to a close at NAS Jacksonville, on 23 September when patrol squadron VP-30 transport (VR) detachment was deactivated. Established in 1977, the VR detachment had conducted global VIP transport missions in support of the Chief of Naval Operations, senior officers and dignitaries using specially configured P-3 Orions. The unit flew more than 32,000

hours and, although it originally had three aircraft, just one VP-3A remained assigned to the VR-Det when it stood down.

Sea King fleet dwindles

The last UH-3Hs operating in Europe were retired when helicopter combat support squadron HC-2 Detachment 1 was deactivated in Naples, Italy, on 15 November. MH-60S Knighthawks operated by Detachment 7 of helicopter sea combat squadron HSC-28 arrived in Naples during October and have assumed the duties previously assigned to the Sea Kings. The 'Ghostriders' of HC-2 Detachment 1 concluded its last major underway period aboard USS *Mount Whitney* (LCC 20) on 14 October. The helicopter was embarked upon the Sixth Fleet command ship which deployed to the Thyrrenian Sea near Sardinia in support of NATO exercise Destined Glory 2005, on 29 September. During the exercise the helicopter detachment and its Sea King made 23 flights carrying 155 passengers and 7,500 lb of cargo. The unit completed a final underway period on 22 October and continued to support operational taskings with its last remaining UH-3H until early November, when the Sea King was disassembled and shipped back to the US.

The last UH-3Hs assigned to HC-2 Detachment 2 at NSA Manama, Bahrain, have also been withdrawn and the detachment deactivated. Known as the 'Desert Ducks', Det. 2 operated in support of the Naval Forces Central Command since arriving in the area of operations in 1976. The 'Desert Ducks' were replaced by a detachment from helicopter sea combat squadron HSC-26. The new detachment, which operates the MH-60S Knighthawk, assumed the responsibility for supporting the Fifth Fleet on 14 September. It has taken the nickname 'Desert Hawks'. Until recently HC-2 operated 12 UH-3Hs that included two in Italy, four with Detachment 2 in Bahrain and six at Chambers Field/Naval Station Norfolk.

The last Sikorsky UH-3H helicopter assigned to NAS Whidbey Island, Washington, departed for the Evergreen Aviation Museum in McMinnville, Oregon, on 28 September. Known as 'Firewood 6', the Sea King was one of three examples operated by the station's air operations department. The other two aircraft departed for storage at the Aerospace Maintenance and Regeneration Center (AMARC) at Davis Monthan AFB, Arizona, on September 13 and 26, respectively. The type first replaced the HH-46 Sea Knight as the air station's search and rescue (SAR) aircraft in July 1981. Since that time the Sea Kings flew 1,228 missions. A pair of Sikorsky MH-60S Knighthawks has replaced the Sea Kings and the new helicopter flew its first SAR mission on 26 September.

Globemaster III updates

The 13th and final C-17A (serial 04-4137) arrived at McGuire AFB, New Jersey, recently. The aircraft is assigned to the 305th Air Mobility Wing's (AMW) 6th Airlift Squadron (AS) but is also operated by the 514th AMW's 732nd AS. The Air Force Reserve Command

wing and squadron operate as associates to the 305th/6th.

Boeing also recently delivered the first C-17A to the Air Force Reserve Command at March ARB, California. Named *Spirit of California*, the aircraft was formally accepted by the USAF at the contractor's Long Beach facility before being flown to the Riverside base. It is assigned to the 452nd AMW's 729th AS, which will have eight Globemaster IIIs assigned by January 2006.

Carrier retired

The amphibious assault ship USS *Belleau Wood* (LHA 3) was decommissioned at Naval Station San Diego, California, on 13 October. The ship was one of five 'Tarawa'-class assault ships and had been named in honour of the historic World War I battle in France in which Marines stopped Germany's last major offensive. It was the second ship to bear the name and followed a light aircraft-carrier (CVL) commissioned during World War II.

Army National Guard changes

The first of six CH-47Ds destined for the Nebraska Army National Guard recently arrived at Central Nebraska Regional Airport in Grand Island. The Chinook was assigned to B Company, 2-135th General Aviation Support Battalion. The initial crews will began training on the heavy lift helicopter in September. Construction of a new $20 million aviation support facility at Grand Island is expected to begin in late 2006.

The Florida Army National Guard's 1-111th General Support Aviation Battalion recently took delivery of the first of six CH-47Ds at the Cecil Commerce Center in Jacksonville. The battalion's three companies, which previously operated the AH-64A, are transitioning to new missions and aircraft including the Chinook and UH-60

Blackhawks. The latter will serve both in general support and air ambulance roles.

Air National Guard changes

In preparation for its upcoming transition to the C-17A, the Hawaii Air National Guard's 154th Wing at Hickam AFB, recently reduced its C-130H fleet from five aircraft to just one. Serial 90-1058 will be the final aircraft assigned to the 204th Airlift Squadron. Although the squadron will convert to the Globemaster III, the new aircraft will be 'owned' by the 15th Airlift Wing's 535th AS and the 204th will become an associate unit to the active-duty squadron. The arrival of the first of eight Globemaster IIIs is scheduled for February 2006.

USAF changes

The 7th Combat Training Squadron (CTS) was redesignated the 7th Fighter Squadron (FS) recently. The 7th, which is located at Holloman AFB, New Mexico, is the formal training unit (FTU) for the F-117A. Besides operating the Nighthawk, the squadron uses

T-38As to carry out its mission.

The 57th Adversary Tactics Group (ATG) was activated on 15 September 2005 at Nellis AFB, Nevada, and was assigned to the 57th Wing. The unit assumed control of the 64th Aggressor Squadron (AGRS), which was reassigned from the 57th Operations Group, and the 65th AGRS which was activated on the same date. Whereas the 64th AGRS flies F-16C/Ds, the new squadron will utilise around eight F-15C/Ds in the aggressor role.

Squadron relocating

As part of US Navy's ongoing transformation efforts, helicopter combat support squadron HC-4, has begun relocating from NAS Sigonella, Italy, to Chambers Field/Naval Station Norfolk, Virginia. The move, which involves five MH-53E Sea Stallion helicopters and 300 sailors, will be completed in summer 2006. The squadron will continue to provide combat support for deployed units through the deployment of rotational detachments.

NASA WB-57 deployment

At 18:14 local on 11 October 2005, WB-57F N928NA (ex 63-13298) touched down at RAF Mildenhall, Suffolk, completing a journey from Ellington, Texas, via Goose Bay in Canada. RAF Mildenhall was selected from among the USAF bases in the United Kingdom as the deployment base of the aircraft because it has plentiful supplies of JP-8 fuel and liquid oxygen (LOX), and adequate runway length. Diversion bases included RAF Lakenheath (if the crosswind over Mildenhall's runway was over 15 knots – the limit for the aircraft) or RAF Fairford. The deployment involved two pilots, up to three System Equipment Operators (SEOs, responsible for the aircraft's payload), six maintenance personnel for the WB-57F, and four other NASA support personnel. The aircraft's deployment to the United Kingdom – only the second time the long-winged variant of the American Canberra had landed in the country – was in connection with the collection of cosmic dust particles, but the opportunity was also taken to acclimatise NASA pilots with flight within the European air traffic communications envi-

ronment. Cosmic dust collectors were attached under the wing, mounted within a faired pylon-like structure. Four collectors are mounted on each structure, retracted inside until the aircraft reaches its operational height, which for the Mildenhall-based missions was 48,500 ft (14783m), low enough not to require the crew to wear a pressure suit. The SEO flicks a switch to deploy the collectors to gather the dust, and later retracts them to stop the samples being contaminated during the decent. On landing the samples are retrieved, labelled and shipped to Houston, Texas for analysis. Four flights were undertaken during the deployment, each of between 4 and 5 hours duration, with the WB-57F returning to Texas on 22 October.

David Willis

The cosmic dust collectors are housed in this installation under the WB-57F's wing. The aircraft is seen in the Mildenhall hangar above.

An F-16CJ Block 50 from the 23rd Expeditionary Fighter Squadron from Spangdahlem, Germany, taxis at Sialuiai air base in Lithuania. The 23rd EFS was operating as part of NATO's Baltic Air Policing mission, providing airspace security for the Baltic states of Estonia, Latvia and Lithuania.

Althea Dragons

Belgian helicopters in Bosnia

From early January 2005, military personnel from various European Union (EU) countries deployed to Bosnia-Hercegovina to launch the Union's ambitious Operation Althea. This operation – which relieves NATO's long-lasting SFOR (Stabilisation Forces) involvement in this theatre of operations – aims to assist, support and boost the Bosnian government institutions in creating a more stable, safer, self-supporting and self-determining structure.

A Belgian Althea crew poses in front of the single A109HA deployed to Bosnia. NVGs are routinely worn, and for operational missions small arms are carried. Both the helicopter and the crew's flightsuits are emblazoned with the EU flag.

To provide flexible transportation to the three EU Bosnia-based Multinational Task Forces (MNTF), a select group of multi-role liaison and transport helicopters was put together by seven European nations: Belgium, Italy, Greece, United Kingdom, Czech Republic, Romania and Switzerland. Some of these nations, and their national armed forces, had already gained substantial Balkans experience during previous UN/NATO-backed missions in Bosnia and/or Kosovo.

One of the first governments to respond to the EU's request to send EUFOR support assets to Bosnia-Hercegovina was that of Belgium. The Council of Ministers decided on 23 December 2004 to send four Agusta A109 multi-role helicopters of the Belgian Air Component. For almost a decade, the Belgian military helicopter pilots, formerly belonging to the Belgian Army Light Aviation, regularly deployed to the Balkans (Bosnia, Kosovo, Macedonia and Albania) as a direct result of the peacekeeping- and humanitarian relief-orientated defence policy of the Belgian Federal Ministry of Defence.

The Belgian helicopter contingent was assigned to Multinational Task Force (North), operating in the northeast parts of the Bosniac and Croat Federation and the Republic of Srpska. Composed of 12 European Union and other European participating countries, the MNTF(North) has its headquarters located at Camp Eagle Base near Tuzla. At Tuzla the Belgian heli-detachment – aka BE HeloDet – was integrated within the Multinational Aviation Company (AvnCoy), co-operating with Greek and Czech contingents flying CH-47D Chinook (Greek Army Aviation) and Mi-17 'Hip' (Czech Air Force) helicopters. In addition to Camp Eagle Base, housing some 750 soldiers, MNTF(North) ground forces also staff 15 Liaison Observation Team (LOT) houses all around its area of operations to act as points of contact between EUFOR and the local population.

Unwilling to jeopardise the smooth deployment of EUFOR forces in Bosnia, as needed to launch the Union's first large-scale operation without delay, the Belgian Air Component (BAC) decided not to fly the helicopters under their own power to

Two of the BE HeloDet's four Agustas are 'slick' A109HOs used for general transport, liaison and training purposes. They regularly fly key EUFOR personnel around the region, often negotiating low cloud and bad visibility in the course of such missions.

Tuzla. A long-range, medium-level navigation through Central Europe in winter may have caused some unfortunate weather-induced stopovers or routing delays, and would have had a negative impact on the maintenance and overhaul cycles of the various helicopters.

At the same time, the BAC lacked the necessary sea or airlift capability to transport the Agusta A109s to Tuzla. During previous peacekeeping operations, A109s had been airlifted one at a time by Belgian Lockheed C-130H Hercules transport aircraft using purpose-built sleds to the out-of-area location. However, the loading and unloading of the dismantled A109 helicopter into the (relatively) confined C-130H Hercules cargo bay always proved to be a time-consuming and meticulous operation. In late 2004, the Belgian C-130H force of 11 aircraft was hit by mechanical problems and various overseas transport commitments.

An A109HO overflies the town of Srebrenica. Much of Bosnia consists of wooded mountains, laced with winding valleys in which the majority of towns and villages are situated. The task of tracking organised criminal gangs in such terrain is challenging.

Consequently, the Belgian Armed Forces were 'forced' to contact the European Airlift Centre (EAC), located in Eindhoven, to look for suitable transport aircraft to airlift the four Agusta A109 helicopters, pilots, support personnel, equipment and spare parts to Tuzla. Since July 2004 the multinational EAC has co-ordinated all requests for aerial transportation from its eight European member nations (Belgium, Norway, France, United Kingdom, Spain, Germany, Italy and the Netherlands).

To be able to adequately allocate all available air transport assets, member air forces need to offer their 'free' transports (and air refuelling tankers) to the EAC. Using the EAC, a 'long-haul' Boeing C-17A Globemaster of No. 99 Squadron, Royal Air Force, landed at Liege-Bierset airport on 7 January 2005 to airlift two dismantled A109s and a vast stock of support equipment to Tuzla. The two remaining helicopters were airlifted by the same method, considerably reducing the deployment time of the Belgian helicopter detachment. Additional support equipment was transported by a French and Belgian C-130H Hercules, and by a Luftwaffe Transall C-160D.

Based alongside a small detachment of US Army UH-60A Black Hawk helicopters, operating independently of EUFOR under a bilateral US-Bosnian agreement, Belgian groundcrew erected a large 'Veldeman' tent, capable of housing all four helicopters during night-time and well-equipped to perform some minor maintenance and overhaul on the A109s.

Initially the Belgian Government committed itself and the Belgian Air Component Helicopter Wing to EUFOR's Operation Althea for one year. Already highly involved during previous Balkans

A crew prepares the 'Angel' medevac A109HO for its day's alert. In addition to the two pilots, the 'Angel' also carries a doctor and paramedic. Note the 'slump' wheel pads used to prevent the aircraft from sinking when landing on soft ground.

operations, the Helicopter Wing planned frequent two- to four-month rotations for the air- and groundcrew of the Wing's three A109 squadrons. Unwilling to overstretch the Wing's personnel in the near future, no formal decision has been taken yet on the prolonged presence of Belgian helicopter crew in Bosnia.

UAV operations

From July 2005, the Belgian helicopter contingent was joined at Tuzla by a detachment of BAC's No. 80 Squadron, operating Israeli-built IAI B-Hunter UAV unmanned reconnaissance aircraft. Although based at Tuzla, this contingent operates its six UAVs and three ground control stations independently of the MNTF(North), and reports directly to COMEUFOR (Commander EUFOR), based at Sarajevo-Butmir. Due to the need to install the laser-guided automatic landing system needed to land the UAV at the end of the runway, the operational use of Sarajevo's Butmir

International Airport, adjacent to EUFOR's compound, was not a viable option. The UAVs are used – weather permitting – on observation missions all over Bosnia.

The participation of BAC's Helicopter Wing in Operation Althea also includes several first-time experiences. For the first time the Belgian helicopter crews are embedded into a multinational aviation company (AvnCoy), initially commanded by a Greek Army Aviation major. From September 2005, the AvnCoy command was transferred to the Belgian detachment commander after the departure of the Greek Aviation contingent. Furthermore, all logistical airfield operations at Tuzla are organised by US civil-contracted companies, being responsible for all local air traffic control, airfield security and refuelling of all Althea helicopters. Finally, all non-flying related logistical support for EUFOR contingents is co-ordinated by the Swedish-run Multinational Integrated Logistical Unit.

The detachment's A109HA (distinguished by the HeliTOW sight above the cockpit) is fitted with an ingenious arrangement in the rear cabin comprising a laptop connected to the HeliTOW system with a hand controller. The operator can use the former missile sight to record reconnaissance imagery.

Although four Agusta A109s, with anti-tank armament removed, are based at Tuzla, only three helicopters are permanently assigned to the MNTF(N). One helicopter, fully equipped for medical evacuation missions, is kept on a 24-hour medical standby at Tuzla. The contingent's five helicopter crews rotate daily to stand alert for medical operations in the MNTF(N)'s Area of Responsibility (AOR), joined by a doctor and assistant medic.

During daytime the ANGEL 01 medevac helicopter needs to be airborne within 30 minutes of receiving a distress call, in the form a 'nine-line' report on the nature and location of the person in need. Once airborne – especially during night-time missions – another pilot will man the contingent's ops centre in the Tuzla barracks. Daily at noon, the new standby crew performs the preflight checks on the medevac helicopter, which is parked in the large 'Veldeman' tent. The first 'live' medevac mission was flown on 2 August 2005 in support of an Austrian EUFOR soldier.

At night, the medevac crew (with night-vision goggles) needs to be airborne within one hour. When used for occasional training or qualification flights by the standby crew, the medevac helicopter needs to remain within five minutes' flying time of the airbase to allow quick refuelling if needed for a real emergency.

Since large parts of Bosnia-Herzegovina are unfortunately still littered with countless known (and unknown) minefields – a sad reminder of the Yugoslavian wars of the 1990s – Althea medevac helicopters are only allowed to land on concrete or 'mine-sweept' helicopter landings sites (HLS). Soon after the start of Belgian Althea missions in January 2005, all Belgian A109s, including the medevac helicopter, were equipped with Swiss-made, wheel-mounted 'slump protection pads' to prevent the helicopter's landing gear sinking in the soft mud or snow of the Bosnian countryside. During summertime, the 'slumps' are still mounted on the medevac 109 due to the extra weight of the medical hardware, doctor and assistant medic during landing operations.

Dragons

The BE HeliDet's three remaining A109 helicopters, using the radio callsigns DRAGON 05-07, are used for a variety of support missions. One of the helicopters is permanently assigned to MNTF(N) and kept on one-hour 'duty ops' standby for liaison and, if needed, observation. Belgian pilots frequently fly EUFOR officials and military personnel to the various EUFOR command sites (especially Sarajevo-Butmir). To prevent bad weather (low cloudbase and reduced visibility) from cancelling these liaison missions, the aircrew quickly 'mapped' terrain-following bad-weather routings through the various valleys and mountain ridges around Tuzla. These low-level routings are frequently practised by the Belgian crews, using the curvy main road between Tuzla and Sarajevo and a winding single-track westbound railroad track from Tuzla to Banja Luka as visual references.

As the detachment is permitted to use 70 hours of the official monthly allocation of 120 for crew-training, one of the remaining A109s is permanently used for pilot-training and qualification flights. Almost daily, A109 crews practise various air traffic patterns and procedures, low-level navigation, landings on the countless helicopter landings sites scattered around the MNTF(N)'s AOR, and (formation) night flying using NVGs.

Finally, the fourth A109 is kept as a spare to allow for periodic maintenance, or as a back-up during unforeseen helicopter unavailability. Spare parts are airlifted from Belgium to Tuzla using the two-weekly supply flight by BAC Airbus A310 or C-130H Hercules transport aircraft, or arrive via the two-weekly Belgian 'Tuzla Express' truck road convoy.

In June 2005, replacement aircrew flew three new A109s from Bierset to Tuzla, making a night-stop at Ljubljana (Slovenia). A week after the arrival of these 'new' A109s, three of the initial helicopters were flown home by the pilots ending their two-month tour of duty. The specially equipped medevac helicopter (s/n H-04) remained at Tuzla, having flown less then its transport colleagues and not yet nearing its next maintenance cycle.

As pre-planned in the Belgian MoD 'Althea roadmap', pilots and groundcrew rotate on a two- to four-month basis to Bosnia. First unit to deploy from Liege-Bierset to Tuzla in January 2005 was the BAC's No. 16 Squadron MRH (Multi Role Helicopter). This helicopter squadron was only re-activated on 1 October 2004 as an operational A109 unit, having been in the past the Army Light Aviation liaison helicopter unit, flying Alouette II lightweight helicopters. In April 2005 No. 16 MRH squadron, having swapped its pilots after months in theatre, was replaced by No. 18 MRH Squadron. No. 17 MRH squadron deployed to Camp Eagle Base in July 2005 for its four-month 'tour of duty', planning to return to Liege-Bierset at the end of November 2005. Pending a possible but yet undecided extension of the Belgian BE HeloDet presence, No. 16 MRH will staff the final detachment and will 'ferry' the A109s home.

Similar to all Althea helicopter crews, Belgian aircrew have to follow an intense preliminary three-week initiation course before deploying to Bosnia. Most importantly, due to the presence of countless minefields in their area of operations, aircrew are given a mine-awareness course, instructed by the Armed Forces Engineering Corps, and a first-aid course to assist their crewmembers in case of accident.

Since all aircrew wear a pistol for self-defence while operating in Bosnia, and during 'tactical' missions an armed gunner/crewmember with a 5.56-mm FNC machine-gun will join the two pilots in the A109, a gunnery course is given to all aircrew. During four days all personnel involved will learn or be re-qualified to shoot the pistol and FNC machine-gun in 'classic' shooting conditions, and also learn and train in self-defence 'responsive' firing techniques, when confronted with 'bad guys' mingled with innocent bystanders.

Other Althea-related subjects include identification of military vehicles, tanks, uniforms worn by all military forces in the theatre and, of course, the rank markings of all European Althea contingents. A two-hour general and intelligence briefing focusses on the geo-policital, historical and religious characteristics of Bosnia-Herzegovina.

All pilots will spend countless hours in the A109 flight simulator at Liege-Bierset to practise flying in visual and IMC conditions, using GPS for pinpoint navigation. Local approach and landing procedures at Tuzla will also be practised in the simulator. Since the Belgian HeliWing frequently deploys to the French Army's mountain flying training centre at Saillagouse, located in the Pyrenées, no preliminary mountain flying is organised prior to an Althea rotation.

Having arrived at Tuzla, the newcomers are familiarised by the 'homebound' contingent by flying several mixed-crew flights in the Area of Responsibility. Pilots and groundcrew rotate every two months. The BE HeloDet's commander however, also being the MRH squadron commander, will stay at Tuzla for four months to guarantee smooth rotation transitions.

For COMEUFOR the fight against organised crime and corruption, unfortunately widespread within Bosnia-Hercegovina, and the maintenance of stability for its multi-ethnic population are the two key objectives of the Operation Althea. To fight these – often cross-border – criminal networks and their illegal activities, which range from illegal fuel and arms traffic, heroin-smuggling and/or illegal logging of trees, EUFOR

regularly uses the multinational helicopter fleet for troop transport and observation.

During these so-called Spring Clean operations, which are often flown at night, the Belgian crews provide various types of support to the ground-troops. Fully able to operate at low level at night wearing night vision goggles, the Belgian A109 might typically be used to lead Czech Mi-17 'Hip' helicopter crews, operating in complete radio silence and without NVGs, who are airlifting EUFOR's tactical reserve forces from Poland, Portugal and Turkey to the landing zone. Once safely on the ground, a Czech groundcrew member will leave the Belgian A109 and mark the landing area for the incoming Czech helicopters using lights. Unfortunately, during insertion simulations, the NVG-capable Greek CH-47D Chinooks proved to be too big and difficult to safely operate in the various confined helicopter landings sites.

Makeshift reconnaissance

The operational anti-tank warfare and target observation background of the experienced A109 air/groundcrew, and their sense of practical creativity, has enabled the MNTF(North) to put a low-key but very effective aerial observation asset into action. The gyro-stabilised, roof-mounted SAAB HeliTOW optical sight, developed to detect and target enemy tanks in Cold War scenarios, is nowadays used to observe the movements of people or vehicles on the ground from a distant, stealthy and out-of-sight position. Based on the BT-49 TOW-simulator, purchased as part of the Agusta A109 package deal in the late 1980s and capable of simulating anti-tank missile shots, the images from the HeliTOW sight are wired to a standard laptop PC mounted in the back of the helicopter. The stick to control the sight has also been mounted in the back to allow the crewmember to operate the daylight and infra-red optics of the HeliTOW.

Initially the BE HeloDet experienced some minor problems in downloading the images. From

July 2005, these problems were resolved by switching to digital recording of the observation pictures or video. After landing, the pictures and additional audio information from the aircrew are screened by the contingent's and MNTF(N)'s intelligence officers. Since only 28 of the 46 purchased A109B's (BA = Belgian Army) were dedicated A109HA anti-tank helicopters, fully equipped and wired to shoot anti-tank missiles during night-time using infra-red optics and carry BT-49 simulators for training, the PC can only by installed in one of the four Althea 109s (s/n H-43). The remaining 18 A109BAs, including Althea's other three airframes, are 'slick' A109HO observation helicopters only capable of operating the daylight variant of the HeliTOW sight.

Although the A109s are not equipped to download in real-time the video imagery or pictures to the operations centre in Tuzla, these low-key but effective observation platforms – very difficult to detect by the 'bad guys' – often generate better

A Belgian A109 lands back at its Tuzla base. The sight of such helicopters, as well as the ground presence, is a reassuring sight to Bosnians as they continue to fight the organised crime that has proliferated in the region since the end of hostilities.

results than the UAVs dedicated to the role. Although they are better equipped for aerial surveillance, the UAVs at Tuzla are frequently hampered by adverse weather conditions. The expanded observation capability of the A109 provides quick reaction time and high-resolution video footage usable for EUFOR operations, and in weathers when other platforms cannot perform.

Stefan Degraef and Edwin Borremans

Below this A109 is the memorial to the 7,000 Bosnian Moslems massacred at Srebrenica and nearby Potocari. In July 2005, on the 10th anniversary of the slaughter, 15,000 Bosnians retraced the steps of the infamous 'March of Death'. As part of Operation Watchful Eye, A109s and B-Hunters provided road-recce and crowd control cover to ensure the event passed off smoothly.

Armenia's Air Force

Hayastani Otayin Udjer

Photographed by Robin Polderman

Surrounded by Azerbaijan, Georgia, Iran and Turkey, the small mountainous state of Armenia takes up 18,000 sq miles (46616 km²) of the South Caucasus. Since its independence from the Soviet Union on 21 September 1991, the country has been trying to build a credible defence. Between 1991 and 1994 Armenia and Azerbaijan fought a bloody war over the Nagorno-Karabakh region. With an uneasy ceasefire in place the air force of the young republic trains for the worst.

Above: One of Armenia's most famous sons is Artem Mikoyan, progenitor of the MiG series. This MiG-19S is on display outside the Victory Park museum in Yerevan as a tribute to the fighter designer.

Left: Bort 83 and 85 are two of the nine Su-25Ks acquired from Slovakia in 2004. Bort 83 has just returned from a weapons training sortie. The UB-32 rocket pods have been removed and the aircraft will soon be towed away from the Gyumri flightline. The 'Frogfoot' is used for attack missions: air defence of the region is entrusted to a Russian Air Force detachment flying MiG-29s from Erebuni.

Above and left: Armenia's air force has two 'Frogfoot' two-seaters. This Su-25UB was supplied with the initial batch, almost certainly from Russia, while a single Su-25UBK arrived with the batch from Slovakia.

Right: The grey/green camouflage identifies this Su-25 as one of the five that were supplied in 1993, the year that the Hayastani Otayin Udjer took over Gyumri air base from the Russians. At the time Armenia was involved in fighting with Azerbaijan over the disputed Nagorno-Karabakh region.

Below: An Su-25K streams its chute on landing at Gyumri. The airfield is also a civil base – in the background are a Yer-Avia Il-78M, and Tu-134s and Yak-40s of the state airline Armenian Airlines.

53661 Shturmovaya Avia Baza, Gyumri

This combined civil/military airfield is located on the outskirts of the city that was destroyed during the 1988 earthquake. During the time of the Soviet Union, Gyumri airbase hosted a regiment flying the Su-7 and, allegedly, MiG-25PDs. The base is strategically located a short distance away from the Turkish border. On 13 April 1993 the base changed hands and has been the 53661 Attack Airbase of the Armenian Air Force ever since. During 1993 a total of six Su-25 'Frogfoot' aircraft was delivered to the Armenian Air Force. Given the country's close ties with Russia the latter country most likely supplied the two-seater and five single-seaters.

During August and September of 2004 the remaining Slovak Su-25s were relocated to Gyumri onboard Il-76 transport aircraft. Nine Su-25K single-seaters and the sole Slovak Su-25UBK two-seater more than doubled the Armenian 'Frogfoot' fleet.

After assembly, the new arrivals initially flew with Slovak roundels and codes still in place. During the summer of 2005 all received the Armenian roundel and a two-digit red code. The special markings the Slovaks applied to two aircraft during 1993 remain unchanged, albeit with a small Armenian roundel added to the fin.

Armament carried by the Su-25s consists of FAB free-fall bombs of various weights, S-5 unguided rockets carried in UB-32 pods, and Kh-25 and Kh-29 guided missiles.

Two Su-25Ks retain the special markings applied when they served with the Slovak air force (above), although the other seven now wear full Armenian markings (left). Standing next to one of the 'specials' is Gyumri base commander Major Igor Martirosyan (below). The Su-25s are operated by the 121 Shturmovaya Avia Eskadrilya (121st Attack Air Squadron).

Below: A reminder of Gyumri's Soviet past is this Mikoyan MiG-25PD 'Foxbat-E' parked alongside three active Su-25Ks.

Left: VIP transport for the government and armed forces is handled by this Mi-8S, operated from Erebuni by the 32822 Helicopter Base but retaining its civil registration ('EK' = Armenia).

Below: Armenia's 'Hip' force consists of relatively modern Mi-8MTVs, equipped with weapon outriggers for free-fall bombs and/or rocket pods, scabbed-on cockpit armour and a rescue winch. Some also have nose guns. One aircraft is maintained at Erebuni on a SAR alert.

32822 Vertoletnaya Baza, Erebuni

Located in one of the southern suburbs of Armenia's capital, Yerevan, Erebuni airport serves a dual civil/military purpose.

When Armenia was still a part of the Soviet Union a helicopter regiment of the 7th Army was based at Erebuni airport. This regiment has since been withdrawn but a number of its helicopters were transferred to the Armenian Air Force. Nowadays, seven Mi-8MTV 'Hips' and 11 Mi-24 'Hinds' are on the strength of the 32822 Helicopter Base, while a single Mi-8S serves as a VIP transport. Of the seven Mi-8MTVs in use, at least one carries a 12.7-mm machine-gun in the nose. Most numerous version of the 'Hind' in use with the unit is the Mi-24P, carrying a twin-barrelled 30-mm cannon on the starboard side of the front fuselage. Both the Mi-24RCh NBC reconnaissance version and the Mi-24K photo-recce variant make up the remainder of the 'Hind' complement.

Interestingly, the 426 Avia Gruppa (Aviation Group) of the Russian Air Force is also based at Erebuni. The unit flies at least a dozen MiG-29 'Fulcrum-Cs' and MiG-29UB aircraft on air defence duties. The 'Fulcrums' replaced an equal number of MiG-23MLDs during 1998.

Below: Mi-24Ps make up the main combat element of the Armenian rotary-wing force, augmented by specialist variants. This Mi-24P has recently been resprayed from its original Soviet-era camouflage to a scheme more in keeping with operations in the southern Caucasus.

Jet training is undertaken in four Aero L-39Cs at the primary fast-jet base at Gyumri. Initially a pair of aircraft (Bort 01 and 02) was received, these having recently been repainted in the standard Armenian camouflage scheme (above). The fleet doubled in size with the arrival of two further aircraft (Bort 03 and 04), which wear a standard Russian two-tone brown scheme (left).

Below: Despite its ancient looks, 'Annushka' ('Little Annie') continues to play an important role with many air arms. The three Antonov An-2s serving at Arzni are used primarily to train paratroops. Like most of the aircraft at the training base, they were taken over from a former unit of the Soviet/Russian DOSAAF training organisation.

In common with other former Soviet states, Armenia relies on Yakovlev's radial-engined designs for its primary trainer and light aircraft needs. A single Yak-18T four-seater (above) is based at Arzni for liaison purposes. The Yak-52 (left) is the standard primary trainer for Russian forces, and Armenia uses part of the fleet handed over in the early 1990s from DOSAAF for the same purpose at Arzni. A single Yak-55 aerobatic aircraft (below) is also at Arzni, but currently only one pilot is qualified to fly it.

56265 Usutsoghakan Baza, Arzni

The 56256 Training Base of Armenia's air arm is located close to the Arzni gorge, just to the north of Yerevan. The base used to house a DOSAAF unit of Soviet origin. All of its aircraft were transferred to the newly established Armenian Air Force during the early 1990s. For basic training 14 Yak-52s are in use. Following roughly 80 hours on the piston-engined trainers, students relocate to Gyumri to fly a further 50 hours on the four L-39C Albatros jet trainers in use. Two brown/green-camouflaged L-39s have been operated since the early 1990s, while two examples carrying the standard Soviet camouflage were recently delivered.

At Arzni a single Yak-18T is in use for liaison duties. To train paratroopers the Military Aviation Institute uses three Polish-built An-2 'Colt' aircraft. The commander of the base is the only pilot qualified to fly the sole Armenian example of the Yak-55 aerobatic aircraft.

Rotary-wing students hone their skills on the Polish-built Mi-2 'Hoplite', of which seven can be found parked outside the hangar at Arzni. They fly the 'Hoplite' for two years before moving on to Erebuni airbase to fly either the Mi-8 'Hip' or Mi-24 'Hind'.

Below: Rotary-wing training is conducted on the Mi-2, all seven of which are seen here parked outside at Arzni. The aircraft display five different colour schemes, and the aircraft on the right retains 'DOSAAF' titles. Although the Mi-2s appear to be in relatively good condition, not all of them are airworthy.

Denel AH-2 Rooivalk

With the exception of Egypt's AH-64 Apaches at the other end of the continent, the Rooivalk is the only dedicated modern attack helicopter in service in Africa. Its development and deployment has been conducted at a leisurely pace, but has been pursued more vigorously in recent years.

The AH-2A Rooivalk is Africa's only indigenous military helicopter to make it into production and service. It is a highly capable, latest generation machine on a par with, and in some respects better than, all other modern attack helicopters, making it the pride of South Africa's local aviation industry and defence force. It forms an integral part of South Africa's peacekeeping and threat deterrent requirements, and provides a powerful force multiplier to the South African Air Force.

In the late 1970s and early 1980s the South African defence forces began expressing the need for an indigenous attack helicopter. United Nations sanctions imposed in 1977 against the apartheid-era country prevented it from acquiring foreign weaponry, so local production was the only feasible route to follow. Fortunately, South Africa had built up a strong local aviation industry over the years, enabling it to pursue this option.

South Africa's local aviation industry was initiated in 1965 when the Atlas Aircraft Corporation (changed to Denel Aviation in April 1996) began licence production of the Aermacchi MB.326M as the Impala Mk I jet trainer, and later the MB.326KC as the Impala Mk II light attack jet. Atlas later produced the radically upgraded Cheetah from the Dassault Mirage III, the Super Puma-equivalent Oryx from the Aérospatiale SA 330 Puma, the indigenous C-4M Kudu light battlefield utility aircraft, as well as the V3B and V3C dogfight missiles.

Before tackling the advanced new attack helicopter project, the government first decided to launch a programme that would give the South African aviation industry the necessary know-how needed to locally produce helicopters. So, in March 1981, Atlas was awarded a contract to design and manufacture the Alpha XH-1 light attack/proof-of-concept helicopter.

First flown on 3 February 1985, the XH-1 incorporated many features and systems of the Alouette III, but had locally produced components in its engine, gearbox and rotor, and had a totally new tandem cockpit fuselage with a 20-mm cannon underneath it. The gunner sat in the nose while the pilot sat behind, both under large, clear-view transparencies that gave impressive downward views. The airframe was of semi-monocoque conventional construction, built mainly of metal with some composites. However, there were almost no battlefield survivability features like armour protection or engine exhaust suppression. The weapons fit consisted of a single

Left: Eight of No. 16 Squadron's AH-2s are seen on the Bloemspruit AFB ramp, home to South Africa's only Rooivalk squadron. All 12 aircraft have now been delivered, but one was lost in a crash in August 2005.

600-round/min GA-1 Rattler cannon, slaved to an indigenous helmet sight that the gunner used to aim the cannon by looking at the target. Stub wings were not fitted, but a production version would have had rockets and anti-tank missiles attached to its wings.

Design of the Rooivalk (Afrikaans for Red Kestrel) began in late 1984 to meet a South African Air Force (SAAF) user requirement specification. In preparation for the Rooivalk, Atlas produced the XTP-1 and XTP-2 (for Experimental Test Platform) concept-proving and systems test-bed helicopters, which were basically modified SA 330L Pumas. The XTP-1, which was modified by Atlas in April 1987, featured a 20-mm GA-1 servo-controlled cannon with 1,000 rounds of ammunition and two stub wings mounted on the fuselage doors, each equipped with two Matra 155 pods with 18 68-mm (2.7-in) SNEB unguided rockets.

The Rooivalk programme soon led to the first prototype, called the XDM (Experimental Development Model), which was rolled out on

15 January 1990. It had a number of features not on the production Rooivalk, including rearward-pointing exhaust pipes, less ammunition and two Makila Topaz turboshaft engines. It was first flown on 11 February 1990 and used to test performance, heat signature from the exhausts and the dynamic systems during flight.

Right: The Rooivalk XDM was rolled out as the Atlas XH-2 prototype for the CSH-2 (Combat Support Helicopter – AH-2 from 1998). Initially it had a taller tailfin, although it was later cropped in line with that of the second prototype (ADM).

Below: The EDM prototype flies without exhaust suppressors over Langkawi island during the 1997 LIMA show in Malaysia. This country remains Denel's most promising export hope, although financial constraints have led to no tangible activity thus far.

Above: Precursor of the Rooivalk was the Atlas XH-1 Alpha. Based on Alouette III dynamics, the Alpha was intended only to prove the gunship concept, but could have been turned into a production aircraft if required. Lacking any defensive systems, the light attack helicopter would have been vulnerable to the increasingly prevalent ground defences in the region, so development of the Rooivalk preferred.

Above: The ADM second prototype retained the Makila Topaz engines of the first aircraft, but had a full weapons system installed. Here it carries the four-round launchers for the Atlas Swift (later ZT35 Ingwe) anti-tank missile, and V3C Kukri air-to-air missiles.

Left: The EDM prototype brought together a full weapons system and the Makila 1K2 engines intended for production. Here it fires the Armscor F2 cannon: cases can be seen falling away. Note the rear-pointing exhaust suppressors.

Although the programme was progressing well, it was temporarily halted later in 1990. The reason why the South African Defence Force expressed the need for the helicopter was to fight guerrillas in Namibia and Angola. So when South Africa started withdrawing from Namibia, the requirement no longer seemed necessary.

South Africa's presence in Namibia began in 1920 when the League of Nations (forerunner of the United Nations) granted South Africa a mandate over the territory after taking it over from Germany in World War I. However, South Africa refused to give Namibia independence. Consequently, in June 1971 the International Court of Justice ruled South Africa's presence illegal, but South Africa continued to govern the country and extract its valuable mineral resources.

As a result, the South West African People's Organisation (SWAPO), a black African nationalist movement led by Sam Nujoma, began a guerrilla campaign against the South Africans. Fighting regularly spread to Angola where SWAPO insurgents were based. South Africa also fought against the Soviet- and Cuban-backed government in the Angolan civil war and attacked many Angolan targets and troops.

Alouette IIIs and SA 330 Pumas were used to give transport and covering fire to South African troops, but were quite vulnerable to ground fire, and picked up many bullet holes – one Puma survived 22 bullet strikes. Five Alouette IIIs and four Pumas were lost to surface-to-air missiles, anti-aircraft artillery and rocket-propelled grenades. Had it been available, the Rooivalk would have played a significant role in suppressing SWAPO insurgents and would have greatly strengthened the South African position.

After around 20 years of guerrilla warfare, South Africa agreed to grant Namibia independence in December 1988. Elections were held in November 1989 and independence was attained on 21 March 1990. But with it, South Africa's need for the Rooivalk evaporated.

Although the government now showed little interest, Atlas continued development, mainly as a company-funded venture in the hope of finding export customers. In 1992 the second prototype, designated ADM (Advanced Demonstration Model), was flown. It had a fully integrated avionics suite and full weapons systems, and was used to test and develop the Denel Mokopa (Black Mamba) long-range anti-tank missile, HOT 3 anti-

tank missile and 68-mm rockets, as well as various other weapons fits. On 21 September 1999 the first Mokopa was fired and in late 2001 it was fully integrated into the helicopter.

The third and final development helicopter was the pre-production EDM (Engineering Development Model), which was fitted with two 1,845-shp (1376-kW) Makila 1K2 engines, digital avionics (which included an automatic flight control system), an inertial navigation system, global positioning system (GPS), two Thales colour multi-function displays, a health and usage monitoring system (HUMS), Thales TopOwl helmet-mounted sight displays, Kentron tracking system and SFIM nose-mounted sight. It also had a reduced IR (infra-red) signature, revised air intakes, a lower maximum take-off weight and could carry up to 16 Mokopa missiles. It first flew on 17 November 1996 and was used to test avionics and sea level performance, as well as certification tests and verification of the Makila engines. By the end of 1997, the EDM had flown 67 hours, the XDM 466 hours and the ADM 414 hours.

Redesignation

The Rooivalk was originally designated Combat Support Helicopter 2 (CSH-2), but was redesignated Attack Helicopter 2 (AH-2) in 1998. There are three basic variants, designated Block 1A, 1B and 1E. Block 1A standard is the basic aircraft with functional mechanical and avionics systems, Block 1B has the TopOwl optical systems and night flying capability, and Block 1E aircraft have a fully operable helmet sight and weapons systems.

In October 2004 Denel stated that the total cost of the Rooivalk programme was R2 billion (about $317 million), not including the R661 million ($105 million) needed to upgrade the helicopters to Block 1E standard. Research and Development (R&D) costs have amounted to R126 million ($20 million), while one-off restructuring costs have come to R124 million ($19.7 million) and production costs R179 million ($28.4 million). A further R305 million ($48.4 million) has been written off.

Although never built, a projected maritime variant of the Rooivalk was studied and shown in model form at the Farnborogh Air Show in

The production AH-2A Rooivalk has a new exhaust system with upturned ducts to deflect hot gases up into the rotor's downwash.

September 1998. This production maritime variant would have had a chin-mounted 360° scan maritime search radar in place of the cannon, an enhanced electro-optical suite in the nose, the tailwheel moved forward 2 m (6.5 ft) to ease ship deck operations, flotation gear on forward sponsons and on the tailboom, manually operated blade-folding, enhanced electronic countermeasures and improved communications. It would also have had shorter stub wings that could carry four Penguin or Exocet anti-ship missiles, as well as air-to-air missiles for self-defence. However, no orders or commitment to build the variant have been forthcoming.

Although not designed as a maritime helicopter, a light brown Rooivalk was used for sea trials, reportedly the same one that was used to test a three-ton (6,600 lb/2994-kg) cargo hook. In October 2003 a R2.162 million ($343,000) contract was awarded to Denel to provide a ship transport capability for the helicopter, allowing it to be transported on ships like the SAS *Amatola*.

Customers

In 1994 the South African Air Force ordered 12 Rooivalks, out of a possible long-term require-

The first production AH-2A (right) was accepted by the SADF on 17 November 1998, and was delivered to Bloemspruit the following January to allow crew training. The pace of deliveries was slow, taking over four years to equip the 12-aircraft squadron. By September 2003 only two of the aircraft were considered weapons-capable, and that clearance was restricted to only rockets and the cannon.

ment of up to 36. The first four aircraft delivered were to Block 1A standard, the next six were to Block 1B standard and the last two are to Block 1E standard. Both Block 1A and 1B aircraft are being upgraded to Block 1E standard, at an additional cost of R661 million.

Only one squadron is being equipped with the helicopter and that is 16 Squadron at Bloemspruit Air Force Base near Bloemfontein. It was originally located at Port Elizabeth, but was disbanded in 1990 and reactivated for the Rooivalk on 1 January 1999. Although deliveries were originally scheduled to begin in March 1996, the first

In conjunction with Marshall of Cambridge, Denel pitched the Rooivalk to the British Army under the name Kestrel. The EDM prototype toured various establishments in the UK, and is seen here carrying dummy Swifts and Kukris. The inability to integrate Hellfire due to a US embargo severely hampered Denel's attempts to market the Rooivalk abroad, and led directly to the development of the Kentron ZT6 Mokopa missile.

Rooivalk was only delivered on 6 January 1999 and entered service in July of that year. Deliveries have been relatively slow – by 2003 only 10 of the 12 Rooivalks had been delivered.

Although no country has actually bought the

The main turret beneath the Rooivalk's nose houses the *TDATS targeting system. It contains a TV, FLIR, autotracker, laser designator and laser rangefinder. The crew are both wearing the TopOwl helmets, which display TV/FLIR imagery and alphanumeric information.*

US refused to supply Denel with the technology needed to make the Rooivalk compatible with American Hellfire missiles required by the British – essential for the new helicopter.

The Rooivalk's next failed export attempt was in 2001 when it competed – as the RedHawk – with the Apache Longbow, Tiger and Mangusta for Australia's AIR 87 requirement for 22 new attack helicopters. However, on 14 August 2001 the Eurocopter Tiger was selected.

In December 2003 Brazil showed limited interest in the Rooivalk. The commander of the Brazilian Air Force expressed the need for an attack helicopter for forest patrolling missions and in countering illegal aircraft carrying drugs and weapons. The Rooivalk would be used for air-to-air and not air-to-ground missions, in conjunction with Embraer EMB-145RS remote sensing aircraft and EMB-314 Super Tucano light attack turboprops. Although there was no budget or timetable to buy an attack helicopter, the Brazilian government has placed a high priority on protecting its Amazon forest. If an attack helicopter could well protect the Amazon forest, the government may fund it.

Malaysian interest

An export order nearly came from Malaysia, but the deal was put on hold after the Southeast Asian financial crisis in 1997. In August 1995 the Malaysian Prime Minister, Mahathir bin Mohamad, visited South Africa and was shown the Rooivalk. He was very impressed and planned to buy eight helicopters, out of a possible requirement of up to 30. Even though the Asian financial crisis put the order on hold, the country still plans to acquire the AH-2s when funding is available.

A number of other countries have shown interest in the Rooivalk and the helicopter's initial prospects looked good. A 1994 survey indicated that 180 Rooivalks could be sold abroad, generating more than R8 billion ($1.27 billion) in foreign exchange, R1 billion ($159 million) in taxes and giving 6,500 people jobs over 10 years.

Algeria, Finland, Singapore, South Korea, Spain, Turkey and Saudi Arabia have shown interest (in August 1997 the ADM was demonstrated in Saudi Arabia), but no orders have been forthcoming.

In order to strengthen the marketing position of the Rooivalk, Denel signed a marketing assistance deal with the Franco-German Eurocopter group in April 1997. The Rooivalk will be promoted to customers looking for a heavy attack helicopter while the Tiger will be aimed at customers looking for a smaller, lighter machine. The two helicopters will become partners, as opposed to adversaries.

The deal also allows Denel to sell its Oryx helicopters around the world. Eurocopter, in turn, got to offer its EC 635 (military version of the civil light twin-engined EC 135) to the South African Air Force. First unveiled in South Africa in April 1998, it was rejected, but went on to sell in Jordan.

The Rooivalk follows the design for a conventional third-generation attack helicopter, with tandem stepped cockpits, stub wings for weapons, and a nose-mounted cannon and sensors. Although the Rooivalk looks like a completely new

---

helicopter yet, the Rooivalk has, and is still, being aggressively promoted for export. It made its international debut at Dubai in 1993 where it was received enthusiastically. Its first big potential sale was in 1994 when the British Army stated a requirement for 91 new attack helicopters, worth an approximate R18.9 billion ($3 billion). The Rooivalk EDM prototype, registered as ZU-AHC, was sent to the United Kingdom to compete with the Eurocopter Tiger, McDonnell Douglas (now Boeing)/Westland Longbow Apache, Bell Cobra Venom and Agusta Mangusta.

Left: The highly manoeuvrable Rooivalk uses some components of the Puma/Super Puma dynamic system. This made the development task easier, and also provides an element of commonality with the Oryx fleet.

Below: An AH-2A is seen during an army exercise. Between delivery of the first Block 1As in January 1999 and the delivery of the first two squadron aircraft in weapons-capable Block 1E configuration, No. 16 Squadron used its aircraft to train in the kind of profiles that it would fly when fully operational. This involved nap-of-the-earth flying, convoy escorts, close support and the whole range of attack helicopter missions.

The Rooivalk was first shown at Middle Wallop international air show in May 1994 and at Farnborough international air show soon after. However, the Rooivalk lost out to the Longbow Apache, built locally in the UK by AgustaWestland as the WAH-64D Apache AH.Mk 1. A total of 67 was ordered on 13 July 1995 at a final cost of £3.1 billion (roughly $5.7 billion, R36 billion).

The main reason why the Rooivalk lost the competition was that it was still in development form and was not fitted with the complete avionics and weapons suite, whereas the Apache was a proven system with more advanced avionics, including a mast-mounted radar. No Rooivalks had yet entered service and had not been fully tested. There were also sanctions against South Africa imposed by the United States, the only country that had not lifted arms restrictions. The

The Rooivalk's two cockpits are essentially similar, and are dominated by the two large multi-function displays that present navigation, weapon system and aircraft systems data, as well as sensor imagery. Many of the important functions can be controlled using the switches on the cyclic control stick top and on the collective control grip located to the pilot's left. Further imagery is displayed on the visor of the TopOwl helmet-mounted sight system. Below the MFDs are a few back-up instruments for use in emergencies, and to the left is a small screen with a data entry keyboard.

AH-2 Rooivalk specification

Dimensions
Main rotor diameter: 15.58 m (51 ft 1.5 in)
Tail rotor diameter: 3.05 m (10 ft)
Length of fuselage excluding gun, with tail rotor turning: 16.39 m (53 ft 9.25 in)
Overall length, rotors turning: 18.73 m (61 ft 5.5 in)
Wing span: 6.355 m (20 ft 10.25 in)
Height to top of rotor head: 4.59 m (15 ft 0.75 in)
Height overall: 5.185 m (17 ft 0.25 in)
Wheel track: 2.78 m (9 ft 1.5 in)
Wheelbase: 11.77 m (38 ft 7.5 in)
Main rotor disc area: 190.6 m² (2052.1 sq ft)
Tail rotor disc area: 7.27 m² (78.23 sq ft)

Weights
Empty weight: 5730 kg (12,632 lb)
Max internal fuel weight: 1469 kg (3,238 lb)
External weapons load with full internal fuel: 1563 kg (3,446 lb)
Max external load with one ton (2,205 lb) of fuel: 2032 kg (4,480 lb)
Typical TO weight: 7500 kg (16,535 lb)
Max TO weight: 8750 kg (19,290 lb)

Performance
(at typical TO weight of 7500 kg [16,535 lb])
Maximum speed: 309 km/h (192 mph)
Maximum cruising speed: 278 km/h (173 mph)
Maximum sideways speed: 93 km/h (58 mph)
Service ceiling: 6100 m (20,000 ft)
Hovering ceiling (In Ground Effect): 5850 m (19,200 ft)
Hovering ceiling (Out of Ground Effect): 5455 m (17,900 ft)
Range with max internal fuel, no reserves: 704 km (437 miles)
Range at max take-off weight with external fuel: 1260 km (783 miles)
Endurance with maximum internal fuel, no reserves: 3 hours, 36 minutes
Endurance at maximum take-off weight with external fuel: 6 hours, 52 minutes
***g* limits:** +2.6/-0.5

machine, it uses some reverse-engineered parts from the Puma and Super Puma, including the dynamic systems of the Puma and engines from the Super Puma. This allows it to be cost-effective, low-risk and easily maintainable as it uses tried and tested components.

It is also very reliable and has a 30-year lifespan, which can be extended, allowing it to remain in service decades after it would normally have been retired. A wide variety of equipment can be added to the airframe, including: pulse Doppler radar, flotation gear, VOR/ILS (VHF Omnidirectional Range/Instrument Landing System), IFF (Identification Friend or Foe) transponder, CRV7 rocket-firing capability, a wire-strike protection system, heated windshields with electrical de-icing/de-misting, video cassette recorder, Mokopa missiles with anti-ship warheads and a crew escape system.

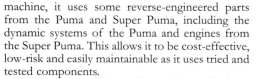

Ammunition for the 20-mm F2 cannon is housed in slab-sided cheek fairings. They are advertised as being able to have two external seats installed on each to transport special forces troops, or possibly to rescue downed aircrew, but they have not been seen in operational use.

The Rooivalk has been designed to survive in an intense combat environment and can operate in all weathers, day and night and at varying altitudes. In order to avoid detection, it has low radar, visual, infrared and acoustic signatures and is optimised for nap-of-the-earth (NOE) flight. Its crash-resistant monocoque primary structure is constructed of aluminium alloy, while the rotor blades and secondary fuselage structures are of composites sandwich, making them light yet tough enough for ballistic tolerance.

The entire landing gear, which consists of two forward-mounted Messier-Dowty main units with two-stage high-absorption main legs and a fully castoring tailwheel at the base of the lower fin, is designed to withstand a landing impact of up to 6 m (20 ft) per second.

Cockpits

The cockpits are stepped in tandem with the weapons system officer (WSO)/co-pilot in the front cockpit and the pilot seated above and behind under a separate canopy. Access to the cockpits is via upward-opening gull-wing window panels, opening on the right-hand side. All transparencies are either flat or have a single curvature, thus minimising sunlight glint. They give the crew forward/downward view angles of 23°. The crew sits in Martin-Baker crashworthy seats with armour protection. An environmental control system provides cockpit air conditioning, giving the crew increased comfort and environmental safety. In order to minimise crew fatigue, the cockpit has been designed to have ultra-low vibration levels. This is largely due to the main engine gearbox being mounted on a vibration isolation system using a tuned beam to insulate the fuselage from rotor vibrations. As a result, it has one of the lowest vibration levels in its class.

FZ90 unguided 70-mm folding-fin rockets are fired from 19-round M159 launchers. The rocket remains an important weapon for the attack helicopter, especially against area and 'soft' targets. Up to four M159s can be carried on the inboard and central wing pylons.

A pair of external seats can be fitted to the Rooivalk's ammunition cheeks, allowing the Rooivalk to transport soldiers or crew. However, this clever system does not yet seem to have been adopted by the SAAF.

Flying controls consist of a duplex four-axis digital AFCS (Automatic Flight Control System) with automatic height-hold, hover-capture and hover-hold, which can be activated by the auto-pilot. Both cockpits are interchangeable and have dual flight controls, which includes hands-on-collective-and-stick (HOCAS). The Rooivalk has a four-bladed fully articulating main rotor (fitted with a brake) that rotates at between 267 and 290 rpm, and a five-bladed starboard mounted tail rotor that rotates at 1,290 rpm. A fixed leading-edge slat is mounted on the horizontal stabiliser, which is mounted halfway up the port side of the vertical stabiliser.

Cockpit instrumentation consists of two 16 x 16-cm (6.3 x 6.3-in) Thales MFD 66 liquid crystal colour multi-function displays (MFDs), two Sextant TopOwl helmet-mounted sights/displays and a position management system in each cockpit. The MFDs show flight control, navigation data (including a moving map on the pilot's display), threat warning and electronic warfare information, weapon controls and imagery from the nose-mounted target detection, acquisition and tracking system. They can also copy information to each other. In addition, there is a basic back-up instrument panel for a 'get home' capability in case of a computer or power failure. All instrumentation is compatible with third-generation night vision goggles (NVGs).

Both crewmembers have Thales (Sextant) TopOwl helmet-mounted sight displays that can cue the sensor turret, the cannon and air-to-air missile seeker heads. In addition to displaying weapons data, they show air speed, power settings, aircraft heading, altitude, rate of climb, turn and slip, collective pitch setting, the helmet line of sight, waypoint bearing, waypoint number, and distance to waypoint.

Avionics

The Rooivalk is equipped with an advanced avionics suite, produced locally by Advanced Technologies and Engineering (ATE) and 15 other sub-contractors. It has dual-redundant all-digital mission computers, MIL-STD 1553B data-

Prime weapon for the AH-2A is the Denel (formerly Kentron) ZT6 Mokopa laser-guided anti-tank missile, which was developed in response to a US embargo on Hellfire and was first revelaed in 1995. Weighing 110 lb (49.9 kg) at launch, the Mokopa is similar in size and capability to the Hellfire, and has a range of launch options including LOBL (lock-on before launch, in which the helicopter designates the target) and LOAL (lock-on after launch, in which the missile flies out using inertial guidance, before acquiring a laser 'spot' from an offboard designator). The missile has a dual-burn motor that boosts it to 5 km (3.1 miles) range in 15 seconds and sustains it to the 10-km (6.2-mile) maximum range in 39 seconds. Up to 16 can be carried on four-round launchers.

buses and a stores management system that conforms to MIL-STD-1760A standard. Other basic avionics include an intercom/audio system and an IFF transponder.

Communications, which are all frequency-agile, consist of two V/UHF transceivers with FM, AM and digital speech processing, that operate between 30 and 400 MHz for normal use, and a single HF transceiver for NOE flights. This radio operates between 2 and 30 MHz and has secure voice and data channels.

The Rooivalk has a comprehensive navigation suite that consists of a Sextant NSS 100-1 eight-channel GPS receiver, Doppler radar velocity sensor, J-band radar altimeter, heading sensor, air data unit, omnidirectional airspeed sensor and a Thales Stratus three-axis strapdown ring laser gyro attitude/heading reference system (AHRS). The navigation and attack system can be programmed by a preloaded cartridge that can hold up to five flight plans consisting of 100 waypoints. These can be edited in flight by either crew member and are displayed on the helmet sights.

Mission avionics

A nose-mounted gyro-stabilised turret contains the Rooivalk's target detection, acquisition and tracking system (TDATS). The turret holds a three-fields-of-view Forward Looking Infra-Red (FLIR) with automatic guidance and tracking, a low-light level TV (LLTV) camera, and a laser rangefinder and laser designator. Related avionics include the crew member's Thales helmet sights, missile command link and tracking goniometer (which measures angles). A Cumulus and Thales Pilot's Night Vision System (PNVS), consisting of a two-axis turret with a 40 x 30° field of view FLIR, is offered as an export alternative to the Sextant helmet sights.

Countermeasures

The Rooivalk has been designed to survive in a hostile environment and, as a result, has a sophisticated suite of countermeasures to help it deal with virtually any threat that it should come upon. Its countermeasures suite is adaptable and programmable in flight, allowing it to match the threat library with the mission's area of operation. The Electronic Warfare (EW) suite incorporates radar and laser warning receivers, and radio frequency, infra-red and laser jammers. Countermeasures are either dispensed manually, semi-automatically or fully automatically.

The radar warning receiver can detect radar signals beyond the source radar's detection range

A Rooivalk launches a ZT6 Mokopa during trials, the first of which was undertaken on 21 September 1999. The missile is only just entering service on the Rooivalk after a protracted development due to intermittent funding. Future developments include the adoption of millimetre-wave and infra-red guidance.

and can pick up most frequencies, including ultra broadband frequencies. These threats are either jammed or chaff is launched.

The laser warning receiver provides broadband laser frequency coverage to detect and display range-finding and missile guidance laser beams. A laser jammer then deals with any threats. Other countermeasures include a flare dispenser and an IR jammer.

Weaponry

The Rooivalk can carry up to two tons (4,400 lb/1996 kg) of advanced weaponry on its stub wings. The wings each have three stores stations, and can carry two or four M159 19-tube launchers for Belgian Forges de Zeebrugge FZ90 70-mm unguided folding-fin aerial rockets (with a variety of warheads) and/or four-round launchers for up to 16 Mokopa anti-tank missiles on the inner two pylons of each wing. Two or four infrared-homing Mistral air-to-air missiles can be carried on the outboard pylons. With the Mistral,

the Rooivalk became the second third-generation attack helicopter, together with the Tiger, to be armed with the projectile.

Equipped with a passive infrared guidance system with trajectory-shaping capabilities and a self-spinning airframe for improved accuracy, the Mistral is powered by a quick-burn solid rocket booster and can reach speeds of up to Mach 2.5, enabling it to achieve its 93 per cent kill rate against highly mobile helicopters and fixed-wing aircraft. The 6-km (3.7-mile) range missile's warhead can either be detonated by impact or by a laser proximity fuse.

Denel's ZT6 Mokopa is a 10-km (6.2-mile) range anti-tank weapon, equipped with a millimetric wave or semi-active laser seeker head that can be locked on to the target either before or after launch. It has a tandem warhead, allowing it to penetrate more than 1000 mm of armour. An alternative anti-ship warhead can also be fitted. A production order for the missile was placed in March 2004.

A No. 16 Squadron crew pulls its Rooivalk to an abrupt stop. The ability to position and re-position around the battlefield with rapidity and at low level is a key requirement for an attack helicopter.

The Rooivak carries a long- or short-barrelled 20-mm Armscor F2 cannon under its chin, fed by two ammunition 'cheeks' holding up to 700 rounds of ready-to-fire ammunition. The cannon is a dual-feed (thus eliminating jamming) gas-operated weapon, able to fire shells at 1100 m/s (3,609 ft/s) at a rate of 740 rounds per minute. Effective range is 2 km (1.2 miles) and maximum range is 4 km (2.5 miles). It can be aimed by the helmet-mounted sight display and be slewed at a rate of 90° per second.

In 2002 the Denel ZT35 Ingwe anti-armour missile was test-fired from the Rooivalk, indicating that it may eventually be fitted to the helicopter. The laser-guided Ingwe weighs 29 kg (64 lb) and has a range of 5 km (2.7 miles).

Engines

The Rooivalk is powered by two licence-built Turboméca Makila 1K2 turboshafts, mounted on each side of the fuselage below the main rotor. They are rated at 1,845 shp (1376 kW) each for take-off and 1,904 shp (1420 kW) continuous power, although in an emergency they can deliver 2,109 shp (1573 kW) for 30 seconds. Both engines have digital control, infrared heat suppressors on their exhausts and particle/sand filters on their intakes, allowing them to operate in harsh climates or bad weather.

Fuel is carried in self-sealing tanks in the fuselage, as well as on drop tanks under the stub wings. Three 618-litre (136-Imp gal)) fuselage tanks give a total internal fuel capacity of 1854 litres (408 Imp gal), while a 750-litre (165-Imp gal) drop tank can be carried on each inboard underwing station. The helicopter can be pressure- or gravity-refuelled and defuelled, allowing it to be quickly refuelled from drums on the ground using an onboard pump system. It is also capable of a 'hot turnaround' with full refuelling and rearming in less than 15 minutes.

The engines drive two independent hydraulic systems, each with a pressure of 175 bar (2,538 lb/sq in). Electrical power is obtained from two 20-kVA alternators providing 200 V of three-phase and 115 V single-phase alternating current at 400 Hz. Two 24 V 31 Ah batteries provide 28 V of direct current power. Other standard systems include a crew oxygen system, fire detectors and fire extinguishers.

Maintainability

One of the most important requirements for the Rooivalk was that it had to be reliable and easily maintainable under the most basic circumstances. It was designed to endure the initial 72 hours of a high-intensity battlefield and then operate for long periods without sophisticated support. A medium-sized transport helicopter with a basic supply of spares plus four personnel is all that is needed to keep the helicopter flying.

The Rooivalk has a number of built-in features

This view shows the diffusing effect of the rotor downwash on the exhaust plume. The SADF has devised a camouflage for the Rooivalk that is effective against widely varying backdrops.

to facilitate maintenance and to detect wear and tear on the airframe. It has automatic fault detection and system status monitoring through its HUMS, it has LRUs (Line Replaceable Units) that incorporate built-in test and evaluation (BITE) monitors, and can download maintenance data to a ground station.

A number of features make the Rooivalk easily maintainable and supportable by ground crew, which include the ability for all systems to operate in silent battery mode, obviating the need for an auxiliary power unit (APU) or ground power unit (GPU). Maintainers can easily access all parts of the airframe without the use of work platforms – the wings and carbon fibre-reinforced plastic engine cowlings double as work platforms and there are integral access ladders in the fuselage. In addition, a large avionics bay door allows easy LRU replacement, the nose-mounted sight is removeable by one man, the batteries are easily accessible and the Rooivalk can be dismantled in the field with an aircraft-mounted portable crane.

Mission

The Rooivalk's main mission is to provide a threat deterrent for South Africa and its citizens and act as a force-multiplier in peacekeeping and enforcement roles. The Rooivalk can easily carry out a number of missions, which include anti-armour, ground suppression, close air support, heliborne escort, counter-insurgency and reconnaissance.

A comprehensive variety of weaponry selected for each mission requirement allows the Rooivalk to carry out a wide range of roles, from anti-helicopter to ferry missions. It can engage multiple targets at short and long range, using the cannon, or rockets and missiles. The Mokopa missile gives it a distinct advantage against competitors as it allows the Rooivalk to attack targets at long ranges.

The Rooivalk's design allows it to be easily deployed across vast distances. With the aid of ferry tanks, it can self-deploy over ranges of more

Two AH-2As exercise with ground armour in open country. A key role envisaged for the Rooivalk is the support and protection of ground forces during the peacekeeping operations that have become an increasingly important role for the SADF.

From the outset the Rooivalk was designed to be very easy and quick to maintain and service in the harsh conditions that can be encountered in the bush.

than 1100 km (680 miles), or longer if refuelling stops are included. It can also be transported in a Lockheed C-130 Hercules or larger aircraft with minimal disassembly.

With its advanced day and night sensors and secure communications, the Rooivalk can quickly and easily gather reconnaissance information that ground troops or other aircraft would find hard to do. Its survivability features allow it to get as close to the action as required without fear of being shot down.

With its excellent mission versatility, self-deployment range and reconnaissance ability, the Rooivalk is ideally suited to provide support for peacekeeping operations. South Africa is regularly involved in peacekeeping, and in the future is bound to use the Rooivalk in these campaigns.

A good example was in 2001 when South African mediators managed to broker peace talks in Burundi after years of unrest. Later that year, a new government was sworn in under the protection of a 700-strong South African peacekeeping force, which included an Air Force element of two helicopters. Sending in Rooivalks was considered, but nothing came of it – probably because the situation was not very threatening, there were not many Rooivalks available (only six in early 2001) and those delivered were still being upgraded.

However, even now the Rooivalk is still not considered ready for peacekeeping duties. Staff departures at Denel mean that the Rooivalk

Without an export order or further SADF requirements the AH-2 Rooivalk faces an uncertain future, but it remains a milestone achievement for the South African defence industry.

cannot be deployed for another year. But soon after that it may be sent to give firepower support to South African peacekeeping troops in the Democratic Republic of the Congo, which is still suffering from sporadic fighting and unrest.

With the cancellation of the state-of-the-art, highly advanced and stealthy Boeing/Sikorsky RAH-66 Comanche in February last year, the Rooivalk has again become one of the most advanced helicopters in the world. Its low price tag, heavy weapon load and advanced sensors make it competitive with other international attack helicopters like the Kamov Ka-50 'Hokum', Mil Mi-28 'Havoc', Boeing AH-64 Apache, Eurocopter Tiger, AgustaWestland A129 Mangusta and Bell AH-1 Cobra. Now that the Rooivalk is a fully operational system, export interest should be forthcoming. But even if it is not, the Rooivalk will remain a proud achievement for South Africa's Air Force and local aviation industry.

Guy Martin

Nevada Airpower

Adversaries, Aggressors, Airlift and advanced tactical training

Photographed by Rick Llinares/Dash 2

Encompassing one of the least densely populated areas of the United States, the 'Silver State' offers huge tracts of unrestricted airspace over high desert terrain. Consequently, Nevada has become a key area for the US military to test its hardware and to forge the tactics and skills that will be used in real combat.

Above: The Air Force rules the skies in the southern part of the state, undertaking a wide range of operational evaluation and advanced tactics training over the vast Nellis ranges. Among the many factors that allow Nellis AFB to provide such an authentic training and test environment is the resident adversary unit, the 64th Aggressor Squadron. The 64th flies 12 F-16C/Ds, represented here by a four-ship painted to simulate the 'Flanker', 'Fulcrum' and 'Flogger'. The newly formed 65th Aggressor Squadron will begin operating F-15C/D Eagles in the aggressor role in 2006.

Above: Headquartered at Eglin AFB, Florida, the 53rd Wing is Air Combat Command's operational test and evaluation organisation, with elements at several bases. The 422nd Test and Evaluation Squadron is at Nellis, operating A-10s, F-15s and F-16s on various trials duties. The 53rd Wing at Nellis was the first service recipient of the F/A-22 Raptor, and is responsible for testing the aircraft and its systems in an operational environment, as well as helping forge a training syllabus and tactics manual.

Around 100 miles east of Reno lies the town of Fallon. A few miles from the town is Van Voorhis Field – better known as Naval Air Station Fallon – which has become the US Navy's principal location for air wing work-up and advanced combat instruction. The latter role is entrusted to the Naval Strike and Air Warfare Center (NSAWC), which combined the tasks of the air-to-air dedicated Naval Fighter Weapons School ('Top Gun') and the air-to-ground Naval Strike Warfare Center ('Strike University'). The main types are the F-16 (above) and F/A-18 (right).

An F-16A and F-16B overfly NAS Fallon (above). The 'Vipers' are the latest recruits to the NSAWC roster, drawn from a batch intended for Pakistan but embargoed and stored for many years at Davis-Monthan AFB. Others from the batch went to the USAF's test fleet at Edwards AFB. NSAWC previously operated the F-14A on adversary duties, but it was withdrawn as part of the rundown of the Tomcat fleet.

Below: VFC-13 'Saints' is a Naval Reserve unit that previously flew Skyhawks on adversary duties from Miramar. Today the squadron operates alongside NSAWC at Fallon, and is equipped with F-5E/F Tiger IIs. The fleet was bolstered in 2005 by the arrival of 32 low-houred aircraft made surplus by the Swiss Air Force.

NSAWC performs a similar role to the USAF's Fighter Weapons School, providing advanced tactical instruction to Fleet pilots. As well as its F-16 adversaries, it also has F/A-18s for the adversary/training role. In addition to teaching advanced air-to-air and air-to-ground techniques, NSAWC is also tasked with providing similarly advanced training for the AEW force, and incorporates the E-2C Hawkeyes that were formerly assigned to the Carrier Airborne Early Warning Weapons School before its amalgamation with NFWS and NSWC to form the current organisation.

Below: NSAWC's only rotary-wing type is the Sikorsky SH-60F, used for advanced training and range support. As well as its training commitment, Fallon provides the Navy's air wings with a site to work-up in weapons delivery and tactical techniques prior to deployment aboard their assigned carriers.

The 'owner' of Nellis AFB is Air Combat Command's Air Warfare Center, which provides an umbrella to both the operational test (53rd Wing) and advanced training (57th Wing) functions. The USAF Fighter Weapons School is an integral part of the 57th Wing, and teaches advanced tactics and techniques to already experienced crews from operational units. When students graduate from the FWS, they take their newly gained experience back to their original units. The FWS works closely with the operational evaluation and intelligence agencies to ensure that new developments in either tactics or the employment of weapons are rapidly taught to their students who, in turn, will take these developments back to the front-line squadrons. The school is divided into type-specific units, with five being represented at Nellis: A-10 (above), F-15, F-15E (left), F-16 (below) and HH-60. Each has its own aircraft assigned. There are two further FWS divisions at Ellsworth AFB and Barksdale AFB to cover the B-1B and B-52H, respectively, but they borrow aircraft from front-line units when required.

The 57th Wing consists of several disparate organisations. The FWS operates separately from the 57th Adversary Tactics Group, which oversees the 64th AS and its aggressor F-16s (above and left), and the 57th OG and its 66th Rescue Squadron that provides HH-60Gs for rescue cover of the ranges. The 64th's busiest time is during Red Flag exercises, but the unit also provides dissimilar types for air combat trials and for FWS work. Another F-16 user parented by the 57th Wing is the USAF Aerial Demonstration Squadron, better known as the 'Thunderbirds'. In recent times the 57th has picked up responsibility for the USAF's growing UAV operations, and has turned the auxiliary field at Indian Springs into the centre for UAV operations and evaluation work. Indian Springs was recently renamed Creech AFB, and houses the USAF's Predator fleet.

Right: Nevada's Air National Guard squadron (192nd Airlift Squadron, 192nd Airlift Wing) operates from May ANGB on the site of Reno-Tahoe IAP. Initially a fighter unit with F-51s and F-86Ls, the 'High Rollers' specialised in reconnaissance from 1961, flying RB-57 Canberras, RF-101 Voodoos and RF-4 Phantoms – including 412 combat sorties during Desert Storm – until re-roled in the 1990s as an airlift wing with C-130E/H Hercules. The 192nd is one of nine ANG units that provide the 15th Air Force with tactical airlift capability. However, it has been put forward for deactivation under the latest base closure proposals, with its C-130Hs to move to the Arkansas Guard unit.

Royal Jordanian Air Force

Silahil Jaw Ilmalaki Lurduni Al Quwwat al Jawwiya al malakiya al Urduniya

Established with RAF help in the aftermath of World War II, the Royal Jordanian Air Force is today one of the most professional and capable air arms in the Middle East region. Re-equipment of the fighter forces with the F-16 is under way, while the RJAF is bolstering its rotary-wing assets, with the accent on increased special forces operations, and revamping its transport fleet.

Left and above: The arrival of the F-16 Fighting Falcon has dramatically improved the air defence capabilities of the RJAF. Despite being high-time airframes, the ex-USAF aircraft were put through a structural life-extension programme. In the near future RJAF F-16s will be modernised to MLU Block 50 standards, with the ability to carry the AIM-120 AMRAAM.

Military aviation in Jordan dates back to 1917, when Arab Forces requested the Royal Flying Corps to bomb Turkish targets. Two air bases were constructed in Jordan by the British, RAF Amman and RAF Mafraq, and some of the buildings constructed in those days are still in use. In 1921 the country now known as Jordan was transformed into a self-governing country under a British mandate. The RJAF was established with major British backing in 1949 as a component of the Arab Legion. Designated initially as the Arab Legion Air Force, the first pilots were RAF officers. The RJAF's first combat jets were de Havilland Vampires, followed by Hawker Hunters that were acquired from the UK but paid for by the US.

In 1964, the Jordanian aircraft went into action for the first time when four Hunters scrambled to fight against Israeli Mirage IIIs. In what became known in Jordan as the Battle of the Dead Sea, one Israeli Mirage was downed and three others were damaged. In 1967, during the opening hours of the Six Day War, the RJAF was virtually wiped out by the Israeli IDF/AF. The RJAF was re-equipped with US-supplied F-104As, but the RJAF played no part in the Yom Kippur war of October 1973.

During the 1970s and 1980s, Jordanian attempts to buy more modern aircraft were blocked by the US: new fighters were needed to replace obsolete material but requests for A-4 Skyhawks and F-16s were blocked. After a meeting of Arab leaders in Baghdad in 1978, it was decided that Jordan should replace its ageing F-104A Starfighter with the financial backing of Saudi Arabia and Iraq.

Today's Royal Jordanian Air Force consists of 15 squadrons based at six main operating airbases around Jordan.

Transport and support

Amman-Marka, home of the transport and most of the helicopter fleet, is located immediately next to the rural areas of Amman, a city that houses almost half of Jordan's population. The transport fleet – consisting of the Air Lift Wing, the Royal Squadron and the Air Police Wing – operates a selection of fixed-wing and helicopter aircraft.

The Air Lift Wing was established in 1971 with the delivery of four C-47s and a number of Alouette IIIs. Soon after its establishment, the wing set about a modernisation programme with the C-47 being replaced by C-130s and a C-119. The Fairchild C-119 was soon withdrawn due to its limited reliability. During 1975/76, four CASA 212A Aviocars were supplied. The wing operated the Alouette III until 1987 when it was replaced by the Super Puma. The wing reactivated No. 8 and No. 14 Squadron following the delivery of 36 ex-US Army UH-1Hs in October 1994.

The Airlift Wing is commanded by Col. Nabil Ababneh and consists of:

■ **No. 3 Squadron** operating four C-130H aircraft and two CN-295s. Until recently another two CN-235Ms were operated on loan from the Turkish Air Force, bridging the gap between order and delivery of the CN-295. The C-212s that the squadron used to operate are stored at the airfield and, if needed, can be brought to flight status in a short period of time, but there are no immediate plans to return them to flying service. The fixed-wing fleet of No. 3 Squadron is also referred to as 'Guts Air Line' and is often involved in international missions around the world.

Right: Jordan operates a mix of helicopter types, of both US and French origin. Here a Bell UH-1H leads a Eurocopter EC 635 and an AS 332 Super Puma. RJAF helicopters support the army, as well as provide civilian emergency cover.

Below: Most of the RJAF's training activity takes place at the Air College near Mafraq, named for the late King Hussein, an accomplished pilot who regularly commanded the Royal Flight TriStar. Screening and primary training is conducted on the T67M Firefly.

Special Operations and Reconnaissance aircraft from the civilian part of the airport, and are equipped with FLIR. Their task is also to perform as an airborne link for data transmission between ground forces. Later this year their equipment will be upgraded to improve the range from 15 to 100 miles (24 to 161 km).

Jordan became the launch customer of the EC 635, the military version of the EC 135, following cancellation of an order for the Portuguese Army. The first of these helicopters were delivered in the second part of 2003 (less than three months after being ordered) and, since then, a total of nine has been delivered to No. 14 Squadron. The EC 635 can fulfil all the roles of a light combat helicopter. Part of its mission is day and night observation, border control, passenger transport, anti-terrorist/special operations and rescue missions. The Jordanian EC 635s have special military avionics systems and all have weapons wiring installed. Though the helicopter could be equipped with weapons, no weapons systems are ever carried; only a FLIR system is used on several aircraft and some are equipped with flotation gear. The RJAF is extremely satisfied with the aircraft and its mission equipment, and is examining options to buy further aircraft.

Training Command

The Training Command of the RJAF is located at the King Hussein Air College near Mafraq, operating the Slingsby T67 Firefly, CASA C-101, MD (Hughes) MD500D and E models, and the Aérospatiale AS 350B-3 Ecureuil.

■ **No. 4 Squadron** operates 16 Fireflies, delivered in 2002. The first step in the career of new pilots is screening and primary training with the unit. The syllabus begins with 14 flights to determine if a candidate has a 'feel' for the job, followed by military and theoretical training for a year. This theoretical part is followed by a move for the successful candidate on to primary flight training with No. 4 Squadron, for a total of around 70 flight hours. During primary flight training, roughly in the middle of this period, a test determines the next step: to continue as fighter pilot or move on to helicopter training.

■ **No. 11 Squadron** operates the CASA C-101 that has been in service since 1987 and takes the students selected for fighter operations after primary training. With No. 11 Squadron the student experiences the jet for the first time and after a total of around 120 flight hours gains his wings. During this phase a total of around 30 sorties are flown to teach students how to make

No. 3 Squadron – 'Guts Air Line' – is the fixed-wing transport squadron, based at Amman's main airport. The backbone of the fleet are four C-130H Hercules (below), two earlier C-130Bs having been donated to the new Iraqi Air Force. The latest type in No. 3 Squadron service is the EADS-CASA CN-295 (above), of which two are in service.

■ **No. 7 Squadron** operates the AS 332AM Super Puma, of which the first of 12 was delivered in 1987. Main task for these powerful helicopters is transport and medevac. Because these helicopters regularly fly international sorties, they wear both military and civilian registrations. Also, because of their humanitarian relief operations, a number of the Pumas wear the insignia of the red Half Moon. Palestinian leader Yasser Arafat was flown in a Super Puma from his HQ to Amman (a well covered event) to receive medical treatment in France before his death. The Pumas are regularly deployed around the country on SAR duties.

■ **No. 8 Squadron** operates around 15 UH-1

Hueys, and recently six Bell 212s have been delivered from an unspecified country. Following the peace accords with Israel, a US MDAP programme saw 18 surplus UH-1H Helicopters delivered in 1994, with another 18 following in 1996. The UH-1 has long been the backbone of helicopter operations with the RJAF, but the number of Hueys is dwindling. Of the 36 that once were operational, 16 have been transferred to the Iraqi Air Force.

■ **No. 14 Squadron** operates nine EC 635 helicopters and two Schweizer SA 2-37A observation platforms. Main task for the squadron is Special Operations. The Schweizers are operated as

Jordan's fleet of CASA 212s is no longer in use, but remains in flyable condition for emergencies. When they were operational with No. 3 Squadron, one of the aircraft had a reconnaissance role.

No. 3 Squadron's CASA connection was maintained by a pair of CN-235s that were operated on loan from Turkey pending delivery of the stretched CN-295s. With the new aircraft in service in 2005, the CN-235s have been returned.

No. 8 Squadron is the Huey unit, operating a mix of single-engined UH-1Hs and twin-engined Agusta-Bell AB 212s. This is one of the latter, parked at its Amman-Marka base.

use of the aircraft tactically. After graduation the new fighter pilot transfers to No. 17 Squadron/OCU operating the F-5 to be trained in the fighter world.

■ **No. 5 Squadron** trains future helicopter pilots on the H500/AS 350. Students receive over 100 hours of flight training before being assigned as co-pilot. After 500 hours pilots are allowed to command operationally.

The F-5 community

The largest contingent of F-5s in the Kingdom can be found at Prince Hassan Air Base. In late 1974, the RJAF obtained 20 F-5As and two F-5Bs from the Imperial Iranian Air Force to replace the Hawker Hunter. They became operational at Mafraq Air Base with 1 and 6 Sqns. The displaced Hunters were donated to Oman. An additional delivery of 10 F-5As and two F-5Bs allowed the formation of 2 Squadron for advanced training.

From 1975 on, the US supplied a total of 61 radar-equipped F-5Es and 12 F-5Fs. This proved to be a big step forward in operations, the aircraft being able to carry weapons like the AIM-9J Sidewinder and Rockeye cluster bombs. They were operated by 1, 6, 9 and 17 Sqns, the latter two units operating from Prince Hassan Air Base, also known as H-5.

RJAF commander

The Commander in Chief of the RJAF is General Al-Bis who was assigned to the job in the autumn of 2004, relieving General Major Prince Feisal Bin Al-Hussein. General Al-Bis has been a long serving pilot of the Air Force, with over 4,000 flight hours, mostly in the Hawker Hunter and F-5. Speaking from his office at the Royal Jordanian Air Force (RJAF) headquarters at Amman-Marka AB, General Hussein Al-Bis told *International Air Power Review***:**

"The main mission of the RJAF is to defend Jordanian air space and to support ground forces. This has been the case since the formation of our air force and the task of defending will remain the same. Throughout history, the Middle East has been plagued with conflicts. Almost every country in the region has at times experienced clashes with its neighbours, Jordan being no exception. These days, however, our country has friendly ties with most of our neighbours. I am a strong supporter of international regional ties and the RJAF is open to learn from other countries and to share information. The RJAF is regularly conducting exercises with its neighbours and allies. These take place both in Jordan and abroad, one example being the major international exercise Bright Star that is being held in Egypt each year.

"The budget for the RJAF is assigned directly from the Armed Forces HQ. Budget restraints hampered our efforts to upgrade our air force – luckily we are now again in the process of boosting our ability to operate with more modern material with the acquisition of a second batch of F-16s that will be upgraded, and with the purchase of new helicopters. This will mean, however, that in the near future the RJAF will be reducing its numbers of fighters, one of the first types to be withdrawn from use will be the Mirage F1CJ aircraft. From this year on, the first aircraft will be withdrawn. They are up for sale and when a customer will buy them, flight operations will immediately cease. The Mirage F1EJ will however remain in service for the foreseeable future.

"After 2008, less F-5 aircraft will be operated by the RJAF. The plan is to have only one operational squadron left. 9 Squadron will remain as an F-5 squadron, while 6 Squadron should be converting to F-16s in the coming years. This squadron should operate the second batch of F-16s that was delivered under the Peace Falcon II deal. Though not in the immediate future, by 2015 a third F-16 squadron is planned to be operational. In the future, the RJAF will probably be an all F-16 force with all the benefits of operating one aircraft type.

"In order to be independent and to increase the budget, the RJAF set-up its own maintenance company called Jordan Aeronautics Company (JAC), formerly known as Marshall. This company will maintain the large fixed-wing aircraft of the RJAF like the C-130, but also aims at being a service centre for these kind of aircraft in the Middle East. In addition to this, a company called Jordanian International Air Cargo has been set up. Though the aircraft that will be operated are civilian registered, they will be flown by Air Force pilots and they will be a big boost in the RJAF's capability to transport soldiers and cargo to faraway locations. The plan is to have two Il-76 aircraft by May 2005: now an Il-76 and a C-130 have been leased. If all goes well, in five years a total of

JY-JIA is the Il-76TD leased by JIAC, a civilian cargo company operated by the RJAF. As well as operating commercially, JIAC provides the RJAF with a readily available strategic transport capability.

five aircraft should be operated.

"Flight safety is an important aspect of our flight operations. Specially for our fighter community, the RJAF acquired a Flight Profile System in 1997 consisting of 24 pods, in order to be able to perform better mission debriefing and to boost situation awareness. The GPS/ACMI system with its Sidewinder-like shape is a real-time system and can be carried on all F-16 and F-5 aircraft. These fighters use the Sidewinder system as weapons and can easily adapt to carry this ACMI system. This system clearly shows how the battle progressed, which aircraft was in the area, which kills were made etc. This proved to be a big step forward in the mission debrief and the RJAF is very satisfied with the system. Unfortunately the Mirage F1 cannot be equipped with the system yet, because it is not able to carry the Sidewinder.

"Special Forces Operations is a very important task of the RJAF these days and has high priority. Though our country has long had to live with terrorism, in the last years special emphasis is put on terror control. We have changed our procedures and are constantly trying to improve our ability to fight terrorists. It is planned to create a separate unit assigned to only perform special operations. The RJAF is looking at buying eight UH-60 Black Hawks in the future to perform these tasks, for delivery as early as next year when plans go ahead. These should replace more UH-1Hs and should significantly boost our Special Ops capabilities. No decision has been made yet, the only Black Hawk type yet flying in the Kingdom are five S-70s operated by the Royal Flight.

"The RJAF has good relations with friendly nations. Since the new Iraqi Air Force was formed, the RJAF has been a major contributor to the new air force. Sixteen completely refurbished and repainted UH-1H Hueys are being transferred to the Iraqi Air Force: the first ones were delivered to Taji Air Base in Iraq on 1 February 2005 and final delivery of this donation should be completed by February 2006. Also two C-130B Hercules were transferred to provide tactical long-range airlift capability. The aircraft, formerly in storage, received an overhaul before delivery and will serve with the 23rd Iraqi Transport Squadron from Al Muthana Air Base. The Iraqi pilots that fly this aircraft received a four-month training course in Jordan.

"Today the RJAF is a well motivated and well trained force of around 12,000 people. The pilots of the RJAF are among the best trained in the Middle East and keep abreast of the latest tactical developments, making them tough opponents. However, our air force has been hampered by a shortage of spare parts and operating funds. Nonetheless, it is fully combat-ready and capable of flexible operations at every level of command."

Above: Jordan's latest helicopter is the EC 635, procured to be operated by No. 14 Squadron in the Special Forces support role. Anti-terrorist and other special operations are increasing in importance, and Jordan is set to acquire eight UH-60s to expand the Special Ops fleet.

Below: Having operated the Sikorsky S-76A for many years, No. 7 Squadron is now equipped with Super Pumas. They are used for general transport tasks, including Red Crescent medevac/emergency duties.

In 1985 the RJAF signed a contract with Smiths Industries to upgrade its aircraft systems and integrate the AGM-65D, of which 60 were ordered in 1983. Modifications consisted of a new HUD/WAC and radar altimeter, along with a BAe LINS 3000 laser INS as part of the Hunter nav/attack suite.

■ **No. 17 Squadron/Operational Conversion Unit (OCU)** After receiving his

wings, the new 2nd lieutenant fighter pilot moves on to the OCU, operating the F-5E/F for fighter conversion. Conversion starts with two months of academics. After that flying starts, although academics remain important throughout the course. After nine months the pilot becomes fully operational and moves on to a front-line squadron.

■ **No. 6 Squadron** As with the other fighter units at Prince Hassan AB, 6 Squadron operates F-5E/F fighters. Though the main task of the squadron is air defence, the aircraft are also operated in the ground attack and recce task.

■ **Fighter Weapons School (FWS)** Headed by Colonel Ziad Al-Qaisi, a very experienced Mirage F1 and F-16 pilot, the FWS is aimed at providing advanced training for pilots in air-to-air and air-to-ground warfare, and they are taught to operate their aircraft in different ways. All instructor pilots are assigned to fighter squadrons and deploy back to their squadron in case of threats or in a wartime situation. All trainees are lieutenants or captains with at least 750 flight hours and section lead status. During the course, a total of 46 sorties is planned: 30 air-to-air and 16 air-to-ground.

The FWS, established in 1985, is the only of its kind in the Middle East. Because of close ties between the RJAF and air forces in the region, the FWS is often visited by foreign students from countries such as Egypt and Oman. Currently one student from the UAE trains at the school.

■ **No. 9 Squadron** The other F-5 unit is based at Al Jafr Air Base/King Faisal Bin Abdul Aziz AB. This air base was built between 1974 and 1976 with Saudi assistance and is a very large facility (over 50 km²/19 sq miles). Only one squadron is based at Al Jafr, though many more could be housed there due to its size. The air base is currently receiving a new runway and infrastructure and its only resident squadron, No. 9 Squadron, is detached to Prince Hassan Air Base.

Muwaffaq Al-Salti/Al Azraq

To honour the first RJAF pilot to die in an air battle, the RJAF decided to name Al Azraq AB after Lt Muwaffaq Al-Salti, who died during combat with Israeli forces on 13 November 1966. The air base, situated in the eastern part of Jordan, was opened in 1981 to house the newly acquired Mirage F1 aircraft that were donated by the Arab leaders.

This quite new air base has good facilities and is located centrally in the country, around 75 miles from Amman. During the 1990s, the air base was often used by US Forces conducting exercises and operating combat aircraft. Several large-scale exercises were conducted with names such as Eager Tiger and Infinite Moonlight. Also, the air base was used to conduct patrols over no-fly zones in Iraq as part of Operation Southern Watch. Recently, a second runway was opened, enabling the aircraft to use their runways in any wind direction. Resident units are No. 1 Squadron operating the Mirage F1CJ and F1EJ fighters, as well as No. 2 Squadron, operating the F-16A/Bs.

Commander of the air base is Brigadier General Aref Al-Daboubi, an experienced aviator with over 3,600 flight hours, mostly in the F-5. Part of his experience was achieved during postings to the FWS as student, instructor, flight leader and commander of the unit.

■ **No. 1 Squadron** was formed in the 1950s flying the de Havilland Vampire. It was the first squadron to be established and is also the only currently active squadron to have participated in

As in many air arms, the Bell UH-1H is the backbone of the transport helicopter fleet, although numbers are dwindling as newer types are delivered. More Hueys are planned for withdrawal from No. 8 Squadron's inventory when UH-60 Black Hawks are delivered, possibly in 2006. While some RJAF UH-1Hs wear the service's standard camouflage, others retain their US Army olive drab finish.

combat operations during the 1967 war. After flying the F-104 Starfighter and the Northrop F-5A, an order was placed with Dassault to purchase Mirage F1CJs in 1980.

Until the arrival of the F-16s in 1996, the 17 Dassault-Breguet Mirage F1CJs and two Mirage F1BJs of No. 25 Squadron spearheaded the air defence assets of the RJAF. In 1981 the squadron was fully operational and more Mirages were ordered, these 17 Mirage F1EJs being delivered to No. 1 Squadron.

The F1CJ model is mainly used for the air defence role, hence its blue/grey colour scheme, while the EJ model was assigned the air-to-ground attack role and was painted in desert camouflage. The blue/grey coloured machines of No. 25 Squadron were amalgamated with the Mirages of No. 1 Squadron but still retain their No. 25 Squadron serials.

To provide training for the pilots the squadron also has two flight simulators, one for each type. These simulators are used to train the new pilots converting to the Mirage and to instruct emergency procedures and instrument flying. The squadron performs its own operational conversion. It usually takes six months to be mission-ready for a new pilot. He will be a wingman and gradually get more responsibility as a pairs leader and later section leader.

Above: A component of the Air College at King Hussein AB, No. 5 Squadron provides rotary-wing training. The advanced part of the course is taught using the MD500E, which has largely replaced the earlier 'round-nosed' MD500D.

Below: Basic training is the principal role of the CASA C-101CC Aviojet, which superseded the Cessna T-37B in the role from 1987. No. 11 Squadron operates the Aviojets, providing elements of tactical instruction towards the end of the syllabus.

Successful RJAF pilot candidates will fly the Slingsby Firefly twice: first during their screening selection process, and again for primary training after an academic break. The Firefly replaced the BAe (Scottish Aviation) Bulldog in 2002.

Partnering the MD500 in No. 5 Squadron is the Aérospatiale AS 350B-3 Ecureuil, a single-engined type that is used mainly for the early stages of the rotary-wing conversion course.

Until the arrival of the F-16, Jordan's most potent fighter was the Mirage F1, delivered in both F1CJ air defence (left) and F1EJ multi-role (above) forms. Both now serve together with No. 1 Squadron. The days of the F1CJ are numbered, the version being due for retirement when the second batch of F-16s enters operational service. The F1EJs are to continue, however, but will probably be withdrawn if a third F-16 batch is acquired. The F1CJ at left is seen with a live Magic 2 missile on the wingtip rail, and an ECM pod under the wing. Despite being operated by No. 1 Squadron, the aircraft retains the serial it wore when flown by No. 25 Squadron (the first one or two Arabic digits of RJAF serials denote the unit).

ing cannot be used on these aircraft. They carry the Matra 550 Magic 2 heat-seeking missiles and Super 530F radar-homing missiles.

■ **No. 2 Squadron** Today, the surviving Mirages of No. 1 Squadron operate alongside No. 2 Squadron operating the F-16. The changes in the Middle East, resulting from the Wye River and Wye II peace accords, have granted the Jordanians one of their greatest wishes – the F-16. After the signing of the peace treaty with Israel in 1994, the Jordanian Government resubmitted its request for up to 42 F-16s to replace its fleet of F-5 and Mirage F1 fighters. On 29 July 1996, a $200 million five-year leasing-deal was authorised, covering the supply of 12 ex-USAF F-16A ADF Block 15OCU aircraft and four F-16Bs as part of the Peace Falcon I programme. The F-16As were

To maintain currency, each Mirage pilot should fly at least one sortie a month. Pilots not being able to fly for a month can only return to flight status after a check flight. This check flight and the conversion to the Mirage F1 today causes an operational problem, since the two F1BJ models the

Successful graduates from the Air College fighter stream proceed to Prince Hassan AB, where they learn the art of fighter operations in the F-5E/Fs of No. 17 Squadron. Jordan's F-5 fleet comprises a mix of silver, grey and camouflaged aircraft.

RJAF used to operate were both lost in accidents, and today Mirage F1 students are taught the tricks of flying the aircraft in France. After a conversion course of about 50 sorties, the students return to Jordan. Because of the problems in training the pilots without a dual-seater, the RJAF is currently looking for replacement aircraft, which will probably be found in Spain.

The Mirage F1 is not able to carry AIM-9 Sidewinders: as a result the ACMI pods that the RJAF operates for situation awareness and brief-

Above: No. 6 Squadron is the operational unit at Prince Hassan. This F-5E has a false cockpit painted underneath to confuse opponents in close-in air combat.

Right: The majority of F-5F two-seaters serve with No. 17 Squadron, the fast-jet OCU.

all high-time ex-USAF aircraft, specially configured for the air defence task, with over 3,000 flight hours on each airframe. An intensive Falcon UP airframe life extension modification was performed at the Ogden Air Logistics Center, and all aircraft were refitted with F100-PW-200E engines. One of these F-16s was lost in an accident in 1997.

Peace Falcon II

A second batch of F-16s was delivered during 2003/2004. These 17 aircraft are USAF Block 15 F-16A/Bs with the Air Defense Fighter modification and were provided to Jordan under the Peace Falcon II Foreign Military Sales programme; 13 single-seat F-16As and four two-seat F-16Bs. In 2004, the RJAF awarded a $87 million contract to Lockheed Martin to support upgrades to these 17 F-16A/Bs. Part of the upgrade is the F-16A/B Mid-Life Update (MLU) that brings the aircraft to Block 50 standard. The MLU consists of enhancements to the cockpit, avionics and weapons. The upgrade also includes the Falcon UP and Falcon STAR structural upgrades. The structural upgrades extend the service life to 8,000 flight hours. This implies that the aircraft could be in service for another 20 years. The letter of offer and acceptance (LOA) was signed in January 2005 and long-lead effort began in February. Modification kits will be delivered from the spring of 2006 through the spring of 2008. All aircraft will be modified between mid-2006 and mid-2009. A brief flight test programme will be conducted in late summer 2006. Apart from better and new systems, a big step forward will be the introduction of the AMRAAM, in addition to the AIM-9 and Sparrow.

The contract was awarded on 30 September 2004. It covers kits, installation instructions, training, and spares and support after the aircraft have been modified. The kit components will be produced in the United States and Europe. The aircraft modifications will partly be performed at Turkish Aerospace Industries (TAI) facilities in Ankara, Turkey, and partly at a base somewhere in Jordan.

No. 2 Squadron also performs the OCU tasks for the F-16 and has a simulator for training purposes. At the moment the squadron commander, Lt Col Ababneh, is looking into the possibilities of splitting the OCU tasks from the operational tasks. Since 1999 the squadron is fully oper-

A view of the flightline at Prince Hassan shows aircraft from all three F-5 units. Nearest the camera is a No. 9 Squadron aircraft, normally based at Al Jafr but using Prince Hassan as a 'bolt-hole' while the Al Jafr runway is renovated.

Royal Jordanian Air Force squadrons

Units not depicted are: **No. 8 Squadron** (UH-1/AB 212 at Amman-Marka), **No. 9 Squadron** (F-5E/F at King Faisal Bin Abdul Aziz AB, Al Jafr), **No. 11 Squadron** (CASA C-101 at King Hussein Air College), and **No. 17 Squadron** (F-5E/F at Prince Hassan AB).

No. 1 Squadron
Muwaffaq Al-Salti/Al Azraq
Mirage F1CJ, F1EJ

No. 2 Squadron
Muwaffaq Al-Salti/Al Azraq
F-16A, F-16B

No. 3 Squadron
Amman-Marka
C-130H, CN-295

No. 4 Squadron
King Hussein Air College,
Mafraq – T67M Firefly

No. 5 Squadron
King Hussein Air College,
Mafraq – MD500, AS 350

No. 6 Squadron
Prince Hassan AB
F-5E, F-5F

No. 7 Squadron
Amman-Marka
AS 332AM Super Puma

No. 10 Squadron
King Abdullah II AB
AH-1F

No. 12 Squadron
King Abdullah II AB
AH-1F

No. 14 Squadron
Amman-Marka
EC 635, SA 2-37A

Left: Three F-16s taxi towards the runway at Al Azraq. No. 6 Squadron is expected to become the second RJAF 'Viper' unit, and the service would like to convert a third (possibly No. 1) to become an all-F-16 fighter force.

Below: Four F-16B two-seaters were supplied with the first batch, and another four with the second. The two-seaters are fully combat-capable, and are fitted with the AIFF 'chip-slicer' antennas ahead of the windscreen.

ational with the emphasis of its task being in line with the motto of the RJAF; air defence. The aircraft are equipped, however, to also perform the ground support role. He is very satisfied with the F-16 and its capabilities. It has proven to be a very stable and reliable platform.

Cobra Town

■ **Nos 10 and 12 Squadrons** In addition to advanced combat aircraft, the RJAF made anti-tank operations an important issue in its future plans. The US supplied an initial batch of 24 AH-1S Cobras, equipped with cannon and TOW systems, that were ordered in 1982 and delivered in 1985. The aircraft were upgraded to AH-1F standard and, in 1995, four additional Cobras were delivered that were mostly used for spare parts.

King Abdullah II AB, the newest air base of the RJAF, was constructed during the late 1990s and became operational in 1999. It was specially constructed for its new inhabitants; the 24 AH-1F Cobras of Nos 10 and 12 Squadrons that first operated out of Amman-Marka. The base is located in a central spot in the Kingdom and in the middle of the Jordanian Army exercising areas.

When rotary-wing pilots leave the Flight School they will get an intensive five-month training course during which they become a tactical helicopter pilot. During the next phase they receive advanced training.

All Cobras have dual flight controls; both crew members are able to operate the aircraft. The crew consists of a pilot, operating from the back seat who will normally fly the aircraft and also is flight commander, and the gunner in the front seat will operate the armament but during flight can also switch to flying the aircraft. Sorties always consist of at least two Cobras: single-ship operations are never undertaken.

The task of both squadrons is to support ground forces and Special Forces. This normally

Above: These No. 2 Squadron F-16As carry ACMI pods on the wingtips, allowing analysis and debriefing of air combat training sorties. As well as its operational role, No. 2 Squadron also acts as the F-16 OCU.

Below: The anti-armour role is considered of great significance to Jordan's defence, and the two-squadron Cobra force is a prized asset. Here a trio of AH-1Fs practises flying using terrain for masking.

consists of anti-tank and armoured vehicles warfare, though in peacetime – apart from exercising with the army – border patrol sorties are also being performed.

The TOW system is the primary weapon, although unguided missiles are used against lighter armoured vehicles. Also the nose-mounted cannon is used for strafing and against lightly armoured trucks. Though an Air Force asset, the Cobras are normally tasked by the army. While each squadron operates its own aircraft, the maintenance branch assigns aircraft for each mission. Crews from a squadron can easily find themselves operating an aircraft from the other unit.

To supplement the fleet nine extra AH-1 Cobras were delivered by the US during 2004 under the US Excess Defense Article Programme.

Eric Katerberg and Anno Gravemaker

Below: Standard armament for Jordan's Cobras are TOW missiles. In addition to the traditional anti-armour role, the Cobras practise operations in support of Special Forces and fly patrols along Jordan's borders.

Sukhoi Su-17/20/22

Swing-wing 'Fitter'

When the Sukhoi Su-17/-22 fighter-bomber and reconnaissance aircraft family appeared in the mid-1960s it was seen as just another improved Su-7B (NATO code 'Fitter-A') derivative. Thanks to a steady improvement effort that lasted into the early 1980s, the swing-wing fighter-bomber's operational capabilities were carefully maintained over two decades so that it can be now considered a textbook example of longevity, reliability and combat efficiency. The war in Afghanistan was the most important event in the aircraft's career.

Illustrating just how far a basic design can be pushed through continuous improvement, a Czech Su-22M4K 'Fitter-K' flies alongside an Su-7BKL 'Fitter-A'. Swing wings improved take-off performance, load-carrying and range.

The integration of a variable-geometry wing with a reliable and fuel-efficient powerplant and structurally sound fuselage, as well as an advanced digital avionics suite and hard-hitting weapons, meant that the Su-7's final derivatives – the Su-17M4 and its export counterpart Su-22M4 – were among the most cost-effective Soviet Air Force and Warsaw Pact tactical strike and reconnaissance aircraft of the mid/late 1980s and early 1990s. Now the type is still soldiering on in

numbers with the Polish Air Force and it is being regarded as a useful NATO asset set to remain in active use until 2010/12.

S-22I swing-wing prototype

In an effort to meet the VVS (Voenno-Vozdushniye Sily, the Soviet Air Forces) requirement for a significant improvement in the Su-7BM's take-off/landing performance, the development of a variable-geometry (also known as 'swing-wing') derivative was initiated in 1963 at the Sukhoi Design Bureau. Chief project designer was N.G. Zirin, and the driving force was to develop a low-level strike aircraft that was well suited to operations from semi-prepared and relatively short airstrips, otherwise unsuitable for Su-7B operations due to the latter's excessive take-off run and landing roll. The then-modern variable-sweep design concept was judged as being the most appropriate solution for improving the take-off and landing performance of the Su-7BM that was, at the time, the main fighter-bomber of the VVS Frontal Aviation, and one which was heavily tasked with low-level tactical nuclear attack.

The swing-wing prototype, known by the internal design bureau designation S-22I, retained the Su-7B's fuselage almost unchanged since this somewhat conservative design approach

In the years that followed the end of the Cold War the former Warsaw Pact nations – faced with new economic realities – embarked on a process to discard their Soviet-era equipment in favour of fewer yet more modern Western types, and the 'Fitter' community in eastern Europe dwindled accordingly. A notable exception is Poland, which has decided to keep its hard-hitting attackers in service into the next decade, suitably upgraded with NATO-compatible avionics.

Above: The S-22I prototype made its public debut at the July 1967 air show at Domodedovo. In the hands of Ye. Kukushyev the S-22I was put through a demanding display routine that highlighted the type's short take-off performance with wings forward, and high-speed manoeuvrability with wings swept. Despite this impressive show, Western analysts believed that the type would not offer sufficient performance increases to warrant being put into production.

promised to save time and drive down development and test expenses. The variable-sweep wing concept adopted for the S-22I was viewed as a sizeable compromise (by comparison with the Su-24 and MiG-23 swing-wing combat aircraft developed in the same period) because the wing roots and the inner section of the wings – representing approximately half the total wingspan – were fixed. The span of the movable wing section of 4.2 m (13 ft 9 in) was restricted by the position of the main undercarriage units that were retained from the Su-7 unchanged. The relatively large fixed centre-wing section panels (known as gloves) contributed to keeping longitudinal stability within limits throughout the entire flight envelope, at any selected angle of sweep. The hydraulically-driven and manually-controlled outer wing sections featured variable sweep angles between 30° and 63° (the former used only for take-off and landing and the latter for the rest of the flight). Later on, a third main wing setting was introduced, at 45° (mid-swept position): it proved useful for cruise and ground attack, while the maximum sweep angle was used for high-speed dashes and smooth ride at low level only. Wing sweep was set using a three-position switch located in the upper left corner of the dashboard.

In an effort to further improve take-off/landing performance, the wings were endowed with a plethora of high-lift devices not present on the Su-7B: the outer wing sections were fitted with three-section slats and slotted trailing-edge flaps, while each panel of the centre section had a two-section double-slotted area-increasing flap.

The S-22I airframe design was frozen in early 1965 and the prototype aircraft was produced through modification of an existing Su-7BM airframe, c/n 48-06. Its roll-out took place in the summer of 1966 and it made its maiden flight on 2 August the same year from Aviaprom's (Soviet Ministry of Aviation Industry) airfield at Zhukovskiy near Moscow, in the capable hands of the Sukhoi Design Bureau's chief test pilot, the famous Vladimir Sergeevich Ilyushin. The S-22I had the distinction of entering the history books as the first Soviet-built swing-wing aircraft to be flight-tested.

Improved performance

By April 1967, the factory tests had been completed and the results met all the expectations of the design team: they proved that the S-22I's swing-wing design provided a reduction in the landing/take-off speed of some 60-100 km/h (42-54 kt). As a result, the take-off run was halved in relation to that of the Su-7BM. Furthermore, the S-22I is said to have demonstrated longer range, relative to its fixed-wing predecessor, despite the fuel capacity being reduced by 330 kg (735 lb) and the gross weight being increased by some 400 kg (890 lb).

The report that was issued by the joint factory/VVS test team cited that '... utilising the swing-wing design in combination with the Su-7BM's airframe would considerably improve the subsonic performance, as well as reduce both the fuel burn per flown distance and the minimum approach speed, which, in turn, will result in further flight safety improvements and further reduction of weather minima [an important considera-

The S-22I was based squarely on an Su-7BM airframe, with little change to the basic fuselage shape. The wing design was not as ambitious as that adopted for the Soviet Union's other variable-geometry programmes (MiG-23 and Su-24), but this was largely governed by the need to retain the Su-7's main undercarriage in the new design. Restricting the sweeping wing panels to around 50 per cent of the span reduced the amount of performance gain, but also reduced the aerodynamic and structural complexities faced by the design team. Furthermore, the wing was initially limited to main settings of fully forward and fully aft, with no intermediate setting (added later). Here the S-22I gets airborne – note the open auxiliary intakes on the sides of the nose.

By the time that production Su-17s appeared a considerable amount of redesign of the fuselage had been accomplished, most notably the fairing which ran the length of the spine in place of the two conduits of the Su-7BM. First-generation swing-wing 'Fitters' were exported under the Su-20 designation, but in two sub-variants. Those for Egypt, Iraq and Syria were Type Bs, with downgraded avionics. Most were fitted with the AL-7F-1 engine. Egypt received its first aircraft in time for them to participate in the Yom Kippur War of October 1973. They were known as Su-17Ks and preceded the main batch of Su-20s. This aircraft is one of 30 that was purchased by Algeria but then given to Egypt.

tion during bad weather landings]'. The only problematic point, according to the test pilots, was the lack of buffeting at high angle-of-attack (AoA) to warn the pilot that the aircraft is approaching departure from controlled flight, a useful protective feature of the Su-7BM.

The first public appearance of the new aircraft was made during the Aviation Day Airshow held on 9 July 1967 at Domodedovo Airport near Moscow. It is noteworthy that Western analysts observing the Soviet air power said immediately after the air display that they see no reason to expect the new Sukhoi swing-wing aircraft – based on the obsolete-technology Su-7B fuselage and powerplant – to form the basis of a new-generation combat aircraft as they predicted that the variable-geometry wing would provide only modest improvements in performance. In the event, the Westerners proved completely wrong in their analysis as the VVS command authorities were highly impressed by the performance of the new aircraft. In November 1967 an order to launch the type's full-scale development and production was issued.

First-generation swing-wing fighter-bomber

The design of the new fighter-bomber that was assigned the VVS service designation Su-17 (internal design bureau designation S-32, NATO reporting name 'Fitter-C') was frozen in

1968. The production-standard supersonic swing-wing strike aircraft introduced a deepened spine housing avionics and systems, while the number of servicing hatches and doors was doubled for ease of servicing (in comparison with the S-22I).

The Su-17's outer wings moved thanks to the use of the GMP-22 hydraulic actuator, consisting of two hydraulic motors (fed from different hydraulic circuits) and four reduction gearboxes. The wings could travel backward in 16 seconds, while movement forward took 19 seconds.

The aircraft retained the AL-7F-1 rather thirsty turbojet of the Su-7B, rated at 94.15 kN (21,164 lb) thrust with full afterburning and 66.7 kN (14,960 lb) at military power. The fuel system was also similar to that of the Su-7BM but with reduced capacity due to the deletion of the wing tanks. Internal fuel capacity was 2790 kg (6,138 lb) and up to four external tanks (550 or 1100 kg each) could be carried to increase the fuel load up to 5620 kg (12,500 lb). The wings boasted two additional hardpoints for weapons or tanks, and the total combat load could reach 3000 kg (6,614 lb). The KS-4S-32 improved ejection seat enabled safe ejection at a minimum speed of 140 km/h (76 kt) at ground level.

The Su-17 introduced a considerably improved avionics suite with the new RSBN-5S tactical aid to navigation, R-832M radio, RV-5 radar altimeter and an all-aspect SPO-3 Sirena-10 radar warning receiver (RWR). Most important was the all-new SAU-22 flight control system featuring automatic/semi-automatic flight modes, as well as automatic/semi-automatic landing capability and a plethora of stabilisation modes. In addition, the SAU-22, working in conjunction with the PBK-2KL sight, enabled automatic toss bombing (useful for the employment of nuclear bombs). Late-batch Su-17s introduced the improved SAU-22-1 fed with altitude information derived from the RV-5 radar altimeter in an effort to provide automatic flight as low as 200 m (650 ft). The follow-on production batches introduced the DUA-3M AoA sensor and an additional pair of overwing fences on the centre wing, while on some early production aircraft the original pair of fences was increased in size. Additionally, late-production aircraft featured a PVD-7 back-up pitot probe relocated to a stretched boom on the port upper nose.

The Su-17 retained the warload capability of its predecessor and its principal weapons were the S-5 series of 57-mm rockets launched from 16- and 32-round packs, as well as S-24 240-mm rockets and various types of bombs and bomb dispensers, complemented by the then new SPPU-22-01 23-mm cannon pod. Every fifth aircraft that rolled off the production line was equipped with an AFA-39 camera, installed aft of the cockpit, and all aircraft featured the AKS-5 gun camera.

Weapons aiming in dive attack and level flight were made possible thanks to the ASP-PFM-B-7 optical sight. The PBK-2KL toss-bombing sight was used to aim the RN-28 or RN-24 nuclear bombs carried on a BDZ-56FNM adapter. The Del'ta-N targeting system for the Kh-23 or Kh-23M (AS-7 'Kerry') radio-command air-to-surface missile with useful range of up to 7 km (3.8 nm) was installed in the nose cone of

Poland was the only export customer for the Su-20 sub-variant A, which was virtually identical to the Su-17M in VVS service. Delivered from April 1974, Polish Su-20MKs were nuclear-capable, and augmented Su-7BMs and Su-7BKLs assigned to tactical nuclear support of Warsaw Pact ground forces. Poland's primary area of concern was northern West Germany, operating against the NATO forces that would have been cut off by a successful Soviet 'Channel dash' advance across Germany. As well as the nuclear role, the Su-20MKs were tasked with conventional attack (primarily against NATO's nuclear-capable assets) and pre-strike reconnaissance.

Egyptian 'Fitters'
Egypt received around 20 very early Su-20s (including a few Su-17MKs) equipped with AL-21F-1 engines, and a quite advanced avionics suite. They entered service with No. 221 or No. 225 Air Brigade, and saw limited action in the October 1973 war, including the use of the 100-kg (220-lb) Dibber anti-runway bomb developed by the Arab Organisation for Industrialisation's Sakr factory. No two-seaters were delivered: conversion training was completed on available Su-7Us. In 1974, there was a break in relations with Moscow, leaving EAF with only poor stocks of spares. Nevertheless, the EAF fleet of Su-20s was reinforced through the addition of some 30 aircraft acquired by Algeria. The next war in which the EAF Su-20s participated was fought against Libya, in July 1977.

By the late 1970s, only some 20 Su-20s remained operational. After establishing good relations with the USA, the Egyptians originally envisaged a comprehensive upgrade for their 'Fitters', including replacement of most of the avionics, the HUD, and a better part of the weapons system. No such plans were realised, however, and the fleet was eventually retired in the late 1980s, when additional F-16s became available. Serials of EAF Su-17MKs and Su-20s were in the range 7701/7799.

This aircraft was one of the very early Su-20s that were delivered in 1972, alongside some Su-17MKs.

all but initial production Su-17s, which retained the SRD-5D radar rangefinder of their Su-7BM predecessor.

Production of the new fighter-bomber was launched in 1968 at the Komsomolsk-na-Amur-based factory (then known as the DMZ factory and now known as KnAAPO/ Komsomolsk-na-Amur Aircraft Production Enterprise, an integral part of Sukhoi Corporation) located in Russia's Far East, and by that time producing all Su-7B derivatives. The first development example assembled there was S-32-1 – it made its maiden flight at Zhukovskiy on 1 July 1969 and was followed not long after by S-32-2. These two pre-production aircraft were used in the type's State Tests, held between July 1969 and May 1971. Based on these results, the new swing-wing fighter-bomber was recommended for service entry; large-scale production commenced at DMZ in late 1970 and it finally superseded the Su-7BM on the production line in 1971. Some 250 AL-7F-1-engined Su-17s were built until 1973, and the last survivors soldiered on with the VVS training units until the mid-1980s. The first VVS unit to receive the Su-17 was the 4th TsBP (Combat Training Centre) in Lipetsk, which took on strength the initial batch of 10 aircraft, while the first front-line unit to be equipped was the 523rd APIB (fighter-bomber regiment) located at Dzemgi in Russia's Far East. The type's operational evaluation and field-testing took place there in 1972/73. According to Sukhoi company official history records, a small number of aircraft in a stripped-out export variant known as the Su-17K (internal designation S-32K, also Su-17MK) were delivered to Egypt to meet an urgent export requirement in 1972.

New engine for improved performance

The Su-17M (internal factory designation S-32M, NATO code 'Fitter-C') was an improved version which appeared in 1972, the most significant difference from its predecessor being the introduction of the Lyul'ka AL-21F-1 turbojet. During the Su-17's initial production run, it was quickly superseded by the uprated AL-21F-3. The new engine was purposely designed by the Lyul'ka Design Bureau (now known as NPO Saturn) for the Su-24 tactical bomber in the mid-1960s, based on the General Electric J79 that powered the McDonnell F-4 Phantom II (acquired as a war trophy in Vietnam). The AL-21F-3's thrust of 76.5 kN (17,200 lb) dry and 110.5 kN (24,700 lb) with full afterburning enabled the designers to remove the Su-7B-style bulges on the Su-17M rear fuselage, which had been dictated by the area rule. The combination of the new fuel-efficient powerplant, improved aerodynamics and the increased internal fuel capacity (to 3630 kg/8,003 lb) resulted in a range increase of around 75 per cent relative to the AL-7F-1-engined Su-17, while the take-off run was shortened by some 15 per cent. The new engine, however, proved very troublesome during the initial in-service period and a number of aircraft were lost due to engine fires. Later on, the AL-21F-3's reliability was greatly improved, and during the 1980s it boasted an impressive flight safety record.

Initially, it was intended that the Su-17M be equipped with the new mission avionics suite similar to that developed for the Mikoyan MiG-23B, but in the event the plan was shelved due to development problems with the advanced avionics systems that delayed their in-service introduction.

The Su-17's two pre-production aircraft were re-engined with the AL-21F-1 and the first of these took to the air on 28 December 1971. The first production Su-17Ms were delivered to the VVS in 1972, and the new 'Fitter' derivative entered service in 1973 (with the 4th TsBP and 523rd APIB) though it was not formally commissioned into VVS service until November 1974. A total of 10 batches rolled off the production line until 1976.

The Su-17M introduced an array of new weapon options, such as MBD3-U6-68 multiple bomb racks (for carrying up to 32 100-kg bombs on eight racks), S-25 250-mm rockets, S-8 80-mm rockets carried in 20-round packs, and KMGU bomblet dispensers. The Su-17M's bombload was increased to 4000 kg (8,818 lb) thanks to the addition of two more underfuselage hardpoints, bringing the total number to eight. Production examples boasted a number of avionics and system improvements such as the Sirena-3M RWR and the ARK-15 ADF. In 1973-1974, an Su-17M was utilised for trials with the then newly-developed air-to-surfaces missile types – the laser-guided Kh-25ML and Kh-29L, the Kh-28 anti-radar missile – as well as the R-60 air-to-air missile. The Kh-28 (AS-9) anti-radar missile was added to the Su-17M's arsenal at a later stage, used with the associated Metel'-A/AV pod-mounted emitter locator/targeting system.

The Su-17M's export derivative appeared in 1972, initially known under the internal factory designation S-32MK, while the new service designation Su-20 was assigned after it had been launched in production (however, the NATO reporting name 'Fitter-C' was retained). The first Su-20 made its maiden flight on 15 December 1972; in the autumn of the same year

A group of Soviet pilots walks past a line of 'Fitters' that comprises both single-seat Su-17M3s and two-seat Su-17UMs. The 'Fitter-H' entered VVS service in 1977, bringing with it important new capabilities, such as an effective SEAD function using the Kh-25MP and Kh-58 anti-radiation missiles, employed in conjunction with the Vyuga-17 targeting pod. Power was provided by the Lyulka AL-21F-3.

Hungary was the only recipient of the Su-22M3 version, this type being much closer in equipment levels to the VVS's Su-17M3 than the Su-22Ms delivered to other export customers. This view highlights the large wing fences of the swing-wing 'Fitter'. Also visible – behind the fuselage pylons – is the ventral strake added during the production run to enhance longitudinal stability.

navigation radar (mounted in a prominent fairing under the nose), IKV inertial reference gyro as well as the new ASP-17S and PBK-3-17s sights, RSBN-6S tactical radio aid to navigation and the SAU-22M automatic flight control system. Working in conjunction with the SAU-22M, the V-144 analog computer provided semi-automatic and fully automatic flight on a pre-planned route with three waypoints and four return airfields. All this equipment underwent trials in an Su-17M in early 1972.

The new version also introduced a nose stretched by 200 mm and a redesigned fuel system with tanks filled with inert gas to prevent explosion of the fuel vapour in the event of a bullet rupturing the tank. In addition, the PTB-800 800-litre drop tanks were introduced during the production run, and they became standard equipment for all following 'Fitter' variants.

The Su-17M2 was able to carry the most modern Soviet guided weapons, already tested on the Su-17M, namely the Kh-25ML (AS-10) laser-guided missile used in conjunction with the Prozhektor-1 laser designation pod mounted under the wing. The powerful Kh-29L (AS-14 'Kedge') laser-guided missile followed in 1975, and in 1976 the Su-17M2's armament was further enriched with the new R-60 (AA-8 'Aphid') air-to-air missile, highly useful for self-defence. In 1979, the S-25L 250-mm modular laser-guided missile was integrated, thus further enhancing the type's guided weapons options. At the same time, the Su-17M2 retained the capability of firing the Kh-23 and Kh-23M radio-controlled missiles, with the Del'ta-NM guidance system housed in a pod instead of being built into the fuselage.

The Su-17M2's first pre-production example was flown for the first time on 20 December 1973, while the type entered regular VVS service in February 1976. It continued in production at KnAAPO between 1974 and 1979.

The Su-17M2D was a derivative powered by the Tumanskiy R-29BS-300 turbojet (originally developed for the MiG-23BN), rated at 78.4 kN (17,248 lb) dry and 112.27 kN (24,794 lb) with full afterburning. The new engine was considerably cheaper than the AL-21F-3 and offered a certain degree of commonality with the MiG-23BN/MF fleets, which were operated by some of the Su-22's export customers. However, the R-29BS-300 had a higher specific fuel consumption and somewhat larger dimensions than the AL-21F-3, thus necessitating extensive redesign of the rear fuselage and the slight enlargement of the dorsal fin. A single example was built for trials at the end of 1974 and it took to the air for the first time on 31 January 1975. Tests demonstrated that the R-29BS-300-engined aircraft lacked any advantages relative to its AL-21F-3 counterpart; moreover, it had significantly reduced performance due to the decreased lift-to-drag ratio of the extensively redesigned rear fuselage section. Nevertheless, the overall performance of this version was more than acceptable for foreign operators and the Su-17M2's export derivative – powered by the R-29BS-300 turbojet and known under the designation Su-22 (internal factory designation S-32MK, NATO 'Fitter-F') – was in production between 1976 and 1980.

'Fitter-Fs' also had restricted equipment and weapons options – for instance, the Fon-1400 was replaced by the SRD-5 radar rangefinder in the nose cone and the guided weapons were represented only by the Kh-23M and R-3S/R-13M missiles. A number of Su-20s and Su-22s were upgraded for carriage of the large KKR-1 photo/ELINT reconnaissance pod, receiving the new designations Su-20R and Su-22R, respectively.

The first Su-22 was built at KnAAPO using a Su-17M2 production-standard airframe and it flew for the first time in February 1976. The type continued in production until 1978 and was sold to Iraq, Peru, Libya, Yemen and Angola.

Two-seat derivative

After having already developed a number of single-seat combat variants for domestic and export use, Sukhoi conceived a dedicated two-seat conversion and continuation training version. It is noteworthy that, prior to the two-seater's introduction in the mid-1970s, all Su-17/M/M2/20/22 pilot

it was demonstrated to high-ranking officers from the Egyptian, Iraqi and Syrian air forces and, not long after, all of them placed orders. The type's flight tests were completed in March 1973 and it was immediately released for export.

Most of the Su-20s were powered by the old AL-7F-1 turbojet and had a downgraded avionics suite, similar to that used in the export Su-7BKL. The Su-20 exported to the Third World countries (sub-variant B) featured a considerably restricted weapons suite, but it was the first 'Fitter' variant to feature provision for two to four R-3S and R-13M heat-seeking air-to-air missiles for self-protection and visual intercepts. The 24 Su-20s sold to Poland (sub-variant A), however, were almost identical to their VVS brethren, retaining the same powerplant, avionics and even the Su-17M's nuclear capability.

The Su-20 continued in production at KnAAPO between 1973 and 1975, and it is noteworthy that its deliveries had begun earlier than the Su-17M, with the initial production aircraft going firstly to the Syrian air force in April 1973, followed by the delivery to the Egyptian air force. The type received a limited baptism of fire during the October 1973 war with Israel.

Improved combat capability

The Su-17M2 (internal factory designation S-42, NATO 'Fitter-D') was the result of the Su-17's second upgrade stage, which called for the integration of an all-new sophisticated nav/attack suite, its most important components being the nosecone-mounted Fon-1400 laser rangefinder and the KN-23 nav/attack suite (originally developed for the MiG-23B). The KN-23 integrated a V-144 analog computer, DISS-7 Doppler

Above and right: Su-22s figured prominently in the Libyan air force's dramatic expansion in the late 1970s and early 1980s. Su-22 'Fitter-Fs' were followed by 'deep-spine' Su-22M 'Fitter-Hs' like these aircraft. As well as attack duties, Libyan Su-22s were used for air defence tasks. Two were shot down by US Navy Tomcats during a confrontation over the Gulf of Sidra on 19 August 1981. For the air defence role Libyan 'Fitters' could be armed with R-3 (illustrated) or the later R-13 'Atoll' Sidewinder-copies, and in the late 1980s received R-60 'Aphids'.

type conversion training was carried out on the obsolete Su-7U. However, the latter's notably different take-off/landing performance relative to the Su-17/17M2 is said to have considerably complicated the conversion process. This was a problem especially for the student pilots who converted to the fighter-bomber just after they had received their two years of initial and basic training on the relatively slow and all-forgiving Aero L-29 at the VVS's Eisk Pilot School (EVVAUL).

Initially given the service designation service designation Su-19U, the two-seater was subsequently redesignated as the Su-17UM (internal factory designation S-52U, NATO 'Fitter-G'). It introduced a completely redesigned front fuselage with

Angolan 'Fitters'

The first batch of some 12 Su-22s (also known as Su-20Ms) entered service with the 15th Fighter Squadron of the Força Aérea Popular de Angola – Defesa Anti-Aviones (FAPA-DAA) in 1984. Four were lost by 1985, when the 15th FS participated in Operation Second Congress – aimed at capturing UNITA-held regions around Mavinga and Jamba. Within two months of operations, only one Su-20M (C507) remained operational: all the others were shot down or lost in additional accidents. A new batch of 12 Su-22s was supplied instead, but further losses were suffered and the FAPA-DAA never managed to keep more than 50 per cent of them in operation. Parallel with Su-20M deliveries, at least five Su-22UM3Ks were delivered. In 1989, a batch of 18 new Su-22M4Ks was delivered from the USSR in order to replace the few battered surviving Su-20Ms. At least six of them are known to have been shot down by UNITA in the period between January 1993 and late 1994.

In 1993 the FAPA-DAA was renamed the Força Aérea Nacional (FAN). Only nine Su-22M4Ks and a single Su-20M (C510) remained operational as of 1999, still serving with the 15th FS. They were meanwhile reinforced by 11 Su-22M4Ks and a single Su-22UM3K from Slovakia. During the 1990s, FAN Su-22s were deployed not only in combat against UNITA, but also during Angolan interventions in Congo-Brazzaville, in 1997, and then in the Democratic Republic of Congo, between 1998 and 2001. In the later conflict they flew combat sorties in support of President Kabila against Rwandan troops and various Congolese rebel factions. The current status of the Su-22 fleet is unknown, especially since much more capable types – including Su-24Ms and Su-27s – were delivered to Angola in recent years and the FAN considerably downsized, to only three flying regiments with three squadrons each.

The Su-22 (Su-20M, above) was from the first batch received by Angola. Serials of this batch were C501 to C512, although the serials were reused by subsequent Su-22 and Su-22M4K batches, which replaced aircraft shot down by UNITA and South African forces. The first Su-22M4K batch used serials in the range C501 to C518 (below), while the former Slovak aircraft carry single letter codes on the fin. The original two-seaters were serialled in the I-31 to I-35 range.

a distinctively drooped nose to improve the pilot's visibility during landing approach and shallow dive attacks – 15° downwards angle of view as opposed to 9° in all preceding variants. The two-seater's airframe was similar to that of the latest Su-17M2 derivative, which was initially designated Su-19 but re-designated as the Su-17M3 in 1974. Other new features common to the Su-17UM and Su-17M3 comprised the distinctively deepened spine housing additional fuel tanks and some avionics/systems. In an effort to save weight in the two-seater, the port wingroot gun was deleted. Both two- and single-seaters also boasted improved aerodynamics thanks to the DISS-7 Doppler radar being mounted inside the fuselage contours instead of being placed in an Su-17M2-style vane. The two-seater introduced the very advanced Zvezda K-36D ejection seat, rated at a minimum safe operating airspeed of 40 kt (75 km/h) at zero altitude.

The Su-17UM prototype made its maiden flight in September 1975 and was in production at KnAAPO between 1976 and 1982. Its export derivative powered by the R-29BS-300 turbojet received the Su-22U (NATO 'Fitter-E') designation and flew for the first time on 22 December 1976. Interestingly, the production examples were flight tested at KnAAPO even before the Su-22M prototype had been flown, as the first production aircraft took to the air in November 1976.

Su-17M3 – the most popular of the 'Fitters'

The Su-17M3 (internal factory designation S-52, NATO 'Fitter-H') was the next single-seat version designed for domestic use. It utilised the two-seater's airframe, with the space originally occupied by the second cockpit now provided for instal-

lation of new avionics components. A ventral strake was introduced onto both single- and two-seater in 1978 and the fin was made taller in an effort to improve directional stability (which had been eroded by the effects of the deeper spine). Two additional pylons for carriage of R-60 missile launchers were mounted onto the non-moving sections of the wing. For guided missile carriage two more hardpoints were added on the fuselage sides, thus increasing the maximum number of the Kh-25 family missiles to four (only two can be carried on the Su-17M2).

Most importantly, the Su-17M3 boasted an improved KN-23-1 nav/attack suite, which added the Klyon-PS laser rangefinder/designator (replacing the Fon-1400) and an improved ASP-17BTs electro-optical weapons sight. In addition, the aircraft received many other new avionics systems, such as the RSDN-10 (A-720) long-range radio aid to navigation, SAU-22M1 flight control system, RV-15 radar altimeter, Parol IFF, SO-69 ATC transponder and the Tester-U3 enhanced flight data recorder. Self-defence was improved thanks to the introduction of the SPO-15 Beryoza-L RWR, and the new 'Fitter' version had the distinction of being the first Soviet tactical aircraft to be fitted with integral chaff/flare dispensers (two six-round KDS-23 in the spine aft of the cockpit firing LO-43 50-mm infrared flares or chaff cartridges). It also featured the option of using the SPS-141MVG Gvozdika active radar jamming pod, or its SPS-142/143 sub-variants covering different frequency ranges.

The M3 version was endowed with much better SEAD capability thanks to the integration of new-generation anti-radar weapons – the Kh-25MP (AS-12) and Kh-58 (AS-11) ARMs together with the associated L-086 Vyuga-17 conformal targeting and launch control pod, receiving the new designation Su-17M3P. The ARM integration made possible a number of Frontal Aviation Su-17M3-equipped squadrons (and later Su-17M4) to specialise in the dangerous Suppression of Enemy Air Defence (SEAD) role.

The Su-17M3 made its maiden flight on 30 June 1976 and the type's test programme was completed in December 1978, but the first production aircraft entered service with the 4th TsBP at Lipetsk as early as September 1977. The first front-line regiment that converted to the Su-17M3 was the 274th APIB based in Kalinin (Moscow Military District), which took on strength the new type in late 1977. The most serious difficulties encountered during the test and evaluation programme are said to have been attributed to the low reliability of the ASP-17BTs sight. The Su-17M3 formally entered VVS service in July 1981 and it soon became the most popular Su-17 derivative, loved by both pilots and technicians. As many as 1,000 aircraft were produced at KnAAPO (including the Su-22M and Su-22M3 export derivatives) between 1977 and 1983.

'Fitter-J' for export

The Su-17M3's export derivative was known as the Su-22M (internal designation S-52K, NATO 'Fitter-J'). Powered by the R-29BS-300 turbojet, it featured – just like its predecessor Su-22 – a considerably downgraded avionics suite (for example, the Fon-1400 laser rangefinder instead of the more capable Klyon-PS) as well as restricted weapons options. The

prototype made its maiden flight in 1976 and the Su-22M continued in production until 1981. Its improved derivative, which boasted the Su-17M3's full-standard avionics suite (although stripped of nuclear delivery capability and with SPO-10 RWR instead of SPO-15L) entered production in 1977 under the appropriate designation Su-22M3 (S-52MK). As many as eight export customers took delivery of the Su-22M, among them being Libya, Syria, Iraq, Peru, South Yemen, North Yemen, Afghanistan and Vietnam, while the Su-22M3 was sold to Hungary only in the Warsaw Pact.

In 1978, an Su-17U two-seater was upgraded with the Su-17M3's avionics suite, receiving the appropriate designation Su-17UM3 (internal designation S-52UM3) – it made its maiden flight in September 1978 and during the same year superseded the Su-17U on the production line. The two-seater's export derivative, powered by the R-29BS-300 (internal factory designation S-52UM3K, NATO 'Fitter-G') and featuring the improved avionics suite, was in production in 1982 and 1983. Export permission for the AL-21F-3 was at last granted in 1983, making it possible for a number of export twin-seaters – ordered by four East European air arms, and later for a number of Third World states – to receive the better power-plant, with the first Su-22UM3 (also known as the Su-17UM3K) deliveries taking place the following year. This variant continued in production until the late 1980s.

The ultimate 'Fitter'

The Su-17M4 (internal factory designation S-54, NATO 'Fitter-K') was the last production variant and the most capable member of the family, successfully integrating the proven Su-17M3 airframe and powerplant with the first generation of Soviet digital avionics. Conceived in 1975 – initially under the new service designation Su-21 – it was to boast a number of new features for improved performance and combat capability, such as the AL-31 afterburning turbofan (developed for the Su-27) and an all-new nav/attack suite similar to that developed for the MiG-23BK (MiG-27K) with the purpose of expanding the weapons options.

Under initial plans, the nav/attack suite was to be built around the Orlan laser/TV targeting system (an improved derivative of the MiG-23BK's Kaira) and an Orbita-series digital computer. The new weapons enabled by the Orlan were to comprise laser- and TV-guided bombs and missiles. The obsolete NR-30 guns were to be replaced by the newer and much more powerful TKB-687 (GSh-301) single-barrelled gun with 1,500 rpm rate of fire (some sources, however, maintain that the GSh-6-30 six-barrelled 30-mm gun purposely developed for the MiG-23BM was also eyed for integration into the Su-17's advanced derivative).

In the event, all these novelties were shelved due to design difficulties and costs considerations – for instance, there was

no space found in the nose suitable for the installation of the heavy and bulky Orlan system. The work on the integration of the new Kh-59 TV-guided long-range air-to-ground missile also failed since, in the early 1980s, the VVS leadership had already decided that the Su-17M4's development potential had been exhausted, and thus cancelled any further costly upgrade works. There was also an idea to significantly improve the Su-17M4's low-speed manoeuvrability thanks to the introduction of automatically actuated slats and flaps but, as might be guessed, this was also shelved.

The Su-17M4 eventually received the 'modest' PrNK-54 integrated digital nav/attack system, consisting of the Orbita-20-22 digital mission computer, an improved Klyon-54 laser rangefinder/designator, the IKV-8 inertial navigation unit, A-321 short-range radio aid to navigation, A-720 long-range radio aid to navigation, RV-21 radar altimeter, ARK-22 ADF, R-862 UHF/VHF radio and a further improved SAU-22M2 flight control system. In addition, the SPO-15LM improved RWR superseded the older SPO-15L.

The PrNK-54 made possible automatic flight on a route with six waypoints and four targets. The newly added bulky hardware required a redesign of the nose section and a fixed intake cone (to save space). This, in turn, resulted in a reduction in maximum level speed at high altitude from Mach 2.09 to Mach 1.7. Nevertheless, the aircraft's overall combat effectiveness was not affected as the capability of reaching Mach 2.0-plus speed at high altitude was considered a luxury for the low-level strike platform (which only rarely flies at supersonic speed). The redesign also reduced the fuel tank capacity to 3770 kg (8,294 lb), the same as in the Su-17M2. Some of the new avionics components were installed in a further stretched spine, while the increased cooling requirements of the new avionics suite necessitated adding an intake at the base of the fin feeding the heat exchanger. The Su-17M4 also introduced the improved Zvezda K-36DM zero-zero ejection seat.

There were no Su-17M4 prototypes, as three development aircraft were produced at KnAAPO by the conversion of standard Su-17M3 airframes. The first of these took to the air on 19 June 1980 and the new derivative's full-scale production took place between 1981 and 1990. The first VVS front-line unit to take the new 'Fitter' derivative on strength was the Kalinin-based 274th APIB. Officially, the Su-17M4 is reported

Poland operates around 15 Su-22UM3Ks to support the fleet of Su-22M4K single-seaters. They were initially distributed around the four operational squadrons, although 6. elt was disbanded in 1997. Late-model 'Fitter-Gs' were powered by the Soviet-standard AL-21F-3 – as here, although early export aircraft had the Tumanskiy R-29BS-300. The original two-seater – the Su-17UM/Su-22U – was the first 'Fitter' to have Zvezda K-36D ejection seats installed.

The Libyan Arab Air Force received a sizeable number of Su-22U/UM3 two-seaters for its three squadrons of Su-22/22M 'Fitter' fighter-bombers. This Su-22UM3K was seen following overhaul at the 558 ARZ (aircraft repair facility) at Baranovichi in Belarus. The long and straight fin fillet identifies it as one of the early export aircraft fitted with the Tumanskiy engine.

Although there were several schemes for continued development of the Su-17/22, the M4 was the final production version. It was widely exported (including to Poland, illustrated).

The M4 was easily distinguishable from earlier models by the intake forward of the fin. Although the Su-17/22M4, like these East German aircraft, was a general improvement on the M3, there were some drawbacks – notably reduced range – and the avionics suite was below the specification of that originally envisaged for the variant.

to have entered VVS service in September 1984.

The Su-17M4's export derivative is known as the Su-22M4 (internal factory designation S-54K, NATO 'Fitter-K'). Powered by the AL-21F-3, its production took place between 1984 and 1990 and it was exported to at least nine states.

The VVS's Su-17M3/M4s and the export Su-22M3/M4s operated by the independent reconnaissance regiments (ORAPs) were capable of using the KKR-1 and KKR-2 series of reconnaissance pods, fitted with conventional wet-film cameras, night wet-film cameras, IR linescan, TV cameras, ELINT systems and other intelligence-gathering hardware. The reconnaissance-capable 'Fitter-H/Ks' featured two additional pod control panels in the cockpit and received the Su-17M3R and Su-22M4R in-service designations, respectively. Late-series, non-reconnaissance-capable Su-17M4s and Su-22M4s saw their weapons suite enhanced in 1987/88 by the Kh-29T TV-guided air-to-surface missile (lock-on before launch weapon with maximum range of up to 13 km/7 nm) with the associated IT-23M CRT display installed in the cockpit, replacing the Su-17M4R's frame-counter in the upper right corner of the dashboard.

In the early 1980s a new version, tentatively designated Su-17M6 (internal factory designation S-56), was proposed by Sukhoi: it was to feature a fixed-geometry wing with a swept angle of 45°, endowed with large leading-edge root extensions. The M6 version was to be powered by the Su-27's AL-31F turbofan engine and to boast a new-generation digital nav/attack suite. However, as the Su-17's airframe was considered too obsolete for any further upgrades the project was abandoned at an early stage.

In total, 2,867 aircraft of the Su-17 family rolled off the production line at the Komsomol'sk-na-Amur factory, 1,165 of which being sold to export customers in at least 15 countries, while 1,702 were produced for domestic use according to official Sukhoi Design Bureau records (prominent Soviet/Russian military aviation researchers Victor Markovskiy and Igor Prihodchenko provide somewhat different figures, namely 771 and 1,095 respectively, making a total of 1,886).

The Su-17 was used to equip no fewer than 32 strike regiments, 12 reconnaissance regiments, one independent reconnaissance squadron and four conversion/advanced training regiments of VVS Frontal Aviation and Naval Aviation. A total of 15 versions was built for export and the unit price is said to have varied between US $2 million for the early Su-20s (exported to Egypt and Syria in the mid-1970s) and $6-7 million for the late-series Su-22M4s acquired by the Warsaw Pact member states in the mid/late 1980s.

Afghan combat experience

VVS 'Fitters' received their baptism of fire in the war in Afghanistan as Soviet forces effectively occupied the country for nearly a decade. The Su-17M3 and M4 rapidly became the principal fighter-bombers and attack aircraft there until the Soviet withdrawal in 1989. First to deploy were two squadrons of the 217th APIB, home-based at Kzyil Arvat airfield in the Soviet Republic of Turkmenia, equipped with the Su-17. Stationed in 1980/81 at Shindand airfield in the south of the country, they saw a lot of use in the close air support role, as well as for strikes against fixed and mobile targets in the territory held by the Mujahideen (Afghan rebel forces). Bombing and rocket attacks were carried out from safe altitude, and in this role the Su-17 proved unsatisfactory due to its poor nav/attack suite and the short mission radius provided by the fuel-thirsty AL-7F-1 turbojet.

In 1982, the Su-17s were replaced by the much more capa-

ble Su-17M3 and, until the end of the war, a number of VVS fighter-bomber and reconnaissance regiments were deployed on a rotational basis for one year to three airfields in Afghanistan (Bagram, Kandahar and Shindand), while others flew combat sorties from their bases in bordering Soviet republics such as Kokadyi in Uzbekistan and Mari-2 in Turkmenistan, on an as-needed basis. The Su-17M4 received its baptism of fire in Afghanistan in 1987. The type – which boasted a very precise navigation system – was often used in the closing stages of the war as a leader for mixed groups of Su-25s and MiG-23MLDs undertaking high-altitude bombing against targets with known coordinates (with no visual contact to be required for the aiming). In such missions, aiming was made possible thanks to the R-720 long-range aid to navigation of the leader's aircraft feeding the necessary information into the Su-17M4's nav/attack suite, and the other aircraft in the group dropped their bombs upon voice command issued by the leader.

An example of the Su-17M3's Afghan experience was the deployment of the 1st squadron of the 156th APIB, home-based at Mari-2. It deployed to Kandahar airfield in Afghanistan in May 1983 and had commenced flying combat sorties within a week of deployment, even without any area familiarisation flights. Usually, strike missions were carried out against targets provided in the combat order issued by the 40th Air Army HQ (40th Air Army, headquartered in Kabul, was the VVS army controlling the aviation assets in Afghanistan), with three to four four-ship flights employed on such missions every morning. Strikes against assigned targets were usually

carried out with 50 seconds' spacing between flights, initially undertaken in several passes from different directions. As the war grew in intensity, and the Mujahideen air defence became stronger thanks to foreign support, the 156th APIB's strike tactics changed to one attack pass only, with a salvo bomb drop or multiple rocket launch.

Squadron deployment

During the first deployment of the squadron – between May and November 1983 – it saw much action in providing air support to ground forces engaged in four large-scale offensive operations, while the second deployment – between January and October 1984 – saw the squadron taking part in two more. During both deployments each pilot logged around 180 combat sorties. As many as five aircraft losses were reported, claiming the lives of two pilots. Two of the losses were attributed to hits from large-calibre machine-guns (mainly the 12.7-mm DShK, a Russian World War II-vintage design manufactured in China, and 14.5-mm units) or small-calibre anti-aircraft artillery (20-mm Oerlikon guns), while the rest are thought to have been due to man-portable SAMs or hardware failures.

A pilot from the squadron – who took part in both deployments – commented that the principal causes for the losses were not related to poor pilot proficiency or the Su-17M3's lack of combat capability. Moreover, this particular Su-17M3 driver with Afghanistan experience, quoted by the Russian aviation history magazine *Mir Aviatsii*, maintained that, in the hands of skilled pilots, the Su-17M3 was a real combat mount.

Old adversaries meet in friendship – a Polish Su-22M4R flies with a pair of Jaguars from the RAF's No. 41 Squadron during a unit exchange. Although the RAF and Polish AF had very close ties during World War II, for much of the Cold War they faced each other as enemies. On the 'front line' in Central Europe both Jaguars and 'Fitters' had tactical nuclear strike commitments, and both were tasked with low-level reconnaissance missions. Today both types still share certain attributes – old yet still viable aircraft that have been updated to be capable of precision attacks with conventional weapons, while retaining an impressive reconnaissance capability.

Nicknamed the 'Strizh' (swallow), the Su-17M was deployed with the 16th Air Army in East Germany from the mid-1970s, equipping the 20th APIB at Gross Dölln (also known as Templin) and the 730th APIB at Neuruppin. There was also a reconnaissance regiment (294th ORAP) at Allstedt. The aircraft above wear the yellow codes of the 20th APIB. The Su-17M4 at top carries the polar bear badge of the regiment's second eskadrilya, while the Su-17UM3 above is adorned with the Guards badge, as worn by aircraft of the third eskadrilya. The 3rd squadron also had a badge of a diving eagle (or later a double-headed Romanov eagle), with the numbers '3' and '20' (or 'XX') superimposed. The first squadron used a bat design. The 20th APIB was one of the last regiments to depart Germany, deploying to Taganrog on 5 April 1994.

The weapons employed the most by the 156th APIB were 250-kg and 500-kg high explosive/fragmentation bombs, which were either dropped in a dive (of between 20° and 40°), or in level flight. Aerial minelaying, using the KMGU mine dispenser (containing as many as 1,248 anti-personnel high-explosive mines), was another routine task for the Su-17M3 pilots, carried out at altitudes between 600 and 1000 m (1,900 and 3,280 ft), and at a speed of 900 km/h (485 kt). Rockets were used on many missions, mainly in the form of the inexpensive 57-mm S-5 fired from 32-round UB-32-57 launchers, but the larger and more lethal 80-mm S-8 and the 240-mm S-24 also saw use during the squadron's two deployments.

Initially, all aiming was made in automatic mode, but in the high-altitude mountain conditions the ASP-17BTs electro-optical sight often failed and this forced the pilots to transition to the manual aiming mode. Between deployments, the 156th APIB's ASP-17BTs were modified in an effort to make them cope with automatic sighting in mountainous conditions. According to pilots with Afghanistan experience, the best bomb precision was achieved during 40° dive attacks, but it was not always possible to enter such a dive in the mountains due to difficulties related to altitude loss after pull-up in the rarefied air. This was particularly true for the strikes made against targets at 3000-5000 m (9,840-16,400 ft) above sea level. In order to be safe, the 40° dive attack commenced from 6000-7000 m (19,680-22,960 ft), but climbing to such levels during hot summer days proved impossible without engaging the afterburner (but this caused excessive fuel consumption that, in turn, drastically reduced the range).

Hot-and-high conditions, a common occurrence in Afghanistan, considerably affected the Su-17M3's airfield performance; as a consequence the weapons load was restricted to three 500-kg bombs. However, the three-bomb carriage imposed handling complexity as the asymmetrical load

made the take-off difficult, especially in strong crosswind conditions. Taking off with a load of four 500-kg bombs at Kandahar was said to have been an impossible task for the Su-17M3 when the air temperature exceeded 35°C because of the insufficient runway length; as a consequence, the usual bombload was most often reduced to two 500-kg bombs only.

After dropping the bombs in a dive attack, the Su-17M3s were required immediately to perform anti-fragmentation and defensive manoeuvring, climbing at a 30° angle and in a right-hand sideslip and all the while pumping flares to fool the seeker heads of the Strela-2M and Red Eye hand-held SAMs that might be launched against the attackers (the Strela-2M originating from Egypt was supplied to the Mujahideen as early as 1983 and the Stinger/Red Eye debuted in 1986). The sideslip is said to have been effective for causing gun-aiming error – as a result the machine-gun and AAA bursts fired tail-on at the Su-17s usually passed by safely to the left.

Visual reconnaissance was another routine task, carried out by the Kandahar-deployed fighter-bombers, usually in pairs. When suspicious activity was detected in enemy-held territory, it was immediately broadcast to the command post using voice messages over the radio.

Tyre bursts were commonplace on take-off, a combination of the hot conditions and poor tyre quality. In such cases, pilots preferred to complete the mission. Brake overheating also led to numerous failures and the aircraft overshot or veered off the runway. In most cases only minor damage was sustained and the aircraft were soon returned to combat-ready status. Pilot comfort was made difficult during hot weather operations due to the low efficiency of the environmental control system, which was incapable of providing sufficient cooling in the cockpit when on the ground.

Reconnaissance operations over Afghanistan

Su-17M3Rs debuted in Afghanistan in 1984, replacing the MiG-21Rs stationed at Bagram airfield with the 263rd ORAE (independent reconnaissance squadron). Their primary task was to search for horse and foot convoys transporting weapons and supplies that regularly entered from Pakistan, heading for numerous enclaves held by the Mujahideen. Other important tasks of the 263rd ORAE comprised frequent road and mountain pass surveillance missions, searches for rebel bases and dugouts, as well as pre- and post-strike reconnaissance and battle-damage assessment missions.

Dangerous visual reconnaissance missions in the deep valleys in the northern part of Afghanistan was usually carried out in pairs at an altitude of 200-400 m (656-1312 ft). The wingman, flying in trail formation behind the leader, slightly higher and often with a certain angle of sideslip, was tasked with looking for air defences and issuing warnings to the leader in case SAM launches or AAA fire was detected. Over flat terrain, the recce pairs flew with lateral separation,

and the wingman was required to stay much higher – up to 1800 m (5,900 ft) – and from time to time to perform a barrel roll for a better look at the ground for threats below the flight path. If signs of suspected activity was spotted, the pair usually executed a second pass to photograph the target.

In addition to the 'pure' reconnaissance sorties, in certain areas the Su-17M3Rs often flew 'search and destroy' sorties, asking the command post for clearance for attacking targets of opportunity. Furthermore, there were certain areas marked on the map – these being located around Kandahar, the Javak pass and Lakray – where the reconnaissance aircraft were allowed to attack convoys without prior clearance from ground control. Typical mission loads for the search and destroy sorties comprised a KKR-1 or KKR-2 pod (equipped with A-39 and A-42-100 cameras) and two 800-litre drop tanks on the leader's aircraft (on short-range missions tanks were often replaced by a pair of 100-kg or 250-kg bombs, or UB-32-57 rocket pods), while the wingman usually carried four to six 100-/250-kg bombs or two-four 32-round rocket packs and a pair of 800-litre drop tanks. A KKR pod was rarely carried by the wingman – it was required only when the targets were of particular importance and a degree of redundancy had to be provided.

A representative example of the 263rd ORAE operations in Afghanistan is the 13-month deployment to Bagram in 1986-1987 of one squadron temporarily drawn from the 101st ORAP, home-based at Borzya in Siberia. This particular squadron was among the units that recorded the highest number of losses during the Afghanistan campaign. In the summer months of 1986 its Su-17M3Rs were heavily used in the strike role, with a total bombload of 1000 kg (most often comprising two 500-kg bombs) or carrying four 20-round packs for firing B-8 80-mm rockets.

As the squadron's pilots recall, the ASP-17BTs sight was mainly used in the manual aiming mode and the minimum altitude for bomb drops was set at 1800 m (5,900 ft), in 25-30° dives, and one attack pass only in an effort to reduce exposure to the (generally strong) enemy air defence. No gun pods are reported to have been used during the deployment while the Su-17M3's built-in 30-mm NR-30 guns were only rarely employed, mostly during rescue operations as last-ditch weapons as the aircraft were required to provided continuous top cover for downed pilots awaiting the arrival of SAR helicopters. Initially, pilots say, manual aiming resulted in low bombing precision, but as thy gained more experience during the later stages of the deployment, precision was considerably improved and CEPs of 30-40 m (98-131 ft) were commonplace (CEP calculation was made utilising battle damage assessment photos).

A strengthening of Mujahideen air defence in the summer and autumn months of 1986 resulted in an increased number of losses suffered by the 263rd ORAE, which in turn led to sorties being performed at higher altitude and increased speed in an attempt to reduce the vulnerability of the reconnaissance aircraft.

Armour protection

All Su-17M3Rs apart from those carrying KKR pods received armour plating on the lower fuselage. The increased weight had an impact on the handling performance, especially in the hot-and-high airfield conditions at Bagram in the summer months. The extreme operating conditions imposed handling technique alterations during take-off, in which the slats and flaps were deployed in the landing position and rotation took place at the end of the runway, just before lift off. Flares were pumped during the initial climb.

Carrying four tanks for a ferry flight, a 20th APIB Su-17M4 taxis past a grass-covered HAS at Gross Dölln. In Germany the two Su-17 fighter-bomber regiments were increasingly assigned to the defence suppression role, especially after tactical strike duties were taken over by MiG-29s and Su-24s. The reconnaissance-tasked Su-17M3s and M4s of the 294th ORAP at Allstedt performed a variety of recce missions with KKR pods. As the 294th was the first unit to leave Germany after reunification, departing in April 1991, correspondingly little is known of its aircraft or activities.

Below: This line-up shows Su-17M4s of the 3rd Eskadrilya, 20th APIB. By 1990 the regiment was transitioning to the Su-17M4, a process completed only when the 730th APIB departed Neuruppin for Russia in April 1991, leaving behind some of its aircraft.

Specification – Su-17M4

Dimensions: wing span 13.68 m (44 ft 11 in) fully spread, 10.025 m (32 ft 11 in) fully swept; wing area 38.49 m² (414.3 sq ft) fully spread, 34.85 m² (375 sq ft) fully swept; aspect ratio 4.862 spread, 2.884 swept; sweepback of the fixed panel 63°; sweepback of the movable panels 30° to 63° at the leading edge; length with probe 19.02 m (62 ft 5 in); length without probe 17.341 m (56 ft 11 in); height 5.129 m (16 ft 10 in)
Powerplant: one Saturn AL-21F-3 turbojet rated at 76.49 kN (17,196 lb) dry and 109.84 kN (24,692 kN) with afterburning
Weights: empty 12161 kg (26,754 lb); normal take-off (with 1000 kg of weapons) 16400 kg (36,156 lb); maximum take-off 19500 kg (43,340 lb);
Fuel and load: internal fuel 3770 kg (8,294 lb); external fuel up to 2402 kg (5,285 lb) in four PTB-800 820-litre drop tanks or 3073 kg (6,761 lb) in two PTB-800 and two PTB-1150 1,150-litre drop tanks; maximum ordnance 4070 kg (8,954 lb)
Performance: maximum level speed at high altitude 1850 km/h (999 kt); maximum level speed at sea level 1350 km/h (729 kt); maximum Mach number at high altitude 1.7; maximum level speed at sea level with full stores 1250 km/h (675 kt); take-off speed (at MTOW) 360 km/h (195 kt); landing speed (at MTOW) 285 km/h (154 kt); take-off distance 600 m (1,969 ft); landing distance 950 m (3,117 ft); practical ceiling 15200 m (49,871 ft); ferry range with four 820-litre drop tanks 2550 km (1,375 nm); range with 1000 kg of weapons at low level 1400 km (756 nm); climb rate at sea level 230 m/s (754 ft/min); maximum *g*-load 7

Sized so that it fills the front windscreen when viewed by the pilot, the ASP-17BTs head-up display/weapons sight was introduced in the Su-17/22M3 (seen here in an Su-22M4), although it suffered from poor reliability in the early days. The sight's pipper can travel through +/-12° in azimuth, and from 6° up to 30° down.

KKR reconnaissance pod family

The Su-17 family was provided with the capability of carrying the KKR series of large underfuselage pods equipped with a wide variety of reconnaissance-gathering hardware.

The **KKR-1** is a 1970s-vintage system with camera equipment and the SRS-9 Virázh ELINT system (or SRS-7 as used on the early production versions such as those fielded by the Polish Air Force Su-20Rs).

The improved **KKR-1TE/2-54** (and its export derivative known as the **KKR-1TE/2-54K**) was fielded in the early 1980s. It weighs up to 800 kg (1,784 lb) fully equipped and 302 kg (665 lb) empty, and imposes a manoeuvring restriction of 3.5 *g* on the carrier aircraft, though there are no speed restrictions. The pod is 6.79 m (22 ft 3 in) long, 0.59 m (1 ft 11.2 in) wide and 0.58 m (1 ft 10.8 in) deep, and is made of two modules. The photo module in the forward part houses one A-39 vertical or oblique camera, one PA-1 oblique camera and one UA-47 night camera. The ELINT module in the rear part houses the SRS-13 Tangazh system (known also as *izdeliye* 33S).

Designed in the 1950s, the A-39 is the most popular Soviet-era camera and can be installed in two positions in the KKR pod – either vertically or forward-facing in the nose. It is used for low- and medium-level operations, between 500 and 5000 m (1,640-16,400 ft), at a maximum speed of 1500 km/h (810 kt). The A-39 is equipped with a 100-mm focal-length lens and uses an 80-mm wide film. Resolution is 44 lines per millimetre in the centre of the frame. The A-39's camera captures a swathe whose width is 0.7 times the altitude of the carrier aircraft when vertically-mounted, and 2 times when oblique-mounted. The length of the photographed swathe in continuous operation is 120 and 180 times the carrier aircraft's altitude, respectively.

The PA-1 is an oblique photography camera used for low-level operations, at an altitude of up to 1200 m (3,937 ft) and at a speed of between 200 and 1000 km/h (108-540 kt), with 80-mm-wide film and 90.5-mm focal-length lens. The PA-1 provides resolution of 20 cm when operating from 500 m (1,640 ft) altitude.

The UA-47 camera is used for nighttime operations in conjunction with the KRF-38 flare dispenser, and is equipped with two lenses inclined at 16° left and right. It is operated between 300 and 1000 m (980-3,323 ft), at a speed of between 750 and 1100 km/h (405-594 kt), and the 80-mm wide film is enough for making 152 pairs of frames. The photo-module of the pod also houses four 38-round KDF-38 flare dispensers, with a dispersion rate of one flare per 0.1-3.5 seconds. The flares are used at altitudes between 300 and 1000 m, providing lighting for 40 microseconds each.

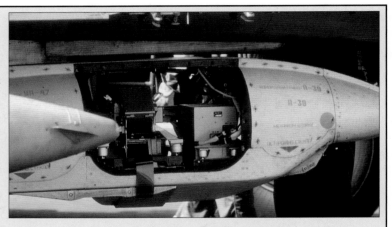

This view shows the open forward compartment of a KKR-1TE/2-54 pod. This section houses the A-39 forward-facing oblique camera and the PA-1 panoramic camera. Both are traditional wet-film systems.

The SRS-13 Tangazh ELINT system is an old-generation ELINT system capable of detecting land- and ship-based radar emissions at a range of at least 100 per cent of the radar's range in its main lobe. Its angular coverage is 25-30° in the horizontal plane (azimuth) and 25-24° in the vertical plane (elevation), recording the carrier frequency, pulse repetition frequency and the side of receipt of the signal (port or starboard). The system's large receiver antennas are located in the rear part of pod, and the received information is recorded on wet film, enough for five or 10 hours depending on the recording speed. Time for processing the gathered information is eight hours per flying hour, and Mean Time Between Failures is 150-200 hours.

The **KKR-1/2** pod features the A-39 and PA-1 camera outfit as well as the Zima IR reconnaissance system (infrared line scanner) and the Tchibis TV system, believed to have been provided with downlink capability to a ground station.

The **KKR-2A** is the version equipped with the A-39 (vertical or oblique) and AP-402 (used for oblique photography) cameras, as well as the Aist-T TV system, Zima IR system and ShRK-1 Trassa datalink system for real-time downlinking of the gathered information to a ground station. The AP-402 is a low-level camera weighing 58 kg, with 80-mm wet film and 90.5-mm focal length lens. It covers a swathe 10 times the altitude of the carrier aircraft, with length of 500 times the altitude of the carrier aircraft.

The **KKR-2Sh** is equipped with the Shtik-2M side-looking radar and the ShRK Trassa datalink system. The Shtik-2M is capable of scanning a 4-24 km (2.1-12.7 nm)-wide swathe, providing a resolution of 5-7.5 m.

The **KKR-2P** is equipped with the Shpil'-2M laser reconnaissance system, covering a swathe four times the altitude of the carrier aircraft, with resolution of 25-400 m depending on the operating altitude.

The **KKR-2E** is a version featuring the Efir-1 radiation reconnaissance equipment.

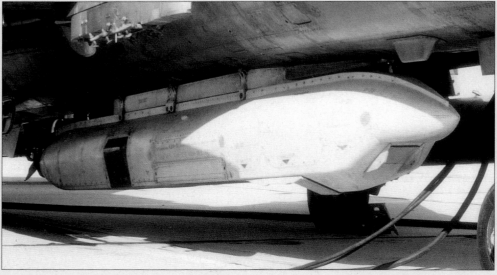
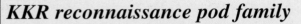

The bulky KKR pod is carried on a special adapter beam beneath the centreline. This is a KKR-1TE/2-54 with the Tangazh Elint system in the rear section, served by side-facing antennas. Note the shutters which protect the window for the forward-facing oblique camera in the nose of the pod. For most reconnaissance missions the 'Fitter' carries a pair of PTB-800 drop tanks under outer wing pylons.

Heavyweight SEAD weapons

Kh-28 (AS-9)

The Su-17M was the first 'Fitter' derivative to be provided with a SEAD capability using the Kh-28 anti-radiation missile (ARM), which formally entered VVS service in 1974. In addition to the Su-17M, it could also be employed by the Su-17M2 and -3 derivatives. Designed by the MKN Raduga company based at Dubna near Moscow, the Kh-28 was intended for destruction of the fire-control radars of the US MIM-23 Hawk, Improved Hawk and Nike Hercules SAM systems. An aircraft-lookalike design dating from the mid-1960s, it weighed 715 kg (1,570 lb) and boasted a 140-kg (312-lb) high-explosive warhead. A nuclear warhead was also developed for the Kh-28. A proximity fuse of the optical type ensured detonation in the event of the missile passing within 5 m (16 ft) of the target; a contact fuse to detonate the warhead in case of direct hit was also provided.

The Kh-28 was powered by one R-253-200 liquid-fuel rocket engine borrowed from the KSR-5 cruise missile, with a launch thrust of 78.5 kN (17,640 lb). The missile had a useful range of between 65 and 80 km (35-43 nm) when launched from high altitude, while when launched at low level (up to 1000 m/3,280 ft) during operations against dense air defences, the Kh-28's range was 35 km (18.9 nm). Maximum speed was 3300 km/h (1,779 kt).

The PRG-28 seeker used by the Kh-28 was capable of covering the A, B and C frequency bands and provided guidance data to the APR-28 autopilot. The autopilot itself featured a prolongation unit for mid-course inertial guidance if the target radar emitter was switched off for up to 10 seconds.

The missile was carried on the PU-28S launcher weighing 100 kg (220 lb), installed on the centre fuselage hardpoint, with the upper fin being recessed into a slot in the rear fuselage, while the bottom fin was folded to avoid scraping the ground. The highly toxic fuel components of the rocket engine are said to have caused considerable difficulties during missile servicing.

Pre-launch target acquisition and launch data calculation was made possible thanks to the Metel'-A/AV pod (A and AV sub variants for coverage of different frequency bands) that housed emitter locator hardware, and the Kh-28's launch control boxes. They provided target bearing detection, bearing and elevation data, and programmed both the flight and terminal attack profiles into the APR-28 autopilot.

Live firing tests demonstrated, however, that autonomous target acquisition using the Metel'-A/AV pod for target search and tracking could be a very difficult task, and thus the system was effective only when employed against radars with known positions. During such missions, the Su-17M pilot was required to reach the pre-calculated launch point and, after the emitting target had been acquired by the Metel'-A/AV pod, to execute the missile launch.

The low level of technologies used in the missile's guidance system made it impossible to achieve sufficient precision in the terminal stage as the Circular Error Probability (CEP) was quoted as 20 m (66 ft), though the low accuracy was partially offset by the powerful warhead.

The Kh-28 was also exported to Syria, Iraq and Vietnam, and saw its debut in real-word combat conditions during the Iran-Iraq war, launched in relatively large numbers by Iraqi AF Su-22M/M4s.

Kh-58 (AS-11)

The Kh-58 was initially conceived as a much-improved Kh-28 derivative. First design works were initiated at MKB Raduga in 1967 and the final design was frozen in 1971.

The solid-fuel engine gives it a high-altitude launch range exceeding 100 km (54 nm). It features two modes of operation – a launch boost of 3.6 seconds duration and thrust of 58.86 kN (13,380 lb), followed by 'sustainer' mode with a duration of 15 seconds and thrust of 9.8 kN (2,230 lb). Maximum flight time is 200 seconds. Launch altitude is between 100 and 22000 m (328-72,160 ft), the speed of the carrier aircraft is between 550 and 1800 km/h (297-971 kt), while the missile's average speed is between 1620 and 2160 km/h (874-625 kt).

This Kh-58 is seen on its loading trolley in front of a Slovak air force Su-25, although the weapon is carried by the Su-22M.

The Kh-58 weighs 490 kg (1,092 lb) and its fuselage is made of steel and titanium to combat the high temperatures in flight caused by the friction between the air and missile, which exceed 500°C. The missile is carried on the AKU-58 or AKU-58U catapult launcher, and two can be carried on the outer underfuselage hardpoints. After launch, the Kh-58 is firstly separated at a safe distance and the then the engine is ignited in an effort to prevent damage from the large rocket plume to the fuselage of the carrier aircraft or engine surge.

The SAU-58 automatic flight control system (described by Russian sources as featuring capabilities comparable to that of the aircraft autopilots of the 1970s) is used in conjunction with the PRGS-58M seeker, quoted to have been capable of operating against various types of radars, including pulse and frequency-agile ones, with selection made by carrier frequency and the pulse repetition frequency of the target radar.

Intended for use against the fire control radars of the Hawk, Improved Hawk and Nike Hercules SAM systems, the PRGS-58M can cover four frequency bands. The A- and A'-bands are designed to cover the continuous-wave radars in the 3-cm wavelength band, as used by the Hawk and Improved Hawk SAMs. Type B band can cover the air surveillance and early warning pulse radars working in the 10-cm wavelength band, while Type C covers the pulse radars working in the 20-cm wavelength band. As Russian Air Force pilots assert, the Kh-58 cannot be used against the Patriot SAM system fielded by the US Army in the 1980s. A prolongation device (mid-course inertial guidance package) embedded in the guidance system is used to determine the target position and continue missile guidance in the event that the target radar was switched off for up to 15 seconds.

The warhead weighs 149 kg (333 lb), including some 58.5 kg of high explosive material, and is detonated by a laser proximity fuse when passing within 5 m (16 ft) from the target; a contact fuse is also available for back up; the warhead's lethal radius is up to 20 m (65 ft).

When launched from low altitude in operational conditions, at 200 m (660 ft) as the carrier aircraft tries to approach the launch point unseen, the missile has a range of 40 km (21.6 nm), while when launched at 7000 m (22,960 ft) its range is up to 70 km (37.8 nm), and at 10000 m (32,800 ft) the range extends to 100 km (54 nm). After launch, the Kh-58 utilises a combined guidance method, firstly using inputs from its own inertial navigation unit and then switching to guidance inputs derived from the PRGS-58M seeker for mid-course and terminal guidance. The Kh-58's CEP is quoted as 5-10 m (16-33 ft).

The PRGS-58M seeker is advertised as being made capable of executing target selection by its carrier frequency and mid-course automatic tracking. After launch, the missile is first stabilised in pitch, yaw and bank before entering into a climb-out until reaching its pre-programmed elevation relative to the target (the elevation is measured by the PRGS-58M's own gyro sensor), then executes a turn in the horizontal plane towards the target (until this moment the autopilot provides inertial guidance) and then proceeds into a shallow dive guided by signals derived from the PRGS-58M seeker using the proportional guidance method. When in the terminal phase, it executes a pop-up with a subsequent steep dive to strike the target from above.

The Vy'uga-17 conformal targeting pod carried on the centre fuselage pylon is used for missile seeker control and feeding target data to the autopilots of the Kh-58, Kh-25ML and Kh-27PS ARMs. It can be used in flight at altitudes from 50 to 15000 m (164-49,200 ft), at a speed of between 600 and 1800 km/h (324-980 nm) and bank angle of up to 20°. The Vy'uga-17 is capable of simultaneously controlling two ARMs, and during strikes against radars with previously known positions it is activated at a pre-planned reference point. After target acquisition, it cues the missile seekers for maximum-range emitter target detection, calculates the minimum and maximum launch ranges and executes cross-cueing of the seeker heads of the two carried ARMs. In the event that the strike is carried out against radars with unknown positions, the PRGS-58M seeker head is used as the search sensor (passive emitter locator unit), receiving search commands issued by the Vy'uga-17. The visual information is provided to the pilot on the Lutch display in the cockpit in the form of several sets of symbols. A functional component of the Vy'uga-17 system, the display is located on the upper right corner of the dashboard, replacing the IT-23 CRT display (on the aircraft configured to fire the Kh-29T ASM) or the recce pod frame counter panel of the aircraft equipped with the KKR recce pod.

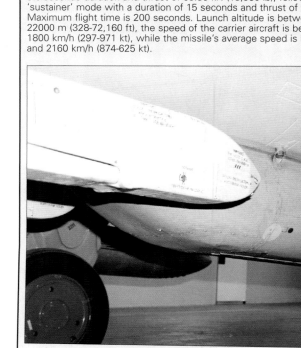

The Vy'uga-17 pod fits snugly beneath the Su-17/22's belly. It can provide targeting information for a range of SEAD weapons, including the Kh-58. It has the ability to control two missiles simultaneously.

'Fitter' weapons

The first guided weapon available to the 'Fitter' was the Kh-23/23M radio command-guided missile (below right), which employed the Del'ta-NM system and was widely used during the 1980s for practice firings. A more capable missile was introduced by the Su-17M2 in the form of the Kh-25ML laser-guided weapon, seen on the Polish AF (7.elt) Su-22M4K above and right (also armed with FAB-250 bombs), followed by the larger Kh-29 (also laser-guided). The Bulgarian Su-22M4Ks below are seen armed with this missile. With the Su-17M3 came the Kh-25MP anti-radar weapon. In the late 1980s some Su-17/22M4s received the ability to fire the Kh-29T TV-guided missile.

This selection of Su-17/22 free-fall weapons comprises OFAB-100 bombs (front row), FAB-250s (back row) and FAB-500-56 on either side.

Armourers cover an S-250FM 250-mm rocket on its trolley. This is one of the most accurate weapons in the Su-17/22's unguided armoury.

This weapons array includes two S-24 240-mm rockets, UB-20-80 rocket pods and a UBK-23-250 twin-cannon pod. Behind are various iron bombs.

Left: In front of this heavily-armed Su-22M4K are two SPPU-2301 ground-strafing pods. The pod in the foreground is displayed inverted, and with its twin 23-mm cannon barrels in the maximum depressed 30° position. The aircraft carries multiple ejector racks on fuselage and underwing pylons.

Right: The R-60M is the standard self-defence missile for the Su-17/22, carried on the intermediate wing pylon.

Landing at Bagram was a particularly difficult exercise as the approach was required to be carried out within the tightly guarded perimeter around the airfield due to the ever-present SAM ambushes. The typical Su-17M3R approach commenced with an overhead pass at 5000-5400 m (16,400-17,700 ft) with wings set at 30° followed by a steep corkscrew manoeuvre with 45-50° angle of bank angle at 500 km/h (270 kt). Landing gear, slats, flaps and speed brakes were deployed at the end of the very short downwind leg and the steep descent continued through the turn onto final approach.

Lining up with the runway centreline before reaching the inner marker (located at some 1,000 m/3,280 ft from runway threshold) at an altitude of 400-450 m (1,312-1,476 ft), the pilot kept the speed below 450 km/h (243 kt), using deployment and retraction of the speed brakes for speed/altitude control. The pilot's aim was that the inner marker was passed at 390-360 km/h (210-194 kt) and 150 m (492 ft), with a rate of descent reduced to 5-7 m/s (16-23 ft). The landing flare was initiated at 10 m (33 ft) altitude, at a distance of 50-100 m (164-330 ft) from the runway threshold. During large formation landings, distance between the approaching pairs was set at 400-600 m (1,312-1,968 ft).

Illumination and countermeasures

From the autumn of 1986, the Su-17M3Rs were often tasked with deploying SAB-100 and SAB-250 illumination bombs, carried out prior to every large-package strike against well-defended targets in rebel-held territory. The heat- and light-emitting SAB-100 and SAB-250 bombs descended slowly, hanging on a parachute usually deployed at 2000-2500 m (6,560-8,200 ft), and during their two- to three-minute descent the powerful heat emissions provided reliable cover for the strike aircraft by fooling seekers of missiles. The heat-seeking SAMs – such as the Polish- and Chinese-made Strela-2M derivatives, Red Eye and the Stinger – are said to have proved vulnerable against this type of countermeasures and, as a rule, aimed at the heat-emitting bombs. Other makeshift countermeasures tried by the Su-17M3Rs called for dispensing the powerful FP-100 photographic flares (launched from the KKR-1 pod and originally designed for use as light sources during low-level photo-reconnaissance missions) but this was not particularly effective.

The SAB-100 also saw use in the target-marking role: with the parachutes being intentionally deployed at a very low level, the bombs impacted the ground still burning and immediately created a gulf of fire in the surrounding vegetation, used as hideouts by the Mujahideen. The strike packages heading to the target area then had no problems of spotting the fire and smoke from tens of kilometres away.

A typical strike package in 1986/87 organised at Bagram involved up to 36 aircraft, and during the fierce Host offensive in March 1987, the 263rd ORAE Su-17M3Rs were tasked with providing anti-SAM cover, deploying SAB-100 bombs for the strikers. Then the main strike package consisting of several four-ship MiG-23MLD groups (another four-ship flight was also employed for top cover when operating in close proximity to the Pakistani border) released their bombs, and the strike was concluded by three to four four-ship Su-25 flights tasked to demolish the remains of the Mujahideen positions.

Twice weekly, the Bagram-deployed Su-17M3Rs were tasked to fly Elint missions along the Iranian and Pakistani borders with the KKR-1/A pod, with the main aim of detecting and tracking the position and changes of the air surveillance radars in the two countries then regarded as hostile. Elint sorties were flown in pairs, in both day and night conditions, sometimes with a refuelling stop being made at Shindand.

Stationed in Bagram between April 1986 and May 1987, the reconnaissance squadron suffered from five losses (another aircraft, a two-seater, was damaged beyond repair in a landing accident), but only two of these are believed to have been due to the Mujahideen. Several other aircraft sustained heavy combat damages but, in the event, managed to return to the base. Most of them were returned to airworthy condition by the squadron's maintenance-repair service.

In general, the Su-17M3R/M4 is reported to have successfully sustained the intense tempo of operations in the rather demanding Afghanistan climatic conditions and harsh operating environment – major difficulties encountered comprising

Streaming its brake chute, an Su-17M4 lands past a row of Su-27s. The 'Fitter' disappeared from mainstream Russian service with unseemly haste in the mid-1990s, a victim of the drive to drastically reduce overall force numbers by retiring all single-engined aircraft. However, a number were retained in regular service until at least 1997, and they saw action over Chechnya. The final front-line Russian aircraft were almost exclusively used in the reconnaissance role in their last years of service. A handful were also retained at Krasnodar to provide a fleet for training foreign pilots.

Su-17s continued to appear at Russian air shows in the 1990s, including this aircraft carrying the Russian flag in place of the traditional star. Aircraft also appeared at the MAKS show to support various attempts to market upgrade proposals to export customers. The first of them involved the integration of French-supplied avionics, but this was too costly for most potential customers. Subsequent all-Russian proposals were cheaper, but drew little interest.

'Fitters' for Yemen

The Yemen Arab Republic Air Force (YARAF – North Yemen) acquired 12 Su-22s (above) in the summer of 1979, forming No. 18 Squadron of the 2nd (Fighter-Bomber) Division with considerable Soviet assistance. Meanwhile, the People's Democratic Republic of Yemen Air Force (PDRYAF –South Yemen) received 20 Su-22Ms (below) and four Su-22UM3Ks in 1986, which served with No. 3 Squadron in the 'Attack Regiment'. In May 1990 the two Yemens were united under Soviet pressure, and a unified Yemen Air Force was created. However, extreme tensions remained between the two factions.

By 1994 Yemens was splitting apart again. A former North Yemeni pilot used an F-5 to shoot down two former South Yemen Shenyang F-7s, and then attacked the former DPRYAF compound at Sana'a, destroying three Su-22Ms. The new North Yemen air force took over many former DPRYAF aircraft, and formed three small Su-22 squadrons within the 1st Air Wing. With Kuwaiti backing the new South Yemen air force acquired 12 Su-22M4Ks in June 1994. In the war that followed, which officially began on 5 May 1995, the North's air force was far better organised and less-reliant on unreliable mercenary pilots, despite the South's acquisition of ex-Moldovan MiG-29s. 'Fitters' were very active during the air fighting. Eventually the North triumphed and the country was united once more. After the war the reunified air force included a number of the almost new 'Fitters' acquired by the South but captured intact by the North at Aden. Serviceability rates fell alarmingly, but have recovered in recent years as aircraft have been overhauled.

Wearing Yemeni markings, this Su-22M was seen at the 558 ARZ overhaul facility at Baranovichi in Belarus.

This Su-22M4 was flown by a defecting Democratic Republic of Afghanistan AF pilot to Pakistan in 1988. Afghan 'Fitters' served with the 355 Air Regiment at Bagram and comprised Su-22Ms and M4s (from 1987). 'Fitters' saw action during the long civil war that followed the Russian withdrawal, flying with both pro- and anti-Taliban forces.

contaminated fuel, foreign object damage and simplified field maintenance due to the high sortie-generation rate of four to six sorties per day.

Survivability improvements

Combat experience during the Afghanistan war in the early and mid-1980s uncovered some unpleasant surprises. Combat damage analysis revealed that the main areas of catastrophic failure in the Su-17M3/M4 (due to bullets, small-calibre projectiles and high-speed fragments of SAM warheads) were the powerplant and its accessories gearbox, as well as the fuel, control and hydraulic systems. Consequently,

in the mid-1980s, a plethora of urgent improvements to the Su-17M3/M4 were introduced in an effort to improve the type's combat survivability, both to aircraft in the field and those on the production line. VVS Su-17M3/M4s deployed to Afghanistan received additional armour plates scabbed on to the lower fuselage to protect the accessories gearbox, electrical generator and the fuel pump, while the fuel tanks were filled with reticulated foam in order to prevent explosion in case of hits by incendiary bullets or missile fragments.

The ever-increasing threat posed by man-portable SAMs prompted a gradual increase in the number of flares, as 32-round ASO-2V chaff/flare dispenser units (launching

LO-56 flares that burn up to six seconds and develop temperatures exceeding 2,000°C) were scabbed on to the upper fuselage – initially four, but later eight and even 12 units (including four on the lower rear fuselage). The ASO-2V chaff/flare dispensers became standard fit on the late-production Su-17M4s and Su-22M4s built in the second half of the 1980s.

As many as 29 Su-17s were reported to have been lost during the Afghanistan conflict, and around half of them are assumed to have been downed by the Mujahideen.

End of the Su-17's VVS story

By 1990, VVS and Soviet Naval Aviation had an Su-17 inventory numbering over 1,000 aircraft, but in the beginning and middle of the decade most of them, together with the entire MiG-23K/D/M inventory, was withdrawn from service and put in long-term storage. Such an abrupt step was justified by the fleet rationalisation and reduction drive that brought about the early retirement of all VVS single-engined combat aircraft, and thus the death of the fighter-bomber aviation branch. By 1997, there were only two Su-17-equipped regiments surviving in Russia, based at Tver-Migalovo and Taganrog, with a total inventory of about 70 reconnaissance-converted Su-17M3R/M4Rs.

Some were used operationally during the first Chechen War between 1993 and 1995. The 43rd OMSHAE – an Su-17M3-equipped independent attack squadron of the Russian Naval Aviation with an inventory of about 35 aircraft – continued to be stationed at Gvardeyskoye airfield near Simferopol in Crimea. In 1998 it converted to the Su-24 'Fencer-C'. A limited number of Su-17/22Ms/M4/UM3s are known to be still in use, or at least are kept suitable for use, with one regiment of the Krasnodar Training & Conversion Centre tasked with type conversion training of foreign pilots. The Russian Air Force test centre at Akhtubinsk (Vladimirovka airfield) still uses a lone Su-17UM3 in the escort and photo chase role.

The Ukrainian Air Force maintained in the 1990s an active fleet of about 40 Su-17M4/UM3s in one regiment based at Lymanskoye near Odessa, and there are some reports that a unit based at Kolomya also used the Su-17M4R until recently

(a 'Fitter' based there is reported to have crashed in 1998), and some aircraft are still kept in storage with the Odessa-based maintenance facility. Until the late 1990s, small numbers of Su-17M3/M4s were operated by the air arms of the former Soviet republics of Turkmenistan and Uzbekistan, but their current status is unknown.

Upgrade plans

In the 1990s, the export versions of the 'Fitter' enjoyed much more active use than their Russian/CIS counterparts. In the middle of the decade, no fewer than 400 aircraft were believed to remain in active service with the air arms of at least 13 countries worldwide. As might be expected, the ready availability of relatively new or little-used airframes, with an average remaining service life of 10-15 years, attracted the attention of the Sukhoi Design Bureau, KnAAPO and several foreign companies specialising in upgrades. However, customers were, and still are, reluctant to appear *en masse*.

Major global upgrade player Israel Aircraft Industries (IAI) was active in the field for the first time in 1993, initiating campaigns aimed at two East European Su-22M4 operators – Poland and the Czech Republic. IAI is reported to have offered them an upgrade package generally similar to that adopted for the Turkish Air Force F-4E Phantom upgrade, the F-4E 2020.

Oozing bulk and menace, an East German Su-22M4 awaits a mission. 'Fitter-Ks' served with Jagdbombergeschwader 77 'Gerhard Leberecht von Blucher' and with Marinefliegergeschwader 28 'Paul Wieczorek', both at Laage near Rostock on the Baltic coast.

There were no realistic plans for the Luftwaffe to use the former East German Su-22M4s operationally after reunification, but a number were employed for tests by WTD 61 at Manching. This example was passed on to the UK, and it now resides alongside ex-East German vehicles and working air defence radars at RAF Spadeadam, the electronic warfare training range in Cumbria.

Czechoslovakia proceeded directly from ancient Su-7BMs and BKLs to AL-21F-3 powered Su-22M4Ks from March 1984. On the eve of the Czechoslovak partiton on 1 January 1993, the air force had a regiment of Su-22M4K/UM3Ks at Namest (20 SBoLP) and a squadron of Su-22M4Rs (47 PLP) at Pardubice, for a total of 57 'Fitters'. After partition, the Su-22s allocated to the Czech Air Force were operated from Namest by 321 Letka until retirement in 2001.

Slovakia's 'Fitter' force was inherited from the break-up of the Czechoslovak air force. Eighteen Su-22M4s and three Su-22UM3s were taken on charge, including some ex-47 PLP Su-22M4Rs. Twelve aircraft were sold to Angola, and the survivors were primarily used in the reconnaissance role until the last was retired in 2002. This Su-22M4R carries a KKR-1 reconnaissance pod.

This package comprised an all-new self-defence suite, glass cockpit and pod-mounted targeting system, plus provision for the use of Israeli-made laser-guided bombs and a wide arsenal of modern Western weapons. In 1994, IAI announced that it had completed definition of its upgrade concept but, by that time, the prospective customers had already changed their minds, thus shelving the Su-22M4 upgrade plans.

In the early 1990s, Sukhoi and KnAAPO promoted their joint upgrade concept. It is noteworthy that, due to a lack of experience and a suitable avionics package, the initial marketing and concept definition steps were undertaken in cooperation with the French companies Sextant Avionique and Thomson-CSF. The French-supplied package incorporated a new mission computer, NSS-100-GPS receiver, Totem inertial navigation systems (for use in the hybrid GPS/inertial nav system), AHV-6 radio altimeter, ADU 300 air data unit, Sherloc radar warning receiver, VOR/ILS/TACAN navigation aids, wide-angle head-up display (HUD), Phantom lightweight multi-mode radar (in the nose cone), Rubis forward-looking infrared (FLIR) navigation and targeting pods, TMV-630 laser rangefinder/designator, Barem pod-mounted electronic counter-termeasures (ECM) system and CINNA 3PN computer-based mission planning system. All of the components of this expensive package had already been proven on French Air Force aircraft and the upgrade price was quoted as being between $2.5 and $3 million per aircraft. In the event, the joint Russian/French proposal also failed to attract any particular interest from prospective customers.

All-Russian upgrade proposal

In 1996/97, a joint team from the Sukhoi Design Bureau, Rosvooruzhenie and KnAAPO continued to pursue prospective customers – this time among the Third World 'Fitter-K' operators. The team promoted an all-Russian multi-level

upgrade plan to the Su-17M5 standard, combined with a comprehensive service life extension programme (SLEP). The proposed upgrade programme was advertised as retaining the existing ground-based infrastructure, spare parts delivery system and technical support intact. The upgrade proposal comprised either the integration of modern digital nav/attack systems or upgrade of the existing PrNK-54 to receive an embedded GPS/GLONASS receiver, plus optional integration of TACAN and VOR/ILS tactical navigation/instrument landing aids. The MK-54 laptop-computer mission-planing system was also included, alongside a new digital interface, making possible the integration of Western avionics and weapons in accordance with specific customer's requirements. Modernisation of the communications system was also offered and provision was made for the integration of a FLIR/laser designation pod, as well as the two-pod Sorbtsia ECM system already proven on the Su-27.

The FLIR pod was advertised as capable of providing night targeting capability for the Kh-25ML and Kh-29L missiles, thus significantly enhancing the Su-22M4's overall combat capability. The cockpit instrumentation was upgraded by the integration of two large-format colour displays, a new wide-angle HUD and HOTAS controls, as well as the Signal-805 radio (or Western types upon customer request). The most extensive upgrade level included integration of the pod-mounted Phazotron Kopyo radar: this would significantly enhance the Su-22M4's anti-ship capability, making possible the use of the Kh-31A Mach 3 sea-skimming ASM with an active radar homing seeker. Other new weapons that could be easily integrated on the modernised Su-22M4, as advertised by Sukhoi, included the R-73E and Magic 2 short-range air-to-air missiles, the S-25L inexpensive laser-guided air-to-ground missile, as well as the KAB-500Kr/KAB-500L TV-/laser-guided 500-kg bombs.

In addition, upgrade of the SPS-141 ECM pod to the SPS-141MK standard was proposed, adding the new capability of jamming CW radars as well as high-PRF and TWS radars; it was achieved through application of digital techniques, high-sensitivity receiver channels with signal analyser, expanded frequency range and new jamming signal forming modes. Additionally, the new KS-418E ECM pod was offered, capable of operating in a plethora of noise and deception jamming modes, in the front and rear hemisphere. The price of the upgrade package was not revealed, but it is believed to have varied between US $2 and $4 million per aircraft, depending on the nature of the equipment and weapons to be integrated upon customer request.

In 2000/01, another Su-22M4 concept was promoted by a Russian company named Gefest i T (which had upgraded the nav/attack system of the Su-24M tactical bombers delivered to Algeria in the early 2000s). The company developed a common nav/attack system upgrade package for both the Su-22M4 and Su-24M (MK, MR, MRK) comprising the SV-24 mission computer, new HUD, GLONASS/GPS satellite navigation receiver integrated into the existing nav system, Obzor-RVB video signal processing unit, TBN-K memory storage unit and the NKPK mission planning/post-mission analysis system. The mission computer was advertised by Gefest i T as being capable of increasing the unguided weapons precision between 3 and 10 times, with CEP in the region of 10-15 m (33-50 ft).

In August 1999, another non-Russian company announced intentions to enter the still unexplored Su-22M4 upgrade market. During the MAKS'99 airshow, DASA (known since 10 July 2000 as EADS-Germany) initiated talks with Sukhoi concerning the possibility of a joint effort to develop and market Su-22M4 service life extension and avionics upgrade packages in eastern Europe. The project was inspired by the successful results of the Westernisation and service life extension of the Luftwaffe MiG-29 fleet, which had been carried out by the German-Russian joint venture MAPS. However, no agreement was reached and then, in early 2000, DASA proposed another Su-22M4 avionics upgrade package: it was to be developed in cooperation with the Israeli company Elbit Systems, based on the package offered for the East European MiG-29s and proven in the air onboard the Romanian AF MiG-29 Sniper technology demonstrator.

The DASA-proposed Su-22M4 upgrade used mostly Elbit-supplied avionics, incorporating the installation of a new digital modular multi-role computer (MMRC), Mil Std-1553B digital databus, wide-angle HUD, HOTAS controls, two colour cockpit displays, new communications system and RWR, as well as an integrated inertial/GPS nav system. Poland, Slovakia and Bulgaria were all targeted as prospective customers in 2000 though with little success; eventually, DASA/EADS-Germany was not able to attract any particular interest.

No market for upgrades

A brief look at the table of Su-22M3/M4 operators outside Europe could reveal that the Sukhoi and EADS Germany market predictions, made in the early 1990s, proved over-optimistic. There are indeed a relatively large number of non-European operators, such as Syria and Yemen, which operate a fleet of little used aircraft, but their 'Fitter-J/Ks' are still considered to be up to date, and their comprehensive (and potentially expensive) upgrade would not represent a cost-effective solution. In fact, significant improvement in the navigation capability of these Su-22M/M4s could be achieved only by the means of acquiring low-cost commercial-off-the-shelf GPS receivers. The Su-22M4's existing range of laser/TV-

guided missiles and lethal yet inexpensive unguided weapons could satisfy the requirements of such air arms.

Countries like Angola (an Su-22M4 operator used its aircraft in continuous operations against the opposition for more than a decade) are known to have been happy over the performance, current weapons suite and the nav/attack system of their Su-22M4s, but need better logistic support to keep the aircraft flying. In 1999, Angola ordered 11 ex-Slovak Su-22M4s and one Su-22UM3, and further small-scale acquisitions from either East European or CIS states cannot be ruled out. The same could be true for other customers considered 'rogue' states, such as Libya and Syria: both of them are long-time 'Fitter' operators and they are likely to benefit from the availability of cheap second-hand, low-houred Su-17M3/4s (currently still held in storage in Russia, Ukraine and other ex-Soviet republics). The possibility, however, that some countries might go ahead with some limited upgrades and integration of new weapons, combined with service life extensions in the near future, cannot be ruled out. Vietnam was the first Su-22M4 operator to adopt this approach, upgrading part of its

A Slovak Su-22M4K is displayed in its shelter with a wide array of ordnance. The aircraft is armed with Kh-25MR (AS-10 'Karen') radio command-guided missiles on the outer wing pylons, with Kh-25MP (AS-12 'Kegler') anti-radar missiles on the fuselage pylons. R-60 (AA-8 'Aphid') short-range AAMs are on the intermediate pylons. On the port inboard pylon is an SPS-141MVG Gvozdika ECM pod, interchangeable with SPS-142 and -143 to cover different frequency bands. In front of the aircraft are various rocket options in varying sizes up to the massive tube-launched 250-mm diameter S-25.

An Su-22M4K taxis out at its base at Malacky, armed with 20-round pods for 80-mm rockets and parachute-retarded bombs. Another aircraft is seen at left dropping retarded bombs while releasing flares from the upward-firing ASO-2V dispensers that are a routine fit on late-model 'Fitters'. Slovakia's Su-22s were initially assigned to 1 Letka of the 3 SBoLP (fighter-bomber regiment), but after the 1 January 1995 reorganisation of the air force, came under the parentage of 33 Stihacie-Bombardovacie Letecke-Kridlo (SBoLK – fighter-bomber wing), which also controlled single squadrons of Su-25s and MiG-21MFs.

With wings forward and with slats and flaps extended, a pair of Hungarian Su-22M3s formates for the camera. While Hungary's aircraft had nearly full 'Soviet-spec' equipment, they were powered by the 'export' Tumanskiy engine.

Another weapons display, this time in front of a Hungarian Su-22M3, shows rockets and small bombs, plus some precision-guided missiles. The aircraft carries Kh-25ML laser missiles on the outer pylons, with Kh-25MP ARMs next to them on the ground. At the extreme left of the display is a Kh-23 (AS-7 'Kerry') radio command-guided weapon.

fleet and seeking to buy second-hand aircraft sourced from eastern Europe.

It is noteworthy that, during the 1990s, some CIS states are known to have provided considerable assistance to the 'hot spot' Su-22 operators – among them Afghanistan, Angola and Yemen (North and South) – in the form of engineering expertise, spare parts and weapons supply, as well as pilots. There have been reports of the transfer of little-used Su-17M4s and spares from Ukraine to South Yemen during the civil war there in the spring of 1994. Currently, the main maintenance centre for the Su-22s operated in the Arab countries is the Baranovichi, Belarus-based 558 ARZ overhaul facility, which offers affordable prices for airframe and engine main overhauls in addition to minor airframe and avionics modernisations.

The only reported 'Fitter-K' upgrade activity in the Third World so far has come from Vietnam. There were several reports in 1997 and 1998 quoting that the Vietnamese Air Force had been considering the possibility of upgrading its Su-22M4 fleet. Sukhoi officials confirmed this fact in private talks with the author in early 1999, but they remained tight-lipped when asked about the details of the upgrade. It is known that as many as 32 Su-22M4s and two Su-22UM3s were upgraded by the Sukhoi/KnAAPO team between 1996 and 1998.

East European upgrade market

Poland was the first country to announce ambitious intentions for upgrading its Su-22M4 fleet. The country was, and continues to be, among the most significant 'Fitter-K' export operators in the world, taking delivery of 100 Su-22M4s and 20 Su-22UM3Ks. In 1999, some 88 Su-22M4s and 18 Su-22M3s were reported as remaining in active service, and in early 2002 the figures were 81 and 17, respectively. The first Su-22M4 upgrade plan was disclosed in the early 1990s, but the loudly stated intentions for the acquisition of a new-generation multi-role fighter by the middle of the decade had greatly contributed to the cancellation of the upgrade effort, as the last 'Fitter-Ks' had been originally been intended to be phased out in 2007. However, a lengthy delay in the selection of the new fighter combined with the country's newly-found responsibilities as a new NATO member revitalised the plans for Su-22M4 modernisation.

In 1999 and 2000, a small number of Powidz-based aircraft had undergone the so-called 'small-scale' upgrade of their navigation and communications systems in order to ensure limited NATO and ICAO compatibility. The work was carried out by the Bydgoszcz-based WZL-2 depot, where Trimble 2101 I/O Approach Plus commercial GPS receivers and Bendix KLU-709 commercial TACAN receivers (provided free of charge by the US but proving not particularly well-suited to hard-manoeuvring aircraft) were added to the existing PrNK-54 nav/attack system. The newly-installed navigation components were integrated within the existing Orbita-20-22 digital mission computer using an interface unit developed by the Czech company Aviation Service. The upgrade was designed in such a way that the cockpit was not affected as information from the new nav components is displayed on the existing NPP 'clockwork' navigation indicator. New anti-collision lights were also installed, while the R-862 UHF radio control panel was upgraded with a continuous frequency selector featuring 8.33-MHz spacing. The upgraded aircraft are operated by the 41st Squadron from Powidz airbase, assigned to the NATO Rapid Reaction Force; it is designated to NATO as a Close Air Support and air-interdiction squadron.

There were some reports that a team of Sukhoi and the Russian arms export agency Rosvooruzhenie (in early 2001 renamed as Rosoboronexport) had also held preliminary talks in 1999 to upgrade an initial batch of about 15 Polish Su-22M4s with new NATO/ICAO-compatible navigation and IFF systems, but that no agreement had been reached. EADS-Germany in partnership with the Israeli company Elbit

Systems was also known to have submitted an unsuccessful upgrade proposal. In June 2000, a surprising announcement by the Polish Ministry of Defence disclosed that it had selected IAI's Lahav Division, teamed with the local WZL-2 depot.

Their comprehensive US $120 million bid was to involve between 16 and 20 aircraft to equip a Quick Reaction Force squadron detached to NATO. The upgrade was to encompass the integration of a new digital nav/attack system, based around a Mil Std-1553B digital databus, comprehensive warning and self-defence suite and, last but not least, provision for Western-made precision weapons. An airframe life extension programme was to be carried out to enable the upgraded 'Fitter-Ks' to soldier on until at least 2013/15.

A mock-up of the upgraded aircraft, Su-22M4 9305, was exhibited with some of the upgrades statically at the Kielce MSPO International Defence Fair in September 2001. In the event, the IAI-proposed upgrade was cancelled by the newly-elected Socialist government, and in January 2003 it was disclosed that Poland at last had agreed a SLEP and mainte-nance rationalisation programme with Sukhoi, intended to be carried out in cooperation with the local WZL-2 depot. The programme is said to have covered 60 aircraft and the upgraded aircraft are to be retained good for use until 2010/12.

The Polish AF received Su-20s from 1975, a total of 24 units continuing in service until 1997. After their retirement, two Powidz-based Su-22M4s were upgraded to carry the KKR-1 reconnaissance pod and other Su-22M4s were made capable of using the indigenous Saturn recce pod, externally similar to the UB-16-57 rocket pod.

In February 2005, it was announced that a privately-owned Polish company named Profus Management will deliver as many as 40 second-hand Su-22M4s in airworthy condition to Vietnam alongside spare parts and weapons before the end of 2005. It was speculated that the aircraft are to be drawn from the Polish AF surplus fleet, but by mid-2005 it was still unclear how it could be fulfilled in time taking into consideration the absence of readily available aircraft in airworthy condition in Poland at the moment. Some 28 Polish AF Su-22M4s and nine Su-22UM3s are to be retired by the end of 2005 and will be

offered for sale, but as their service life has been exhausted, future operation will be possible only after a major overhaul. Other options for rapid delivery, if the announcement is correct, will be to acquire used Su-17M4s/22M4s from Ukraine, Russia or other East European states.

East European Su-22M4 operators faced tight defence budgets in the mid- to late 1990s, and the most important issue for them has been the extension of the service life and the rationalisation of fleet maintenance in order to reduce the excessive direct operating costs (DOC). In 1998, the Su-22M4's DOC were quoted as being about US $5,540 per flying hour, estimated at 180 per cent that of the Su-25K, according to information released by the Bulgarian Air Force in 1998. The original service life of the Su-22M4 was set by Sukhoi in the late 1980s at only 1,500 hours and 19 years (whichever is reached first). Engine TBO is 450 to 500 hours.

The aircraft was originally designed to undergo one major overhaul throughout its service life, at 800 hours and 10 years, and priced at around $1.8 million, but this is believed to have been considerably lower than the actual airframe and equip-ment endurance. Further life extensions could be granted by application (with payment) from Sukhoi – initially to 2,000 hours and then to 2,500-3,000 hours, with the extensions depending on the airframe and equipment condition.

In 1997, the Bulgarian AF, in an effort to save funds, had independently introduced an extension of the time between

Bulgaria's 18 Su-22M4s and three Su-22UM3s served with 2/26 Eskadrila, 26th Razuznavatelna Aviobasa at Dobritch. Although they received new transponders in the 1990s, there were insufficient funds for a full NATO compatibility upgrade.

A reconnaissance technician carries an A-39 camera which has just been removed from the KKR-1TE/2-54 reconnaissance pod carried by the Bulgarian Su-22M4R in the background. The film will be removed and processed before analysis can begin – a time-consuming and laborious process.

An Su-22M4K from 40.elt lands on the Kliniska DOL (Drogowy Odcinek Lotniskowy – highway strip) during the annual Polish Air Force exercise. Dispersed operations were a key part of Warsaw Pact strategy during the Cold War, and the variable-geometry 'Fitters' were easily able to use the short highway strips that would have been available near their peacetime bases. The DOL exercise strip is just 16 m (52 ft) wide and is 2000 m (6,560 ft) long.

This Polish Su-22M4 from 8.elt carries a pair of SPPU-2301 pods. The barrels of the twin 23-mm cannon can be depressed for ground strafing from level flight. Poland's Su-22 fleet has been pared down, but the survivors are undergoing a significant upgrade process to allow as many as 60 aircraft to operate within the NATO framework to around 2010/12.

overhauls (TBO) of part of its Su-22M4 fleet to 12 years (these were all low-houred machines, well below the 800-hour limit) and there were no technical problems experienced during this time.

Expensive maintenance and the high cost of spare parts offered by Russian and Belarussian contractors were among the main reasons for the considerable reduction of serviceable numbers in the Czech, Slovak and Bulgarian Su-22M4 fleets, and for the withdrawal from use of the Hungarian Su-22M3s in 1998. Czech Su-22M4/UM3s remained in active service until early 2001, and the type saw its eventual replacement by the Aero L-159 light combat aircraft later the same year, while Slovakia managed to sell 11 of its Su-22M4s and one Su-22UM3 (more than half of the total 'Fitter' fleet), to Angola in two consignments, concluded in 1999 and 2001. A handful of reconnaissance-tasked aircraft were retained until early 2002. Currently, as many nine Su-22 are in storage or in museums.

The Bulgarian Air Force was among the most enthusiastic Su-22M4 operators in the 1980s but was badly hit by tight budget constraints in the following decade. All aircraft received Becker Avionics ATC 2000-(2)-R transponders in the mid-1990s, and Trimble 2021 I/O Approach Plus commercial GPS receivers were installed (but not integrated into the aircraft's nav/attack system) in early 2000 on the two-seaters. The last two to three airworthy Su-22s were operated until final retirement in May 2004.

Hungary received 12 Su-22M3s and four Su-22UM3s, which entered service in March 1984. Until its retirement in February 1998, the 'Fitter' fleet amassed a total of 16,863 hours. A fleet

rationalisation programme due to tight defence budgets and base closures prompted the early withdrawal of the Su-22s, although five single-seaters and two two-seaters had been fresh from their main overhauls, and each airframe had some 1,000 hours left on it, enough for 10 more years of regular service. The Su-22's withdrawal left the Hungarian AF without a dedicated strike and reconnaissance capability.

Peruvian 'Fitters'

The Fuerza Aérea del Peru took delivery of 32 Su-22s and four Su-22Us, with the first of these taken on strength with the 111th Air Squadron in June 1977 (contract value was $250 million). Three years later, 16 Su-22Ms and three Su-22UMs followed, taken on strength by the 411th Air Squadron, based together with the 111th at La Joya. The new strike aircraft were delivered along with a wide array of unguided ordnance, R-13M air-to-air missiles, KKR reconnaissance pods and SPS-141 ECM pods. Soon after delivery, the new strike aircraft received their baptism of fire in the 1981 war with Ecuador, amassing some 79 flying hours in 60 sorties.

Peruvian Su-22s were also used in the air-to-air role in a clash which received high publicity. On 2 April 1992, an unidentified USAF C-130 transport from the Panama-based 310th Airlift Squadron was intercepted by a pair of Peruvian 'Fitters', some 56 km (35 miles) off the Peruvian coast. After the C-130 had ignored the visual and radio warning issued by the Peruvian pilots, the Su-22s were ordered to open fire with their NR-30 guns, aiming at 'non-vital' areas of the target. After receiving 30-mm hits, which killed one crewmember and caused some damage, the C-130 made an emergency landing at El Pato. After landing it was discovered that the C-130 was an Elint aircraft gathering intelligence on drug trafficking.

In 1992, the surviving Su-22s and Su-22Ms were upgraded with Western navigation aids such as VOR and ILS, and with the Russian-made ARK-15M ADF. The fleet was also adapted to use French-made BAP-100 runway-denial bombs. In an experimental programme a pair of Su-22s received French-made air refuelling probes, but fleet-wide introduction was cancelled due to lack of funds. In January 1993, the 411th Air Squadron was decommissioned and all Su-22/22Ms were gathered together with the 111th Air Squadron. In 1994, a second upgrade stage called for the installation of GPS receivers and Collins ARC-182 radios.

During the 1995 conflict with Ecuador for control over the Cenepa river valley, the 'Fitter' fleet is reported to have flown 61 hours in 40 attack sorties, dropping in the process 80 tons of bombs (principal weapons during the interdiction missions were 250-kg and 500-kg bombs). Two Su-22s were lost during

Left: Vietnam acquired 32 Tumanskiy-powered Su-22M/UMs in 1980, assigning them to the 923 'Yen The' Trung Doan (regiment) at Tho Xuan to the south of Hanoi. Here groundcrew load one of the aircraft with the massive Kh-28 first-generation anti-radiation missile, which first saw combat action in Iraqi hands during the Iran-Iraq war.

the conflict after being intercepted by two Ecuadorian Mirage F1 fighters with Magic air-to-air missiles (the Peruvians, however, still maintain that the Su-22s were downed by the Peruvians with Igla shoulder-launched SAMs).

After the war, the fleet was reduced in size as 12 aircraft were phased out in September 1996. The survivors (believed to number 18) received some survivability enhancements in the form of locally-made chaff/flare dispensers and upgraded SPO-10 RWRs. In 1997/98, Israeli-made SAMP chaff/flare dispensers were added together with the Elisra SPS-20 RWR and other improvements. Some sources say that a number of Peruvian 'Fitters' were upgraded with assistance from the Belarussian 558th ARZ Company with the nav/attack suites borrowed from the Su-17M4. In 1997 all surviving aircraft were relocated to Talara (El Pato) and are now being operated from there alongside Su-25s. In 2001/03, reports cited that the 'Fitter' inventory numbered 12 Su-22s, eight Su-22Ms and three Su-22UMs. In 2004, the 111th Air Squadron celebrated 38,000 flying hours in the 'Fitter' since it had entered service in 1977.

Middle East 'Fitters'

The Syrian Air Force is another major 'Fitter' operator which received its first Su-20s during 1973, a total of 12 units. Additional numbers of Su-20s were delivered, followed by the much improved Su-22M in the late 1970s and early 1980s. Syrian KKR-equipped Su-22Ms took part in the 1982 war in Lebanon, flying numerous recce sorties to map Israeli force deployments and the air defence in the border areas of Israel and Lebanon.

In the mid-1990s, Syria was reported to have maintained an inventory of 50-60 Su-20s, Su-22M3s and Su-22M4s (delivered in the mid/late 1980s), and was looking to upgrade and life-extend these, with some sources alluding that a contract was signed (or at least negotiated) with Sukhoi/KnAAPO in 1996.

Other Arab 'Fitters' that have seen combat are those of both Yemen states, who fought each other in a bloody civil war in 1994/95. The depleted 'Fitter' inventory of the unified Yemeni air arm is believed to have been strengthened by at least four ex-Ukrainian Su-17M4s, having previously served with the Limankoye-based reconnaissance regiment.

African 'Fitters' still soldier on in two countries – Libya and Angola – whereas the third major 'Fitter' user in the continent – Egypt, the Su-17's and Su-20's launch customer – no longer flies the swing-wing fighter. After breaking up the friendly relations with the Soviet Union in the mid-1970s, the serviceability of the Egyptian Su-20 fleet worsened from year to year, and

Pilots from the 923 Trung Doan discuss a mission in front of a line of Su-22s. The nearest aircraft is an Su-22UM two-seater. This first batch for Vietnam is also thought to have contained a few Su-22MRs with the capability to carry KKR-1 reconnaissance pods.

the type was eventually withdrawn from use in the early 1980s. In 1980, the inventory numbered 54 aircraft and, four years later, two of the still airworthy Su-20s were sold to West Germany for testing with WTD61 at Manching.

The Libyan AF took delivery of the 'Fitter' in the mid-late 1970s, initially Su-22s and later the improved Su-22Ms. They received their baptism of fire during the military operation in Chad and recorded several losses. The most publicised Libyan use of the Su-22M was for patrolling the Gulf of Sidra armed with R-3S and R-13M missiles, and two were downed by USN F-14s on 19 August 1981 during a brief dogfight. In the 1990s, the Libyan Su-22/22M inventory numbering about 50 aircraft was hit badly by the UN-imposed embargo and its current state is unknown. Several aircraft, for instance, were noted held in storage at the 558 ARZ company in Belarus.

Left and above: These Su-22M4Ks are operated by the Vietnam People's Air Force, serving with the 937 'Hau Giang' Trung Doan at Phan Rang in the south of country. Vietnam acquired around 50 Su-22M4/UM3s in 1988-90 to bolster the Su-22M fleet. It is understood that Vietnam is buying further aircraft from Poland.

During the 1990s, when tensions erupted into open fighting with Ecuador, Peru's 'Fitters' were repainted in this dark green/grey camouflage. In more peaceful times they have also adopted colourful tiger nose markings. This Su-22 is undergoing maintenance at SEMAN Industries.

Below right: Peru's second batch of 'Fitters' comprised 16 'deep-spine' Su-22Ms (illustrated) and three Su-22UM two-seaters. They were initially assigned to Escuadrón 411 at La Joya, but were merged into Escuadrón 111 at El Pato in 1991 as an economic measure. The unit was heavily involved in the 1995 Cenepa war with Ecuador. Today it flies around a dozen Su-22Ms as its primary equipment, with the surviving Su-22s held in reserve.

Below: Two Su-22s were modified with bolt-on refuelling probes for trials, although a lack of funds prevented a fleet-wide introduction.

The Angolan air arm is among the latest 'Fitter-K' operators receiving 18 Su-22M4s and two Su-22UM3s in the 1990s, just before the end of Russian support for the Marxist regime. The 'Fitter-Ks' were soon in use against UNITA opposition forces, operated by the one of the squadrons of the 26th Fighter-Bomber Regiment at Namib airfield. They are highly prized for their speed and most of their combat missions were flown with bombs dropped from safe altitude, well outside the reach of UNITA's shoulder-launched SAMs and AAA – however, the precision of such operations is low. In 1999 and 2001, 11 Slovak Su-22M4s and one Su-22UM3 strengthened the already depleted 'Fitter-K' inventory, and most of the operations were carried out with mercenary pilots. Some sources allude that Israeli mercenary pilots have been heavily involved in training local pilots for the Su-22s, but there are no confirmation that they have flown combat missions.

A major user of the 'Fitter' family in the 1980s was Iraq. In the mid-1970s its air arm firstly complemented its Su-7B inventory with a small number of Su-20s before switching to the procurement of the much better equipped Su-22M in the early 1980s, followed by the Su-22M4 by the middle of decade. Some 60 Su-22Ms are believed to have been taken on strength, together with 24 Su-22UM two-seaters, while the Su-22M4

deliveries numbered as many as 120. The swing-wing 'Fitters' became the main ground-attack types during the war with Iran between 1980 and 1988, carrying a wide array of unguided ordnance as well as some precision-guided weapons and ECM pods. The close air support and battlefield interdiction missions were flown mainly armed with the S-5 rockets carried in 32-round packs, while other missions were carried out with free-fall bombs. The 'Fitters' were also the main SEAD assets of the Iraqi air arm, carrying the anti-radar Kh-28 missile. During the prolonged 'cat-and-mouse' game with the Iranian MIM-23 Hawk SAM crews, the Iraqi air crews gained extensive anti-radar experience. In addition to the Kh-28, the more modern and flexible Kh-25ML was also used according to some sources, but confirmation is not available; the same is true for the Kh-58 ARM.

During the 1991 Gulf War, the Iraqi Su-22 fleet was idle, just like most other fighter and attack types in the inventory, and were not used to repulse the offensive of the coalition forces. The only sorties carried out were defections to Iran in an attempt to avoid the fleet's destruction on the ground by the coalition's repetitive bombings of the hardened aircraft shelters at the main airbases (an unknown number of Su-22s were destroyed in these raids). In the event, some four Su-20s and 42 Su-22M/M4s fled to Iran, and three more were shot down by the USAF as they attempted to flee. In March 1991, just after the end of the hostilities, two more Su-22Ms that violated the no-fly zone imposed by the coalition forces fell victim to USAF F-15s. As many as 10 Su-22Ms and two two-seaters sent for overhaul to the Baranovichi, Belarus-based 558 ARZ repair facility on the eve of the Iraqi invasion of Kuwait were retained there due to the UN-imposed arms embargo.

The Afghan Air Force was also another major 'Fitter' operator throughout the 1980s. The three dozen or so Su-22Ms delivered in the early 1980s to the Bagram-based 355 Air Regiment were used in anger against the opposition forces, and in 1987 the survivors were complemented by a number of Su-22M4s. Afghan AF Su-22M/M4 never participated in coordinated strikes with Soviet AF units, and the willingness of the Afghan pilots to fight was described by Russian sources as 'having not been particularly high'. The Afghan Su-22s suffered not only from attrition due to pilot errors, opposition air defence and sabotage on the ground – as many as three aircraft were shot down by Pakistani Air Force F-16s in the border area while another aircraft, a brand-new Su-22M4, was flown by its disaffected pilot to Pakistan in 1988.

Su-22M/M4s continued flying in combat in Afghanistan long after the Soviet withdrawal in 1989, firstly supporting the government and then the opposition forces. 'Fitter' pilots of the Afghan AF found themselves fighting on both sides after the Taliban movement had seized power in the country. There was some information that the Uzbekistan government provided air support to the Northern Alliance during the battles in the late 1990s, in 2000 and 2001, and in the course of operations both the Uzbek AF and Northern Alliance air arms flew numerous sorties from bases located in Uzbekistan.

By the beginning of US-led operation to remove the Taliban from power in October 2001, some four to six Su-22s were reported to have survived but took no part in the fighting, and were probably destroyed on the ground.

Alexander Mladenov;
additional box material by Tom Cooper

Iraqi and Iranian 'Fitters'

The first batch of 12 Su-20Ms arrived in Iraq in 1974, and entered service with No. 1 Fighter-Bomber Squadron, then based at Rashid AB, in Baghdad. Some 40 Su-20Ms and then around 60 Su-22Ms and Su-22M3Ks were delivered by the summer of 1980. When Iraq invaded Iran, in September 1980, it had the following Su-20/22 units: No. 1 FBS at Kirkuk with Su-20Ms; No. 5 FBS at Kirkuk with Su-22M3Ks; No. 44 FBS at Kirkuk with a few Su-20Ms (equipped for recce and also called 'Su-20R' by the Iraqis) and Su-22M3Ks; and No. 109 FBS at Shoibyah with Su-22Ms. Two additional units, including No. 69 Squadron, were formed during the war, and equipped with Su-22M4Ks.

The Su-20/22 was the IrAF's workhorse during the war with Iran. They were deployed from the first until the last day of the war, and used for all possible combat-related purposes. From 1986, IrAF Su-20/22s were equipped with 2200-litre (484-Imp gal) spray tanks for chemical weapons (before this time, their main means of chemical weapons delivery were bombs). Su-22M3Ks and Su-22M4Ks were also deployed as 'Wild Weasels' during the war with Iran using the Kh-28e (designed for use against HAWK SAMs) and Kh-28c (for use against C-Band radars).

According to USN intelligence reports, the IrAF still had a fleet of 119 Su-20s and Su-22s at the begining of the Gulf War in January 1991. The actual figure of operational airframes was between 85 and 90, distributed between six units. Four Su-22s were claimed as shot down by USAF F-15 fighters during and immediately after the fighting; 14 were destroyed on the ground Finally, 46 (including four Su-20s) were flown to Iran, of which two (including one Su-22UM3K) were shot down while arriving there. This should have left the IrAF with some 25-30 airframes as of mid-1991. However, by the late 1990s, concentrating on challenging the US and British aircraft flying control operations over the northern and southern no-fly zones over Iraq, and thus emphasising interceptors, the IrAF concluded that the type was useless for its purposes. All the advanced equipment – including their chaff and flare dispensers, as well as SPS-141 ECM pods – was removed from surviving airframes and mounted on the remaining IrAF interceptors instead, mainly MiG-23MLs. Only around 20 Su-20Ms, Su-22M3/4Ks and Su-22UM3Ks are known to have been captured by US and Australian troops in 2003.

As originally delivered from the USSR, the Iraqi Su-20Ms were camouflaged in sand and dark olive green over, and 'Russian light blue' underneath. Over the years, while going through overhauls, some aircraft received a slightly darker camouflage, consisting of light earth and dark green. Iraqi Su-22Ms, Su-22M3/4Ks and Su-22UM3Ks were camouflaged in a number of completely different patterns. The most widespread was earth and dark olive green on top with Russian light blue underneath, but there were plenty of variations of this pattern, especially those including sand. Also, a number of Su-22M3Ks and a few early Su-22UM-3Ks were camouflaged in sand and dark olive all over.

After the war with Iran, the whole fleet was re-serialled in a manner similar to that of the Yugoslav air force. According to the new system, each aircraft type received its own designation and this was used for the first two digits of the serial. The third digit of the serial represented the variant, and the last two the individual number of aircraft. Correspondingly, Su-20s were serialled in the range 205xx, while Su-22M3/4Ks and Su-22UM3Ks were serialled in the range 225xx.

Of the 44 Iraqi 'Fitters' that landed safely in Iran in 1991 – mainly at Nojeh AB, near Hamedan – four were Su-20s. The Islamic Republic of Iran Air Force (IRIAF) tested a number of Su-22M4Ks for some time, and several were seen still wearing their camouflage pattern as used by the IrAF – but with full IRIAF markings. Eventually, however, the IRIAF found them unsatisfactory and dropped the idea of introducing them to service. Reports about possible delivery of several ex-Iraqi Su-22s to the forces of the Northern Coalition/United Front, led by Ahmadshah Massoud, in northern Afghanistan in the 1990s, were never confirmed. Nevertheless, Iran was a staunch enemy of the Taliban and the IRIAF is known to have flown dozens of combat- and reconnaissance sorties over Afghanistan at the time.

Most of the former IrAF Su-20/22s were 'stored' in the open at Nojeh AB. A number of them – still in their full IrAF markings – were used for filming the movie *H-3 Attack* (released in 2004), about the legendary mission flown by eight IRIAF F-4s over the whole of Iraq to strike the Iraqi H-3/al-Wallid AB, in the west of the country. In recent years reports surfaced that the Islamic Republic Guards Corps Air Force (IRGCAF) might be interested in putting them into service, but no confirmation of this being the case has been forthcoming.

A reminder of what aircraft were like in the days before composites and computers took over aircraft design completely, the Su-22 is a popular and still capable warhorse. While many former Soviet types have rapidly fallen by the wayside, the 'Fitter' is scheduled to remain in service well into the next decade, thanks to countries such as Poland and Vietnam.

Captured at Kirkuk by US troops, these Iraqi Su-20s had not only survived the long and bloody war with Iran, but also two full-scale aerial onslaughts by coalition forces in 1991 and 2003. It was from this base that Iraqi 'Fitters' launched the first air attack against Iran, hitting Nojeh AB on 22 September 1980. A few days earlier, on the 14th, an Su-22 on a reconnaissance sortie probably became the first victim of the Phoenix missile, shot down by an Iranian F-14 Tomcat.

HT-18
'Vigilant Eagles'

For prospective US Navy, Marine Corps and Coast Guard helicopter pilots – and many from overseas air arms – the first step to becoming versed in the art of rotary-wing flying is taken at Whiting Field, home of Training Air Wing Five. Here two squadrons fly the Bell Sea Ranger to produce over 300 qualified helicopter pilots every year.

Based at Naval Air Station Whiting Field near Milton, Florida, Helicopter Training Squadron Eighteen (HT-18) was established on 1 March 1972 and is a component of Training Wing Five. As one of only two squadrons (the other being HT-8 'Eight Ballers') charged with training Navy, Marine Corps and Coast Guard helicopter pilots, HT-18 flies in excess of 36,000 hours and produces over 300 new helicopter pilots annually. HT-18 also trains foreign students and instructs Italians, Germans, Danish, Spanish and Saudi Arabians. On 10 March 2005 the squadron celebrated the completion of 1,000,000 Class 'A' mishap-free flight hours in the TH-57 helicopter. It took 35 years to accomplish

the million-hour mark, which started at the squadron's conception back when it operated the TH-1 Huey.

All Student Naval Aviators begin their flight training with approximately 100 hours in the T-34 or T-6 training aircraft. Following selection to the advanced rotary-wing pipeline, the aspiring aviators are introduced to helicopters and complete 40 flight hours in the TH-57B. This stage of training familiarises the student with hovering, take-offs and landings, auto-rotations and VFR navigation. Later stages of training include 75 hours of instruction in Basic Instrument Flight, Radio Instruments/Airways Navigation and Helicopter Tactics, including formation flight, search and

rescue operations and external load operations. Initial shipboard qualifications are conducted in Pensacola Bay aboard the Helicopter Landing Trainer (IX-514). At the completion of the syllabus, the newly designated Naval Aviator will have accumulated approximately 215 hours of fixed- and rotary-wing flight time.

Each TAW-5 squadron gets approximately 37 helicopters a day to train with, and about 22 are C-models and the remaining 15 are B-models. After a student gets about 15-20 flights under their belt in the TH-57B, they start flying the more complicated C-model. V-22 students may also come through HT-18. Osprey students have a reduced syllabus since they do a lot of training at Corpus Christi in multi-engined aircraft and already possess their instrument rating. Their main goal is to learn hovering and how to fly helicopters for those coming out of fixed-wing communities. Student 'winging' ceremonies are undertaken every two weeks, and a typical graduating class ranges between 22-30 aviators that receive their 'Wings of Gold'. HT-18 has an average of 70 instructors on hand.

Above: The early stages of the HT-18 course are undertaken in the TH-57B, which is less well-equipped than its TH-57C stablemate. For the first 40 flying hours students learn the rudiments of helicopter flying – controls, emergencies and simple visual navigation. Once students are competent in the art of controlling a helicopter, they begin to learn how to employ helicopters operationally.

A TH-57B hover-taxis along the HT-18 ramp at Whiting (above). The airfield is extremely busy, requiring care for all operations from the base, a solid grounding in procedures for the students. The two Sea Ranger squadrons also share the field with three fixed-wing training squadrons, while nearby NAS Pensacola is also a busy training base. Below a pair of TH-57Cs flies over the Pensacola Beach area.

Sea Ranger

HT-18 flies both the TH-57B and TH-57C Sea Ranger, which are all new-build aircraft manufactured by Bell Helicopter Textron located in Fort Worth, Texas. Based on the popular Bell 206B, the aircraft has a maximum gross weight of 3,200 lb (1452 kg), a maximum forward airspeed of 130 kt (241 km/h), and can remain airborne for over 2.5 hours. Some aircraft have an external hook mounted on the bottom that is used for external load training. There are currently about 120 TH-57s on the Whiting Field ramp, with the majority being C-models.

Sea Rangers have two pilots, space for three passengers in the back, and are powered by a single Rolls-Royce 250-C20J engine. The TH-57Bs are basic 'hands-on' day-VFR trainers that allow students to learn the essential flight characteristics. The first B-model was received in October 1981 and the last example was delivered in December 1985. The TH-57C is an FAA-certified single-pilot IFR aircraft and has additional NAV aids. It is used for advanced training and has dual attitude gyros, heading indicators, VOR, TACAN, ADF, ILS and GPS systems. The C-model weighs around 150-200 lb (68-91 kg) more than a

TH-57B (depending on the particular airframes) due to the additional instrumentation on board. The first C-model was received in December 1982 and the last example was delivered in February 1985. Vertex Aerospace at NAS Whiting Field performs contract maintenance.

Future potential TH-57 upgrades may include a stick (collective) shaker to help prevent overtorque situations, and crashworthy attenuation seats. Also, there may possibly be a 'glass' cockpit added, but that could be a costly upgrade and would take years for it to materialise. As with many communities, the upgrades are at the mercy of funding dynamics.

The Commanding Officer is Commander Gerald Briggs and XO is Lieutenant Colonel Joe Richards. Lt Col Richards flew CH-46Es, joined the Marines in 1985, served on both west and east coasts, and has over 3,000 flight hours to date. When asked about the squadron and the latest in the community, he said, "As far as changes for the syllabus, we are starting to integrate NVG night flying into training to give the students a taste of what they will experience in the fleet. Currently we have six aircraft that have been modified with NVG-compatible lighting, and ultimately we will end up with 15 NVG-capable TH-57s. We use

The US Navy began its long association with the Sea Ranger in 1968, when the TH-57A (based on the Model 206A) entered service. It was superseded by the Model 206B-based TH-57B and TH-57C, of which 47 and 77 were purchased, respectively. Two Sea Rangers are assigned to NAWC-AD at Patuxent River, while the remainder are divided between HT-8 and HT-18 with TAW-5 at Whiting Field.

ANVIS-9 NVGs and our goal is to introduce NVGs to all students coming through here.

"The Bell TH-57 is such a proven design that it is the only helicopter in the inventory that does not have a slated future planned replacement – the Sea Ranger will be around for a long time. Training Wing Five flies 11 percent of the total hours flown by all Navy and Marine Corps aircraft during peacetime, and we are huge players in Naval Aviation. That is a lot of flying and our squadron alone flies about 32,000-34,000 hours a

For the time being the TH-57 remains more than adequate for the task, and no replacement is in sight. Upgrades – particularly to the TH-57C operational trainer – will continue to be applied to enhance flight safety and its usefulness in the trainer role.

year, compared to a typical fleet squadron flying in the neighborhood of 4,000-5,000 hours in a year.

"Whiting Field is one of the busiest airfields in the nation and, while the aircraft here are small, the numbers of take-offs and landings are some of the highest. The product that we put out every two weeks is great. A highlight occurs every other Friday when the students graduate in the presence of their families. Seeing the ceremony and that personal moment held every two weeks – it doesn't get old. It is a fun job and I always look forward to coming to work every day."

The HT-18 Operations Officer is Major Michael Marko who has 1,700 hours and came from the west coast CH-53E community. He added, "As long as the basic 'skill set' that is

Carrying TAW-5's 'E' tailcode, a pair of TH-57Cs flies over a runway marked out with helicopter landing spots. As well as the main field, Whiting controls a number of auxiliary fields nearby in Florida and Alabama, which can be used by TAW-5's training fleet for pattern practice.

required to check into fleet squadrons remains the same, we will fly the TH-57 and give the students the necessary skill that enables them to progress into their assigned fleet helicopter. The instructors put in 10-12 hour days and fly two to three times a day on hot days. I am proud of all the talented instructors we have here that work so hard every day and who send well-trained aviators on their way to fleet."

Ted Carlson/Fotodynamics.net

Philippines
Air Force and Navy

Photographed by
Rogier Westerhuis/Aero Image

Once a modern air force, today the Philippine Air Force (PAF) faces a constant struggle to keep its aircraft and helicopters flying. Over the last few decades it has had little opportunity to strengthen its forces and to develop capabilities similar to those of it neighbours. Factors that have contributed to this include political instability, corruption, the 1997 Asian currency crisis and the ongoing operations against communist and Moslem insurgents. Despite modernisation plans and improvements through working more efficiently, today's PAF is seriously lacking the hardware and finances to meet its requirements. With many aircraft mothballed over the years, the PAF has recently launched the 'Aircraft Recovery Program'. The aim is to make these aircraft airworthy again, which is considered to be the most cost-effective way of increasing the capability of the air force. In general, the PAF has to go to great lengths to get the most out of what little it has.

A unique and symbolic formation: with the insurgent problem in different parts of the country, increasing troop transport capability and close air support (CAS)/ counter-insurgency (COIN) capability have become the main priorities within the Philippine Air Force. Troop transport capability is provided by the Lockheed C-130, whereas the CAS and COIN effort is entrusted to the Rockwell OV-10 Bronco.

The Sikorsky S-76A is a legend within the Philippine Air Force as it played an important role in the 1986 EDSA 'People's Power Revolution' that led to the fall of the dictatorship of Ferdinand E. Marcos. The first of these Sikorsky helicopters arrived in 1983 with the armed version – the AUH-76 Eagle (locally known as 'Gunbirds') – being assigned to the 20th Air Commando Squadron. The unarmed S-76s (locally known as 'Whitebirds') were allocated to the Search and Rescue task.

Previously used for training with Air Education and Training Command, the SIAI-Marchetti SF-260TP has adopted a more aggressive role. The fleet received this appropriate colour scheme and was assigned tasks that included light attack, forward air control, close air support and limited maritime patrols. The 17th Attack Squadron 'Jaguars' was reactivated in July 2001 to fly the SF-260TP, of which six are assigned. It previously flew the AT-28D 'Tora-Tora' and the OV-10, but the squadron was decommissioned in 1996. The unit is part of the 15th Strike Wing, based at Maj. Danielo S. Atienza Air Base, Sangley Point, Cavite.

Below: After a fatal crash of a Northrop F-5A on 2 May 2001 the remaining fleet of eight F-5A/Bs was grounded, and has not flown since. A 'Task Force Freedom Fighter' was created in the autumn of 2004 to conduct a comprehensive Non-Destructive Inspection of the fleet to determine whether it was feasible to bring the F-5 back into service. Although the results were satisfactory, the decision to retire them was taken in October 2005. An aircraft performed a ceremonial 'taxi-in and shut-down' to mark the end of the F-5's career. One of the F-5As was painted in this attractive scheme and kept in excellent condition, just in case.

Above: The Rockwell OV-10 works closely with the MD520MG Defender in the close air support (CAS) and counter-insurgency (COIN) roles. The OV-10A entered PAF service in 1992 and is assigned to Nos 16 and 25 Attack Squadrons, part of the 15th Strike Wing based at Sangley Point.

Left: The Thai government donated eight OV-10Cs to the Philippines. The first four arrived in September 2003, followed by the final four in May 2004. Except for a few minor changes the aircraft still fly in their Thai paint schemes. This one carries the new 20-mm cannon pod.

Below: To reduce the risk of mid-air collisions the air force painted part of the upper side of the Broncos' wings white to increase visibility. The top side of the vertical stabiliser of the MD520 tail and part of the wings of the SF-260TP are also painted white for the same purpose.

Above: Specialists in finding uses for their grounded equipment, the PAF has decided to strip the 20-mm cannon from the Northrop F-5s and mount them into locally developed pods. With this gun pod the fighting power of the Bronco is further boosted. When this photo was taken the system was still being tested, although it had seen limited operational action.

Right: This Bronco entered PAF service as one of the 24 OV-10As delivered in 1992. It was placed in storage due to a lack of funding and spare parts. As a result of increasing requirements for CAS- and COIN-capable aircraft for Internal Security Operations, and because of a lack of money for new equipment, the PAF decided to introduce the Service Life Improvement Program (SLEP) for the OV-10A. Initially, six OV-10As will be refitted with overhauled and modified Garrett T76-G-10/12 engines and the Hartzell four-bladed propeller system.

Left: This aircraft was delivered in 1981 as one of an order for three Fokker F27 Maritime Patrol Aircraft (MPA), but in recent years its radar equipment was removed (the auxiliary fuel tanks were retained) and the aircraft is now used in the transport role by the 221st Transport Squadron.

Below: The PAF's main transport is the C-130. Due to a lack of funding and parts, the PAF faces a constant struggle to keep these aircraft flying. With the designation of the Philippines as a major non-NATO US ally, the Philippine Armed Forces have access to the Excess Defense Article Program. This will hopefully result in the acquisition of surplus C-130s to further increase the heavy transport capacity. This heavily used C-130H is flown by the 222nd Heavy Transport Squadron, part of the 220th Airlift Wing based at Benito Ebuen Air Base, Mactan Island, Cebu.

Above and right: *The first McDonnell Douglas MD520MG Defenders entered service in 1990 and proved an ideal asset in counter-insurgency (COIN) operations. The agile Defender has seen considerable combat in various parts of the country, and usually operates in close co-operation with the OV-10s. The helicopter can be armed with a gun pod and rockets and, to make them suitable for night flying missions, the PAF recently tested forward-looking infrared (FLIR.) Two squadrons under the 15th Strike Wing are equipped with the Defender.*

Below: *Success in a guerrilla war is greatly dependent on the capability to airlift troops from one area of engagement to another in the shortest possible time. The UH-1H Huey is the ideal helicopter for the job and for this reason the PAF has around 40 of these helicopters in its inventory. The PAF is in constant need for more tactical troop transport capability and therefore the expansion of its fleet of UH-1 Huey helicopters forms one of the main priorities. Recently, 20 Hueys were acquired through Singapore Technologies Aerospace (STA) with deliveries completed in the second half of 2005. The helicopters delivered by STA were overhauled and modified before delivery, making them night-fighting capable. The training of pilots in flying with night vision goggles was conducted in conjunction with the USAF's 6th Special Operations Squadron.*

The 205th Tactical Helicopter Wing is based at Benito Ebuen Air Base at Mactan Island in Cebu. It is heavily involved in internal security operations.

Above: This UH-1H is fitted with an M60 7.62-mm machine-gun in the door for protection, a routine fit for Philippine Hueys. The 205th Tactical Helicopter Wing consists of three squadrons: 206 THS 'Hornets' at Edwin Andrews Air Base, Zamboanga, 207 THS 'Stingers' (badge left) at Col. Jesus A. Villamor Air Base, Pasay City, Metro Manila, and 208 THS 'Daggers' at the wing's HQ base at Mactan. These squadrons have deployments all over the country.

Right: For years this Sikorsky S-76 has been the sole example used for VIP transport by the 252nd Helicopter Squadron, based at Col. Jesus Villamor Air Base, Pasay City, Metro Manila.

Below: As part of the Aircraft Recovery Program, it was decided to restore this Sikorsky AUH-76 and to bring it back to flying condition after it had been in storage for some time. This particular helicopter is seen while on its way for a deployment in the Palawan area to protect the offshore gas rigs from possible terrorist attacks.

Flying training is conducted by the Air Education & Training Command (AE&TC) based at Fernando Air Base near Lipa city. The 101st Primary Flying Training Squadron 'Layangs' is equipped with the Cessna T-41D and Cessna 172. This T-41D is passing the crater of Taal volcano, located on an island in Lake Taal, south of Manila. The AE&TC, previously the 100th Flying Training Wing, moved in 1998 from Fernando Air Base to General Santos Air Base in South Cotabato, where the weather is more favourable for flying training. Due to the increase in insurgent activities in that area, the AE&TC returned to Fernando Air Base in 2000.

Left: To meet the requirement for more flying training capacity, the PAF recently acquired two Cessna 172s on the civilian market. These two aircraft were delivered in 2004 and are still not painted in air force markings.

Below: The SF-260MP is flown by 102nd Basic Pilot Training Squadron 'Musang' for basic and advanced flying training. The low number of aircraft available however (just one in April 2005) means that the T-41D is often used instead.

As part of the Air Force Modernization Program of 2000 several major commands were scheduled for downgrading. As a result, the Air Defence Command was deactivated on 31 March 2005 and downgraded to an Air Defence Wing. As a consequence, the 5th Fighter Wing (patch above) was deactivated and downgraded to the 5th Tactical Fighter Group with the 7th Tactical Fighter Squadron, flying the SIAI-Marchetti S-211, being the only squadron equipped with aircraft.

Eighteen SIAI-Marchetti S.211s were delivered to the PAF, but today only four remain in service. Another four are expected to re-enter service after completion of a Major Structural Inspection and repair work carried out by Philippines-based AEROTECH Industries. This is part of the PAF-initiated Aircraft Recovery Program. After completion of a Major Structural Inspection and repair, the S-211s are painted in this light grey colour scheme.

The PAF bought two different versions of the S-211. Those aircraft with serials (painted on the nose) starting with '0' are the 'warrior' version equipped with pylons, gunsights and provisions for an auxiliary fuel tank. Those aircraft with a serial starting with an '8' were initially delivered as jet trainers but later modified for the light attack role. The S-211 is mainly used for maritime patrol missions and training, but can also be armed with 0.50-in calibre (12.7-mm) machine-guns on the centreline hardpoint, multi-purpose bombs and rockets pods for counter-insurgency (COIN) and close air support (CAS) missions. Although not of much use as a true fighter aircraft, while awaiting the arrival of a 'modern' jet fighter to replace the stored F-5s, the PAF is also using these aircraft to keep the 'fighter spirit' alive. The aircraft are flown by the 7th Tactical Fighter Squadron 'Bulldogs', part of the 5th TCG at Basa Air Base (formerly used by the USAF and known as Clark Air Force Base).

Philippine Navy – Naval Air Group

Despite the Philippines having around 12,500 nautical miles (23148 km) of coastline and 7,107 islands, the air component of the Philippine Navy – the Naval Air Group – has always been surprisingly small. Aerial operations are considered to be a task of the air force. Therefore, the navy was only permitted a small force of light observation and utility aircraft and helicopters to provide air assets for naval operations. A turbulent political history and the Asian currency crisis of the late 1990s have hindered development and modernisation. With a small and ageing fleet but growing demand, the Naval Air Group is desperately awaiting better financial times in order to modify and reinforce its assets. The upcoming upgrade from Naval Air Group to Naval Air Wing could be a first step to what will hopefully be a more prosperous future.

Left: The current commander of the NAG, Captain Willie A. Ilustre, has plans to restore this Piper J-3 Cub to keep part of the history of the NAG preserved for future generations. In 1947 the Air Section was created within what was known at that time as the Off Shore Patrol (later to be the Philippine Navy). The first aircraft to enter service were a Stinson L-5 Sentinel and Boeing PT-13 Stearman, both on floats.

Left: In 1975 the first two Britten-Norman BN-2 Islanders were delivered to what was then called the Naval Air Unit. The NAU was renamed the Naval Air Group only a few months later. Presently, seven Islanders are used by the Naval Air Group and all are assigned to Naval Air Squadron MF-30 (MF standing for Multi-purpose Fixed-wing.) Australia-based Hawker Pacific Ltd recently re-equipped one of the Islanders with a Honeywell weather and sea search radar, new avionics and new communication equipment, making the Islander IFR-capable. In an ideal situation four aircraft would undergo this modernisation.

The Philippines is a country prone to natural disasters, with an average of 20 typhoons threatening the nation each year. The four squadrons of the 505th Search and Rescue Group are regularly called upon to support civilians in distress. The 505th SRG uses the Bell 205 (illustrated), UH-1H, Huey II (a modified UH-1H) and Sikorsky S-76 for disaster relief operations as well as other SAR duties.

Left: The NAG is based at Naval Base Cavite at Sangley Point and shares facilities with adjoining Maj. Danielo S. Atienza Air Base. The mission of the NAG is to prepare and provide air assets for naval operations. The five remaining BO105Cs belong to Naval Air Squadron MH-40 (MH stands for Multi-purpose Helicopter.)

Above: The Bolkow/MBB BO105 entered service in 1974 and is still the only helicopter in service with the NAG. The NAG has only two BO105Cs operational and three others are awaiting overhaul and modifications. With the cost for complete overhaul estimated at around 15 million pesos each (US$268,000), it may be more economical to buy refurbished BO105s or other used helicopters instead. With no funds available a decision is not likely to be made soon. The BO105s work closely with an anti-terrorist Ready Force called 'Task Force Stingray'. One of the tasks of 'Task Force Stingray' is the protection of tourist destinations against terrorist activities. The small fleet of BO105s has been supplemented during the years with aircraft previously used by the Philippine Air Force.

Naval Aviation Training Squadron-50 (NATS-50) is responsible for the training of pilots. In the basic flying training phase the NAG cadet flies the Cessna 152 (above), followed by the Cessna 177 (right). After successful completion of the basic course the young 'naval aviator' will start the Equipment Qualification Course on either the BN-2 Islander or BO105. The Cessna 177 was delivered in the early 1990s and is also used for liaison.

Left: The 5052nd Search and Rescue Squadron is one of the four squadrons under the 505th Search and Rescue Group. SAR crews wear Dayglo flight suits as a matter of routine.

Below: The PAF decided to upgrade two UH-1Hs in accordance with the Bell/Textron Huey II upgrade programme. This Huey II is the first example that entered the PAF on 17 February 2004, the first to receive this update package in Southeast Asia, and the first to be modified outside the Bell facilities at Forth Worth, Texas. The helicopter is equipped with parts that have a longer service life and an engine that performs better in hot-and-high environments. Because of its improved performance and the need for this capability in the SAR role, the Huey II was assigned to the 505th SRG.

Desert Testers:
586th FLTS 'Roadrunners'

Operating on and over White Sands Missile Range – the largest military installation in the United States – Air Force Materiel Command's resident unit, the 46th Test Group, offers a wide variety of truly unique test capabilities, from indoor evaluations to actual flight-testing. The latter is the core business of the 586th Flight Test Squadron, providing agile, responsive and expeditionary flight tests of the most advanced equipment in the world to ensure that the US military stays ahead of global threats.

The 46th Test Group is the largest associate unit based at Holloman AFB and reports to the 46th Test Wing headquartered at Eglin AFB, Florida – itself part of the Air Armament Center (AAC), Air Force Materiel Command (AFMC). The Test Group performs a unique test and evaluation mission for which it operates world-class test facilities for high-speed sled track testing, navigation and guidance system testing, radar signature measurements and weapon system flight testing. It is the liaison office for all Air Force programmes tested at White Sands Missile Range (WSMR). Some 450 military, civil service and contractor personnel staff the Group, of whom two-thirds are scientists, engineers and technicians.

The 46th Test Group directs the activities of the 746th Test Squadron, 846th Test Squadron, National Radar Cross Section Test Facility,

Directed Energy Office, Detachment 1 and the only Air Force Flight Test Squadron based at Holloman AFB, the 586 Flight Test Squadron.

The 746th Test Squadron, also known as Central Inertial Guidance Test Facility (CIGTF), was established in 1959 to test and validate intercontinental ballistic missile guidance systems. Today it is the DoD's 'Center of Expertise' for testing and evaluating GPS user equipment, inertial guidance systems for aircraft, missiles and spacecraft, as well as Doppler and stellar-aided inertial navigation systems and navigation subsystems such as sensors, gyroscopes and accelerometers. It is considered the leader in GPS open-air combat threat environment testing and the centre of excellence for testing all aircraft and missile guidance systems. The 746th Test Squadron is the Responsible Test Organization (RTO) for GPS user equipment tests, being

responsible for test, evaluation and verification of GPS user equipment and GPS integrated systems.

The 846th Test Squadron operates the Holloman High Speed Test Track (HHSTT) and its capabilities are unique in subsonic through hypersonic velocities. It is the world's premier rocket sled test facility and its 50,788-ft (15480-m) sled track is the longest in the world. Payloads up to full-scale aircraft can be tested at selected portions of the flight environment under accurately programmed and instrumented conditions. The test capabilities provided by the Test Track fill a distinct gap between laboratory investigations and full-scale flight tests. The squadron is also DoD's 'Center of Expertise' for all ejection seat testing using the Multi-Axis Seat Ejection (MASE) sled and is the lead facility for all supersonic tracks.

The National Radar Cross-Section (RCS) Test Facility (NRTF) is the primary DoD facility for outdoor static cross-section measurements. Located on White Sands Missile Range, this one-of-a-kind NRTF consists of two separate but complementary outdoor test sites, combining the best of monostatic and bistatic radar cross-section measurements. The NRTF Main Site facilities are capable of radar cross-section amplitude and phase measurements, antenna pattern measurements, glint and near-field measurements. A wide variety of targets can be tested, from models to full-size aircraft or ground vehicles. Targets can be

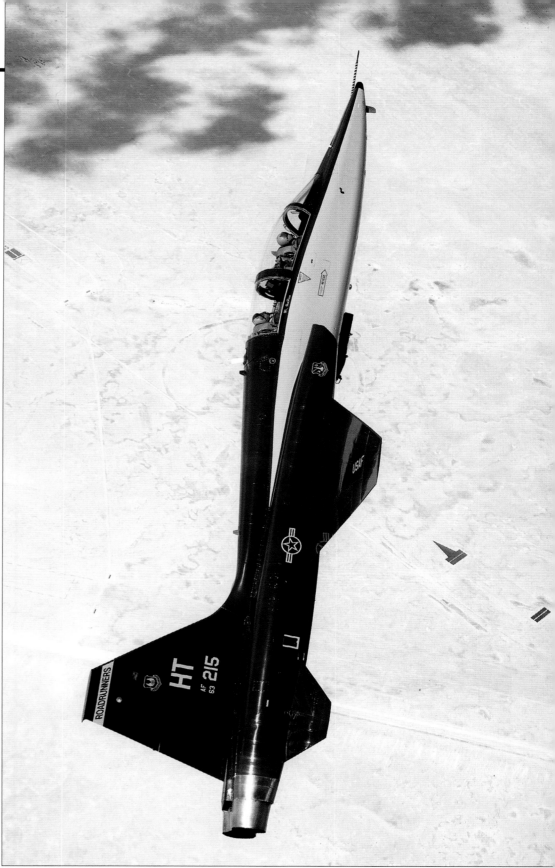

The 586th FLTS (above left) is a component of the 46th Test Group (above), itself a part of the 46th Test Wing – the USAF's primary armaments test organisation. The flying operations of the 'Roadrunners' are just one element of the 46th Test Group's work at Holloman – it also conducts ejection seat and other sled tests, and co-ordinates work on directed-energy weapons and advanced navigation/landing systems

Left and right: One of the 586th's three AT-38Bs cavorts over the White Sands National Monument close to the base at Holloman. The unit's Talons are very busy, not only acting as platforms for a myriad of equipment tests, but also being employed as simulated targets and as target-tugs.

mounted on pylons or columns at a variety of heights, orientations and locations to meet virtually any measurement requirement. The other site contains the RCS Advanced Measurement System (RAMS) that is designed for monostatic RCS measurements of RCS targets, and is equipped with a fully – in the ground – retractable pylon.

The 46th TG Directed Energy (DE) Office was formed to assist the Air Force DE Program Office in designing meaningful and effective test programmes. DE weapons are devices using electromagnetic waves or sub-atomic particles to achieve military objectives. The 46th Test Group's unique test capabilities, with its remote and secure location adjacent to WSMR's airspace, make the Test Group the ideal test organisation to conduct passive, active or lethal DE weapon tests against airborne targets using the two High Altitude Test Sites at North Oscura Peak and Salinas Peak, working with the Air Force Research Laboratory Directed Energy Directorate at Kirtland AFB, New Mexico, on test and evaluation of the Air Force's new YAL-1A Airborne Laser.

Last but not least, Air Force Detachment 1 (DET 1) provides a liaison function for coordination of all Air Force test and training operations at White Sands Missile Range.

'Roadrunners'

Originally constituted as the 586th Bombardment Squadron (Medium) on 15 February 1943, the unit was known as the 'Bridge Busters'. Stationed at MacDill Field (USA), Ardmore Army Air Field (USA), Kellogg Field (USA), Boreham (UK), Holmsley (UK), Tour-en-Bassin (France), Bricy (France), Cambrai-Niergnies (France), Venlo (Holland), Kitzingen (Germany) and Bolling Field (USA), operating B-26 Marauders and A-26 Invaders, it was disbanded on 31 March 1946. It was consolidated on 1 October 1992 with the 6586th Test Squadron based at Holloman AFB, which was designated and activated on 15 December 1982. The 6586th was re-designated as the 586th Test Squadron on 1 October 1992 and renamed 586th Flight Test

Squadron on 15 March 1994, acquiring the name 'Roadrunners' – New Mexico's state bird.

Currently commanded by Lieutenant Colonel Dawn Dunlop, the 586th Flight Test Squadron is the only Flight Test Squadron based at Holloman AFB and is responsible for a wide variety of test and test support missions that are executed on the White Sands Missile Range. The unit supports 46th Test Group programmes as well as Air Armament Center (AAC) objectives for Air-to-Air and Air-to-Ground weapons development, and, as such, it plans, co-ordinates, analyses and conducts flight tests of avionics, guidance, navigation and laser systems, as well as advanced weapon systems.

Weapons testing includes ordnance being fired or dropped with live warheads.

In addition, the unit provides operational support to those Department of Defense test aircraft staging out of Holloman AFB. The unit provides target, chase and photo support for numerous Joint test programmes, and is involved with the oversight of UAV and commercial testing at Holloman. The unit is also responsible for range recovery, collecting the different type of test articles dropped on the range. This can vary from target tows to actual weapons such as, for example, a JASSM that may need to be recovered and sent back to the manufacturer for further analysis.

Above: Many of the test fixtures are carried on the centreline pylon of the AT-38B, but in this case the aircraft carries an ALQ-167 'Squeaky' programmable ECM pod. Jamming is required for many of the tests performed by the 586th FLTS. Inflight controls for centreline stores can be fitted in the rear cockpit.

Left: Current commanding officer of the 586th Flight Test Squadron is Lieutenant Colonel Dawn Dunlop, pictured here with one of the unit's AT-38Bs.

To perform these test and support missions, the unit owns three Northrop AT-38B Talons, one Beechcraft C-12J and, until recently, one McDonnell Douglas F-15D Eagle. All aircraft have been extensively modified to execute their specific test missions.

Although the Talon is still operated in large numbers in the USA, not many of the fighter training version remain in service today. Nevertheless, it is not difficult to recognise those operated by the 'Roadrunners', as their AT-38B aircraft are painted in a unique glossy two-tone grey colour scheme, portraying the Test Group insignia on the right air intake and the Test Squadron badge on the left. For their new mission the Talons have been modified with the indigenous Fighter Instrumentation and Navigation System (FINS). This system relies on inertial navi-

gation and global positioning input to develop a reference for the required Time-Space-Position-Information, essential for testing, and is capable of providing an accuracy of less than 1 m² (10 sq ft).

Each aircraft also received radar altimeters and moving map displays, giving it a 200-ft (61-m) AGL low-level navigation and target support capability. An on-board instrumentation system is integrated with the navigation/display system to capture real-time display information for the pilot. When required, additional clip-on GPSs are added. For specialised tests, oversized test equipment can be rack-mounted and installed in the rear cockpit, replacing the ejection seat.

Dual C-band radar beacons are permanently installed for ground-air radar tracking, whereby the frequencies are switched based on requirements. In addition, the aircraft are equipped with Mode 4 IFF and dual UHF radios, and all three aircraft have been modified to provide multiple format photographic and film coverage when they are used for chase missions. For this purpose DC power connections have been installed in the rear cockpit for hand-held film and video cameras, but the unit also utilises a specially constructed helmet-mounted video camera. Video from either the front or rear cockpit can be encrypted and downlinked in real time, or stored on the internal high-speed 8-mm video tape recorder.

Located in the spine of the aircraft aft of the cockpit, a test equipment bay has been configured to allow installation of additional test support systems and GPS/navigation test articles. Data from test equipment can be stored by the instrumentation system, along with the time-space-position information from the navigation system, to permit post-flight evaluation of test equipment performance, but can also be downlinked together with the video images in real time.

Externally, the aircraft has a single modified centreline pylon to enable the carriage of different types of test and operational stores ranging from pods for chaff, flares, Global Positioning System (GPS) navigation, precision data recording and telemetry, to electronic countermeasures (ECM) and Air Combat Manoeuvring Instrumentation (ACMI). For some particular stores, electronic control needs to be executed by the operator in the back seat.

In addition, all three aircraft are capable of carrying the ALQ-167 ECM pod – affectionately dubbed 'Squeaky' – which is a modular jammer programmable with a wide variety of electronic jamming techniques. Additionally, an ALE-40 chaff and flare pod is available for carriage.

Towed targets

Another test capability is offered by the Low Radar Cross Section Tow Target (LRCSTT) system that supports many different types of towed targets. Based at Redstone Arsenal, Alabama, the US Army Simulation, Training and Instrumentation Command (STRICOM) Targets Management Office (TMO), a division of STRICOM's Project Manager for Instrumentation, Targets and Threat Simulators (PM ITTS), provides most of these targets for the Tow Target Program, such as the Patriot Omni-directional Training Aerial - Tow (POTA-Tow), Very Low RCS Tow Target, Infrared Tow Target and the Aerial Gunnery Tow Target.

Once airborne, the target is lowered by cable from the pod and trailed behind the aircraft. Upon completion of the test, the target is dropped at WSMR for recovery, after which the test aircraft returns to Holloman. Depending on the type of test article and its characteristics, these are dropped at a pre-determined location on the range at a speed of around 250 kt (463 km/h) and 500-ft (152-m) altitude, to make a controlled drop possible.

Another new and unique capability offered by the squadron is the Low Observable Instrumented Tow (LOIT), being an instrumented low-observable tow target used for signature evaluation. For this capability a specially modified pod is used, covered with laser reflective material allowing the cross-section to be varied in flight by extending or

retracting the fins of the pod. Upon completion of the test, this pod is also dropped for recovery when towed by an AT-38. Other airframes like the F-16 have the ability, however, to retract the pod and bring it back to base. Finally, standard travel pods have been configured to allow carriage and operation of GPS position measurement equipment.

To mark the modifications made to the aircraft, the non-standard systems and wires have been painted orange, except for those non-standard components that should not be painted because of their functionality (like the TACAN and C-band radar beacons). The aircraft themselves are not involved in any new developments related to the aircraft type itself, but purely serve as host platforms and, besides their capability to carry the aforementioned pods, the Talons are intensively used to fly cruise missile profiles, photo safety chase and target missions.

Test restrictions

However, the Talon has some restrictions and can therefore not cover the full spectrum of test requirements. As the engines for the T-38 are not as robust as the engines of an F-16 or F-15, for example, above 35,000 ft (10668 m), on a standard day or cooler, the aircraft has minimum Mach restrictions. This means that, above that altitude, the pilot has to maintain the minimum Mach speed by moving the throttle only 1 in (2.5 cm) every 3 seconds. It gets worse when the temperatures are lower. Then the operational ceiling of the aircraft also goes down. This not only restricts the

The 586th's AT-38Bs have an equipment bay in the spine that can house various test instruments. This example carries an unusual test fixture.

The Talon provides fighter-like performance at a fraction of the cost, making it ideal for many of the test duties assigned to the 586th. The squadron can employ a helmet-mounted camera system (right) to allow the pilot to record chase imagery without having to physically operate a camera. Note the co-located QF-4s of Det 1, 82nd ATRS in the background.

aircraft's manoeuvrability, but moreover does not leave a lot of room for corrections or, worse, errors. Consequently, safety requirements sometimes limit the type of profiles that can be flown and some specific high-altitude test requirements cannot be met with the Talon. Also, the Talon is limited in its endurance, and longer or repetitive missions on occasions can become real challenges to support adequately.

At this moment, no requirement exists to upgrade the three Talons to C-model, as this modification would result in losing the pylons and thus pod capability. The electronics of the C-model would occupy valuable space currently storing test equipment and there would be no place to put FINS back in. Moreover, the accuracy of navigation would be downgraded, as FINS is a more sophisticated and precise navigation system than that installed in the C-models.

However, due to the many modifications made to the aircraft, the test Talons are approximately 300 lb (136 kg) heavier than a standard T-38, which increases the landing speed by about 5 kt (9 km/h). Therefore, to have the aircraft undergo the engine modification similar to that of the C-model is an option currently being studied.

Right now the aircraft continue to be supportable in the sense that maintenance and spare parts are readily available, and in the short term there are no concerns over having to replace the Talons. However, the unit is keeping an eye on what the next aircraft could be for the time when the Talons become unsupportable. The difficulty here is to find an aircraft that would meet all the existing and foreseeable mission requirements, but also has the capability to chase slow assets, something

problematic to accomplish with the Talons as well. Last year, a total of some 400 flight hours were accumulated by the AT-38Bs.

Beech 1900

The squadron also owns and operates a highly modified C-12J (Beech 1900 Airliner) with multiple antenna and receiver modifications for a variety of guidance/navigation, avionics and electronic countermeasures tests. This is the only C-12J in US military service used for test purposes, and many of its missions are flown to provide test support for the 746th Test Squadron, which is the GPS & Navigation test squadron at Holloman.

With a maximum gross weight of 16,600 lb (7530 kg), the C-12J can accommodate a maximum of four test stations or equipment pallets in its spacious cabin. As the aircraft acts as a host platform, no development is done for the aircraft type itself. Missions include tests for GPS jamming environment characterisation, for GPS development, for non-GPS systems development, and many other – mostly classified – projects.

Other test projects involving the C-12J are LADAR, the laser equivalent of radar, in which the aircraft is targeted by the LADAR tracking

Formerly an Edwards machine, this F-15D was assigned to the 586th until August 2005. It is now part of the 46th Test Wing's 40th FLTS at Eglin AFB, but remains available to the Holloman unit for its requirements. A key programme it has been involved in is infra-red measurements using the ATIMS podded sensors.

system and Joint Precision Approach and Landing System (JPALS). In the latter project, a next generation landing system is being developed and tested to facilitate pilots on their approach to aircraft-carriers, fixed bases, tactical airfields and forward operating bases.

JPALS is considered a key system for US military forces and will contribute to increasing mobility and rapid response capability on a global basis. JPALS is similar in concept to the civilian Local Area Augmentation System (LAAS) and is based on differential Global Positioning System (GPS) technology. It consists of modular avionics and ground/shipboard components to provide a range of landing minima and system configurations. As well as coping with the weak signal power from GPS satellites, JPALS-equipped aircraft are expected to operate in the presence of significant radio frequency interference (RFI), and one of the studies is focussing on the use of Controlled Reception Pattern Antenna (CRPA) technology.

UAV Systems Operations and Validation programme

From November 2005 the 586th will also be involved with UAV flight testing. As the US government recognises that actually no rules and regulations exist for flying UAVs in commercial airspace, the Air Warfare Office of the Under Secretary of Defense (Acquisition, Technology & Logistics) signed a Memorandum of Understanding with the 46th Test Group to support See and Avoid Technology for Unmanned Aerial Vehicles (UAVs). Also involved is the Unmanned Aircraft System Planning Task Force, Defense Systems. This see-and-avoid system is required to meet the Federal Aviation Administration (FAA) equivalent level of safety requirement to integrate UAV operations in the National Airspace System (NAS).

As none has been fully integrated or approved, the FAA is not accepting the use of these air vehicles by, for example, law enforcement agencies. Called the UAV Systems Operations and Validation Program (USOVP), this new test programme started in November 2005 and will evaluate a 'see' sensor developed by Air Force Research Lab Sensors Directorate for Unmanned Aircraft Systems (UAS). This will be the first 'see' sensor tested on a UAS and is to prove to the FAA that a UAS equipped with a see-and-avoid system can fly throughout the NAS demonstrating the equivalent safety of a manned aircraft, able to 'file and fly' a flight plan within 24 hours like a manned aircraft.

Being a commercial venture, the New Mexico State University, Physical Science Laboratory (NMSU PSL) in Las Cruces, New Mexico, is the prime contractor for this programme and, based on an in-house market analysis, NMSU PSL selected the Israeli designed and built Aerostar UAV – a commercially available off-the-shelf tactical UAV made by Aeronautics Defense Systems – to meet the programme's objectives.

Having a strategic alliance with Aeronautics Defense for the duration of the programme, General Dynamics Information Systems (GDIS) will lease two

Aerostar Tactical UAVs to NMSU PSL in support of UAV operations, training/certification and sensor technology development, and the UAVs will be deployed at Las Cruces. In addition, General Dynamics Ordnance and Tactical Systems (GDOTS) was awarded a contract by NMSU PSL for a UAV Ground Control System.

NMSU/PSL, the 46th Test Group at Holloman Air Force Base, and White Sands Missile Range have combined their UAV efforts to produce a joint regional UAV Test and Evaluation Center (UTEC) to help develop a 'file and fly' capability for UAVs. Over the coming months, while the UAVs are operated by the NMSU PSL, the 586th Flight Test Squadron is to maintain the overall safety and risk management oversight, as well as being responsible for the test plan and how the users develop that. At the same time, the squadron plans to test the 'sense and avoid' system as developed by the Air Force Research Laboratory from the WSMR Stallion Army Airfield in the north-west corner of the range. The programme provides the DoD with a unique short- and long-range UAV operations capability to support testing of UAV platforms and systems in civil airspace on a routine basis.

In addition to testing UAV components, the 46th Test Group hopes to purchase the two Aerostars in 2006 under the UAV Systems Operations and Validation Program follow-on contract.

With the UAV's payload capability, the Test Group will be able to test airborne jammers, sensors, and other equipment at a fraction of the normal cost. The 46th Test Group is already working with the Edwards AFB Flight Test Center on developing the procedures to fly the Boeing X-45C and Northrop Grumman X-47B Joint Unmanned Combat Air System (J-UCAS) developpment aircraft in the NAS from Edwards AFB to Holloman AFB.

This solution has to meet performance requirements of a highly mobile force at all locations, including operations at austere forward operating areas worldwide, potentially being hostile, where the aircraft will be receiving ranging and navigation data from the satellite constellation and differential ranging data or corrections from a ground/shipboard station via a datalink. This, in turn, will allow the aircraft to terrain-follow, evade threats and then safely land.

In another programme, a cost-effective Innovative Global Positioning System (GPS) jamming programme called JAMFEST is organised to broaden both the operational and test communities' awareness of GPS vulnerabilities. Organised for the first time in May 2004, for the duration of one week, several ground-based, mobile and aerial jamming systems were employed, offering some 59 jamming scenarios to some 12 customers, including multi-service Department of Defense (DoD) organisations, several defence contractors and civil organisations.

At JAMFEST, participants are able to walk, drive or fly through GPS jamming environments to test the effectiveness of advanced anti-jamming technologies while exposed to real-life signal jamming scenarios. For this programme, the 746th Test Squadron sets up and characterises the jamming environment and provides a 'quick-look' report stating when jammers are activated and describes the active interference signals.

In support of this programme, the 586th Flight Test Squadron characterises the jamming field using its C-12J, to which purpose the aircraft is configured with Controlled Reception Pattern Antenna (CRPA) ports and Fixed Reception Pattern Antenna (FRPA) ports on the top and bottom of the fuselage. For this mission, the C-12J carries 746th TS-owned test equipment designed to collect airborne reference measurements of the GPS jamming environment, flying data-collection sorties that span the airspace and altitudes used by the different systems under test. Part of the equipment installed in the cabin is a rack-mounted Central Inertial Guidance Test Facility (CIGTF) Reference System, consisting of navigation sensors/subsystems and the Data Acquisition System (DAS), the latter performing the primary functions of data-collection and real-time control for the Embedded Global Positioning System/Inertial Navigation System (EGI), GPS receiver/receivers, Standard

Six Raytheon (Beech) C-12Js were bought by the USAF, initially to serve in the operational support airlift role with the Air National Guard. This example is now assigned to the 586th FLTS to undertake navigation/landing equipment tests, principally on behalf of the 746th Test Squadron – the GPS specialists. The empty cabin can quickly be reconfigured with various test loads (right).

Navigation Unit (SNU) INS and Cubic CR-100 Range/Range Rate Interrogator/Transponders System (RRS). JAMFEST IV took place in the first week of November 2005.

A study is currently being undertaken to fit two pylons underneath the C-12J's fuselage (same pylon as used by the AT-38) to enable the aircraft to carry podded test equipment. Over the last seven years, the 586th averaged some 155 hours per year with the C-12J, but, as with all aircraft in use, test programme hours can vary significantly with the users' need for test support.

F-15 for test

Until the end of August 2005, a single F-15D Eagle was assigned to the 586th Flight Test Squadron. This specific aircraft was previously an Edwards test plane about to be de-commissioned when the 586th managed to convince the 46th Test Wing to acquire it for its test purposes. This was because of its specific operational and representative radar capabilities, and some other capabilities that it possessed.

As a consequence of performing different types of testing, both the Edwards and Eglin test wings have two different instrumentation requirements and, prior to its transfer, the aircraft needed to be de-modified at Edwards AFB. Its instrumentation equipment had to be removed as it was test-specific, but the aircraft retained its Suite 5 software loads, several of the components for the Joint Tactical Information Distribution System (JTIDS) and other upgrades. Each of the 46th Test Wing F-15s is in a test-specific configuration, some having instrumentation capabilities or having been modified for developmental weapon testing. In order to make its F-15 fit for its own tests, the Test Wing at Eglin made its own basic set of modifications, installing its own instrumentation equipment for data collection and real-time monitoring before it was transferred to Holloman.

These are basic modifications allowing test parameter recording of data and/or voice, and data telemetry of parameters to a ground control

room. This particular aircraft also received other test-specific modifications for testing with the Airborne Turret Infrared Measurement System (ATIMS). This is a multi-sensor data collection pod used to determine the operational effectiveness of infrared countermeasures to deny acquisition or decoy the seeker head of captive missile seekers, and can, among other capabilities, evaluate the effectiveness of countermeasures such as flares. At the end of October 2005, the F-15D temporarily returned to Edwards, to perform further tests flying the ATIMS III pod.

Prior to this, at the end of August 2005, the aircraft was flown back to Eglin, as currently the entire fleet of Operational Test and Evaluation (OT&E) and Developmental Test and Evaluation (DT&E) Eagles is based there. Currently the aircraft does not have FDL and the 586th is working with Eglin to get that modification accomplished as soon as possible, as it is needed to support many of the unit's test/target requirements. In addition, in the February 2006 timeframe, this specific aircraft will receive another set of modifications for some upcoming testing. To avoid any misunderstanding, although the aircraft

Above: The 586th FLTS has been very busy supporting the F/A-22 programme, providing chase and target functions. A typical mission is flying chase on weapon tests and recording them. The photo shows a recent supersonic JDAM drop, recorded from a 'Roadrunners' AT-38B.

Below: As well as airborne chase, the 46th Test Group also has a complex suite of ground recording equipment for weapons tests. These images capture a live JASSM trial.

is no longer on the books of the 586th, it will continue to be used by the 586th FTS.

As Lieutenant Colonel Dunlop explains: "A sharing agreement with the 46th at Eglin is now being worked out, orchestrating the use of this particular airplane and thus determining what particular test is done when. With this single aircraft, the 586th is providing essential support to a significant number of test programmes of the Air-to-Air Wing, Air-to-Ground Wing, Operational Flight Program Combined Test Force (OFP-CTF), Edwards Test Wing, US Navy at Point Mugu, US Army at Fort Bliss, F-35 Program Office, Joint Forces Command, 746th Test Squadron, Air Force Test Pilot School and Air Force Research Laboratory. Not to forget that the F-15 Eagle is an ideal aircraft for high-speed, high-altitude and long-endurance chase, something the 586th Talons cannot do."

Test programmes

No other range in the USA has so much test work going on from the different branches of the US armed forces as the White Sands Missile Range. The US Army, for instance, is very active with Patriot Advanced Capabilities tests. The US Navy has its in-ground Desert Ship on the range and the US Air Force does the majority of its overland testing on WSMR. With its current mission providing agile, responsive, expeditionary flight test for the development of weapons and sensors, the 586th test work is extremely diverse. As well as being involved in the big 'limelight' test programmes, the unit is also very much part of the many smaller developmental test programmes as they potentially may grow into something significant as well.

Being the only Flight Test Squadron at Holloman, the 586th is in a unique position not only able to support the different branches of the armed forces and their test activities on WSMR, but more specifically has the capability to provide these branches with test capabilities to support network-centric initiatives, joint service testing, and multi-service command and control testing. As Dunlop explains: "From June 2004 to June 2005, we supported 800 sorties for a total of 52 different test programmes of the Air Force, Army, Navy, Marines, Missile Defense Agency and the FAA, in which we either were the Responsible Test Organization (RTO) or the Participating Test Organization (PTO).

"Looking at the history of the unit, over the years the squadron mission and scope have grown significantly beyond the initial charter of the 586th Flight Test Squadron. The increasing and diverse demands of simultaneous flight test, programme management, joint programme test support, UAV development, commercial liaison to WSMR, and research and development needs to get back in balance with the squadron's manning and resources. We need to better align our mission within the Air Force and Air Armament Center guidelines to then tailor our resources and manning to best meet those objectives."

As the flight test squadron has become more integrated into the test process, the need to have a larger echelon of qualified test pilots has grown as well, as it is the skill of a test pilot which allows proper understanding of the developmental test process and then mitigate the risks and evaluate the test. Better understanding of the test itself will result in a better interaction with the owner of the test, resulting in better coordination regarding the

actual requirements versus the performance of the platform.

Being part of the 46th Test Wing, the 586th in principal is a developmental flight test unit, however a lot of support is provided to Operational Testing (OT) taking place at WSMR, which can be weapon support, ECM support, control room support and range recovery for these operational tests. Some examples of this OT involvement are the F-15 avionics development (currently Suite 5), JASSM chase and recovery (F-16, B-52, B-1) and providing support for the F/A-22 FOT&E missile shots at WSMR.

The flight test squadron typically acts as a PTO, providing weapons and photo chase, and range recovery in support of Eglin's based Air-to-Ground Wing, which undertakes a significant part of its operational tests at WSMR, testing the Direct Attack, Area Attack and Long Range Attack systems such as the Joint Direct Attack Munition (JDAM), Joint Air-to-Surface Standoff Missile (JASSM), Small Diameter Bomb (SDB), Joint Standoff Weapon (JSOW), Sensor Fuzed Weapon (SFW), Wind Corrected Munitions Dispenser (WCMD) and many other specialised programmes. In the same sense, the 586th supports the AMRAAM operational test and evaluation shots at WSMR of the Air-to-Air Wing.

Development Test roles

On the developmental test side, the unit participates in classified programmes like Bright Eyes, Big Safari, HEL-LATTS and QWIP Quick Climb. Joint Development programmes supported by the 586th are, for example, the Complementary Low Altitude Weapon System (CLAWS) and Counter Rocket, Artillery, and Mortar (C-RAM). Tested by the US Marines Tactical Systems Support Activity at WSMR, CLAWS is a mobile, rapidly deployable, high-firepower, all-weather, stand-off air defence system aimed to extend the Marine Air-Ground Task Force (MAGTF) commander's 3-D defences against attacks by cruise missiles, fixed- and rotary-wing aircraft, and unmanned aerial vehicles. For these tests the AT-38s typically mimic incoming cruise missiles.

Developing a capability to defeat rockets, artillery, and mortars, C-RAM must successfully identify, track, and engage the incoming threat munitions and is being developed by the US Army's Air Defense Artillery branch working with the Field Artillery School. The 586th provides target support simulating incoming rocket profiles.

Last but not least, the 586th is involved in Roving Sands, Global Strike Exercises, Homeland Defence Exercises and the recently held Weapon System Evaluation Program (WSEP). In the latter case, it was the first time in approximately 20 years that the 53rd Weapons Evaluation Group conducted an off-station WSEP, and in the period between 25 April until 6 May 2005, the 586th hosted Combat Archer 05-10, supporting the deployment of 18 aircraft and a total of 300 personnel to Holloman AFB.

The flight test squadron planned, managed and executed 21 air-to-air missile shots at the WSMR, providing the Combat Air Forces with many 'first-ever' shot profiles accomplished with operational aircraft and aircrews. The data gained from the missile firings will provide invaluable information regarding weapons systems performance, in turn enabling the 53rd WEG to document weapons systems performance in an over-land environment as opposed to the normal over-water tests.

Marnix Sap/MIAS

Desert Falcon flight test

In an exclusive agreement with the Department of Defense, Lockheed Martin's most advanced F-16 to date – the Desert Falcon – is undergoing its development testing at Holloman AFB, making use of the adjacent White Sands Missile Range.

During a hotly contested competition that spanned six years, UAEAF pilots made over 90 evaluation flights in the F-16 as part of the new fighter selection process. Lockheed Martin Aeronautics Company eventually won the competition and signed a commercial contract worth $6.4 billion with the government of the UAE in March 2000. The contract called for both the development of the Block 60 Desert Falcon (formal designation F-16E and F) and production of 55 single-seat F-16Es and 25 two-seat F-16Fs, scheduled to be delivered to the UAE by 2007. This contract is an entirely commercial programme with no Foreign Military Sales (FMS) support from the US government, the only government involvement being the release of encryption codes related to some five systems of the aircraft. As the contract covers a complete package, the deal also includes commercial training, support and flight-testing.

In this sense, the Air National Guard training base at Tucson IAP, Arizona, first had to compete against two other locations in the US prior to the 162nd Fighter Wing being allocated the Desert Falcon training mission. The first of 15 F-16 Block 60s arrived in September 2004. Another part of the training component is the supply of an F-16 Training System, comprising Unit Level Trainers (ULT) and Weapon System Trainers (WST) that provide pilots with real-world operational training capabilities for day and night missions in all weather conditions. Developed by Lockheed Martin in Akron, this training system will interface via local and long-distance networks, and will interface with the UAE's existing Mirage 2000-9 training systems. On the technical side, since 2001 senior engineers and Block 60 programme personnel from the UAE have been in the Lockheed Martin factory in Forth Worth, Texas, and training of engineers and maintenance technicians continues in Texas.

More importantly, considering that the Desert Falcon is such a substantial evolution compared to the 4,000-plus F-16s delivered thus far, with so many new capabilities, a completely new test programme had to be launched to validate the type. Although many of the Block 60 features have either been previously qualified on Block 30 and 50 aircraft, such as the conformal fuel tanks (CFTs) and uprated General Electric F110-GE-132 engine, or have been tested at a component level (like the Northrop Grumman APG-80 radar tested on board the company's BAC 1-11 testbed based in Baltimore), in the UAE's Desert Falcon the new structure, engine, avionics and flight controls were combined for the first time.

AESA radar

One of the many hardware features that have been integrated with the Desert Falcon is Northrop Grumman's active electronically scanned array (AESA) APG-80 agile-beam-radar (ABR), requiring the elimination of the pitot tube to improve radar performance. Also included is an upgraded environmental control system (ECS), designed for the UAE's harsh desert conditions, that provides cooled air to the cockpit and avionics. Another novelty is Northrop Grumman's AN/AAQ-32 Integrated Forward-looking infrared and Targeting System (IFTS), for which the wide-area FLIR navigation sensor is housed above the nose and the targeting pod is mounted under the port-side of the intake. Also new is the integration of the ALQ-165 electronic countermeasures system, also known as the Airborne Self Protection Jammer (ASPJ).

Providing active and passive jamming is the newly designed Falcon Edge Integrated Electronic Warfare Suite (IEWS). Another important new feature is the Combined Intermediate Automatic Test Equipment (CIATE) programme, capable of automatically testing all three Northrop Grumman sensor systems – the APG-80, IFTS and IEWS – and detecting faults to allow subsystem repair down to component level.

With the Desert Falcon selected as a platform, the UAE additionally purchased $2 billion in sophisticated armament, including AIM-120 AMRAAM, AIM-9M Sidewinder, AGM-88 HARM, AGM-65 Maverick, AGM-84 Harpoon, laser-guided bombs, 20-mm ammunition and other weaponry. To support their purchase, the UAE ordered a complete development, test and validation programme for the F-16E/F and, apart from the typical range of tests (high-angle-of-attack evaluations, departure/deep-stall testing, load surveys, flutter and overall aircraft performance tests), precision ground collision avoidance and weapon-systems testing was also planned.

As the US government is not officially involved, and supersonic and weapons testing cannot be performed at the Lockheed Martin facilities at Fort Worth or Marietta, this flight testing also had to be run commercially. A competition was run between the different US Air Force test ranges and, based on time, infrastructure and cost, Lockheed Martin announced in February 2004 that Holloman AFB and its adjacent White Sands Missile Range was to become the principal site for developmental testing of the Desert Falcon. In a

The second F-16F flies over the White Sands range during a test sortie. On its port intake pylon the aircraft carries the targeting pod for the Northrop Grumman AAQ-32 IFTS. The pod houses a long-range targeting FLIR and laser designator.

separate agreement, the USAF also provides aerial-refuelling support on a commercial basis.

For the Desert Falcon test programme (and contrary to previous F-16 models), standard production aircraft are used in the form of the first three production F-16F aircraft. The first test aircraft is the two-seat F-16F that made its maiden flight at Forth Worth, Texas, on 6 December 2003 with Lockheed Martin F-16E/F chief test pilot Steve Barter at the controls. The second aircraft, RF-2, flew in April 2004, followed by RF-3 joining the test programme in June 2004. All three are in common flight test configuration with Lockheed Martin's own instrumentation equipment installed, and these test aircraft rotate back and forth to Fort Worth on a regular basis for upgrades or the installation of test-specific instrumentation. All three have received modifications and fortifications to the aft upper and under surface for the installation of the emergency spin/departure recovery system for high angle-of-attack tests.

The Block 60 capabilities are being gradually developed, tested and delivered in blocks or Standards. Standard 0 was the initial release and, for instance, only included basic air-to-air and air-to-ground radar modes of the APG-80 radar. Upon verification of all the contractually defined functionalities, the first production aircraft were delivered in this Standard for training in Tucson. Standard 0 testing started in mid-2004 and ended in early 2005.

Standard 1 is the full hardware configuration and provides capabilities basically equivalent to the F-16 Block 50/52. The following Standards are mainly software upgrades to extend the aircraft's capabilities in stages, reaching its full operational capability in Standard 3. Currently the development is at Standard 2 and projected to be completed in February 2006, adding additional system and weapons capabilities to the Desert Falcon. A new contract is currently drawn to stay at Holloman until Standard 3 is reached, scheduled to be achieved by mid-2007.

The Desert Falcon Flight Test Team at Holloman encompasses some 34 people, pilots not included. The contract with Lockheed Martin stipulates to provide the Test Team with buildings and general support, but also that the 586th Flight Test Squadron is the Participation Test Organization (PTO) for the Block 60 tests performed by the Desert Falcon Flight Test Team operating from Holloman's north ramp. The Flight Test Squadron has the government oversight for security, safety review and risk management boards fulfilling a government requirement allowing Lockheed Martin to test the Desert Falcon at WSMR.

As a consequence, Lockheed Martin is also able to use the range telemetry system and coordinate with the 46th Test Group all its safety and risk management boards prior to testing, as the military have accepted some level of accountability. However, Lockheed Martin also has its own telemetry ground station for real-time time-synchronised data collection, and this data is not shared with the Air Force. Being fully independent, Lockheed Martin does not require any other support from the Air Force such as chase aircraft.

Test Group approval

The risk management board approves the risk level (whether it is low, medium or high). Lockheed Martin requires formal approval from the Test Group to do high-risk level testing at WSMR. The board reviews what loading and configuration Lockheed Martin is testing, what envelope is being pursued, what are the 'knock-it-off' criteria and as the manuals (Dash 1) are progressively drafted, how the obtained test result is weighed into the next tests by discussing the decision tree and test cards before going to the next test.

A typical example of the tests undertaken by the Desert Falcon Flight Test Team at Holloman are the Flutter Excitation Tests for the Block 60 running from September 2005 to March 2006. The F-16E/F is 70 per cent different compared to its predecessors, and all envelopes and configurations have to be newly defined and validated.

These tests are aimed to clear an envelope in a specific configuration and a flutter excitation system is used, entailing a special control panel to be installed in the cockpit. During the flight a programmed mode is excited to monitor the aircraft's response and verify whether the response is within the safety margins. The parameters usually measured and recorded during a flutter excitation test are airspeed, pressure altitude, outside air temperature, flutter exciter, elevator position, aileron position, rudder position, wing tip accelerations and wing strain gauges. Real time analysis is necessary to assess the results from each individual excitation in order to decide whether to proceed to the next test point.

A single flutter excitation test will identify the frequency and damping data for one structural surface at one flight condition of Mach number. The flight test engineer will establish a table of flight conditions where a series of flutter excitation tests are desired. The test series will start at a moderate airspeed, well below the expected flutter boundary, in order to establish a baseline damping level for the surface. The tests will progress in small increments toward the predicted boundary, with data analysis occurring between each excitation. This is then performed for all the different possible configurations.

Upon completion of the test schedule in the USA, all three aircraft will be handed over to the UAE with the instrumentation equipment still installed. The gun pack, which is currently replaced by instrumentation, will be delivered in a box for installation *in situ* if required.

Above: The sides of the second F-16F are covered with photo-calibration marks to allow the accurate recording of weapon drops. The three test aircraft will be delivered to the UAEAF after the programme is complete, but will retain their test equipment so that the UAEAF can continue with its own tests if it decides to do so.

Below: The third F-16F sits on the ramp at Holloman AFB. Although the Lockheed Martin test programme is formally discrete from the US DoD activities, the fact that it uses the White Sands range and some of its facilities necessitates close co-ordination with the DoD's test organisations – principally the 586th FLTS.

The second F-16F has two forward-facing, podded Schneider Kreuznach Cinegon C-mount high-speed video cameras mounted just inboard of the tailplanes to record weapon separations. Three further cameras are carried in each AIM-120C AMRAAM digital camera pods on the wingtips, facing inwards.

A-26/B-26 Invader

By most measures the Douglas Invader is one of the world's great aircraft. Fast, powerful and sleek, the Invader certainly has the aesthetics. Its combat record is unequalled by any other aircraft, flying combat missions in the last year of World War II, during the Korean War and the long US involvement in southeast Asia, in addition to a multitude of smaller conflicts. The Invader was also widely used by foreign air arms, the last example being retired from military service in the early 1980s, while it also had a parallel civilian career that spawned many executive conversions.

Preliminary design studies for a new light bomber began during the autumn of 1940 at Douglas's El Segundo factory in California following a review of the Douglas A-20 Havoc that had highlighted some inadequacies with the aircraft. By November 1940 these had been quantified as a lack of interchangeability of the crew in flight; insufficient defensive armament; a 6-g design strength that limited it to shallow glide bombing instead of a moderate dive bombing capability; long take-off and landing distances incompatible with operations from unimproved fields close to the front-line; and a lack of speed, especially at intermediate altitude.

To improve all-round defence it was suggested that a new aircraft should have powered turrets with 0.50-in (12.7-mm) calibre weapons and more ammunition. Offensive armament was to include a 75-mm (2.95-in) cannon, requiring a new fuselage so that the weapon could be accommodated in the nose with an armourer to service the weapon in flight. To improve strength and reduce weight a new aluminium alloy, 75S, was to be used. Sophisticated two-piece aerofoil-shaped wing flaps were designed that had a higher lift coefficient than those used by the A-20. The National Advisory Committee for Aeronautics (NACA) recommended using a laminar flow aerofoil section and smooth skin surface to improve speed.

By the end of January 1941 a Douglas team led by Edward H. Heinemann and Project Engineer Robert Donovan submitted a proposal to the US Army Air Corps for the manufacture of two prototype aircraft, one configured as a twin-engined light bomber and the other as a night-fighter. With a mid-mounted wing with a laminar

The epitome of a wartime light bomber, the Invader design drew heavily on that of the A-20 that preceded it. This A-26B exhibits the early nose-gun arrangement with four to starboard and two to port. The inner surfaces of the engine nacelles were painted black to reduce glare in the cockpit.

First of the many: the XA-26 Invader prototype runs up on the ramp at Mines Field, Los Angeles, in 1942. The type's remarkable longevity can be summed up by its warpaint – it spanned the era from early wartime olive drab to three-tone tactical camouflage in Vietnam.

Above: A-26Bs and Cs from the 386th Bomb Group head for a target in Germany in April 1945. All are fitted with the four twin-gun fairings that could be scabbed under the wings to augment nose and dorsal turret armament, the latter often being fixed to fire forward.

Below: A trio of B-26Bs from the 731st Bomb Squadron, 452nd BW, returns to its base at Itazuke, Japan, after a raid on North Korean targets in May 1951. Night-camouflaged aircraft served alongside natural metal examples throughout the Korean War.

flow aerofoil and electrically operated double-slotted flaps, the new design was to be powered by a pair of 2,000-hp (1492-kW) Pratt & Whitney R-2800-27 radial engines in large nacelles. Much work was undertaken on the nacelle design, including the construction and wind tunnel-testing of mock-ups to reduce drag and increase engine cooling. A large bomb bay able to carry up to 4,000 lb (1814 kg) of bombs or two torpedoes was located below the wing spars, while external racks on the wings increased the ordnance that could be carried. Remotely operated dorsal and ventral gun turrets, each with a pair of 0.5-in (12.7-mm) machine-guns, were controlled by a gunner located in a rear compartment in the aircraft.

The US War Department inspected a mock-up of the aircraft on 11 and 22 April 1941, resulting in an order for two prototypes under contract number W535 AC17946 at a cost of $2,083,385.79 (plus a fee of $125,000) on 2 June 1941. The first prototype, the XA-26-DE, was a three-seat light

bomber with a glass bombardier's nose, while the second aircraft – the XA-26A-DE – was a two-seat night-fighter. A third prototype was added three weeks later. Known as the XA-26B-DE, the last prototype was to be a solid-nose version armed with a 75-mm (2.95-in) cannon.

Production contract

On 28 February 1941 Douglas put forward a proposal to produce 500 examples of the new aircraft at Santa Monica at a cost of $142,250 each. The first aircraft were expected off the production line 20 months after contract signature. However, the Army Air Corps' Material Division saw the cost as excessive and negotiations continued until agreement was reached on 31 October 1941, when Douglas was awarded contract AC21393. In order not to interrupt production of the A-20 Boston, C-47 Skytrain and C-54 Skymaster at Santa Monica, the USAAF decide that its new aircraft would be built at new

Douglas factories at Long Beach, California, and Tulsa, Oklahoma. Time was lost while these facilities tooled up for production of the Invader, while Douglas was criticised by the Army Air Force for not assigning a sufficient number of engineers to the programme.

The XA-26-DE made its maiden flight from Mines Field (near the site of Los Angeles International Airport) with Ben Howard at the controls on 10 July 1942, although it had originally been scheduled for 15 January 1942. The flight had been pushed back because of delays obtaining landing gear struts, self-sealing tanks, the turrets and government-furnished equipment. The XA-26-DE undertook the majority of the type's flight test programme, although it was joined during the year by the XA-26A-DE and XA-26B-DE.

Meanwhile, the USAAF was finalising which versions of the aircraft it wanted. Eight days after the XA-26-DE made its maiden flight, it

The red trim on this B-26C identifies it as a 13th Bomb Squadron 'Grim Reapers' aircraft, seen parked next to an AD Skyraider of VMA-121 at Pohang. The 13th's sister unit in the 3rd Bomb Wing, the 8th BS, flew the first and last bombing missions of the Korean War.

instructed Douglas that the 500 aircraft in the initial production batch would have the 75-mm (2.95-in) cannon installed, while 200 interchangeable gun noses with six 0.50-in (12.7-mm) machine-guns would also be built. This decision was soon abandoned, however, and the company was told to speed up development of the bombardier-nosed version while investigating

Above: The Invader had an important career in the US Navy, albeit in second-line roles. Most of the JD-1s wore brightly coloured tails for their primary role as a target-tug for training naval gunners and fighter pilots.

alternative armament to the 75-mm (2.95-in) cannon for the solid-nosed version. These included various combination of 0.50-in (12.7-mm), 37-mm (1.46-in) T20 cannon and 75-mm (2.95-in) cannon, although all work on the T20 stopped in August 1943.

On 17 March 1943 a further order was signed for 500 A-26Bs to be built at Tulsa, Oklahoma, primarily armed with the 75-mm (2.95-in) cannon. The designation A-26B was originally allocated to a production version with the 75-mm (2.95-in) cannon, while A-26C was the solid-nose aircraft with the alternative, smaller calibre armament. This was later changed with all solid-nose aircraft, irrespective of armament, becoming A-26Bs while the production version of the XA-26-DE with the bombardier's nose became the A-26C. These two versions were the only examples to enter production. Development of the armament for the A-26B finally resulted in the adoption of six (later increased to eight) 0.5-in (12.7-mm) M2 machine-guns in the noses of the majority built, plus two more of the weapons in both the ventral and dorsal gun turrets.

The A-26 Invader was originally planned to replace all of the US Army Air Force's medium bombers in service from 1942. Indecision as to the exact role the aircraft would undertake – mirrored in later years by subsequent designation changes – meant that production of the aircraft was delayed time and time again. The estimated

date for the first production A-26s was to have been July 1943, but creation of production tooling took a lot longer than anticipated. Personnel were transferred from the C-74 Globemaster programme, while the first six production aircraft were built using the tooling on which the prototypes were assembled.

In January 1943 Douglas informed the USAAF that production of the Invader would slip until October. The USAAF had accepted only 21 examples by March 1944. Famously General Arnold wrote about the slow delivery of the Invader to the USAAF that "I want the A-26 for use in this war and not the next war". While initially both Long Beach and Tulsa were to build both the A-26B and C models, this was changed so that the former concentrated on producing A-26Bs and the latter factory A-26Cs. Again, the changes added more delays to the programme, especially for the A-26C.

Flight tests with the XA-26A-DE demonstrated that the aircraft was a capable night-fighter. Unfortunately for Douglas the aircraft did not demonstrate a significant advantage over the Northrop P-61 Black Widow and was quickly dropped from USAAF plans. The only Invaders to serve as night-fighters were a handful of aircraft converted for the French forces for use over Algeria in the early 1960s.

At war

The Invader had its combat debut in July 1944 when four A-26B-5-DLs were evaluated under low-level combat conditions in New Guinea by crews from the 13th Bombardment Squadron, 3rd Bomb Group. Having transitioned from A-20 Havocs, the crews found the visibility from the cockpits poor and the forward armament had a limited punch in strafing attacks. Elevator forces encountered during low-altitude pull-outs were judged to be heavy, range was considered inadequate, the life raft stowage was unsatisfactory while the lower turret was little used and believed unnecessary.

This experience resulted in a redesign of the canopy, raising its profile to give the pilot a better view to the sides, and creating an easier method of escape for the pilot and navigator in the front office. The aircraft's basic armament was increased, with two more 0.5-in (12.7-mm)

In the form of the DB-26 the Invader played an important role in the development of unmanned aircraft. Aircraft like this DB-26C carried hundreds of target drones for missile tests and practice shoots. Here the aircraft carries a Ryan Q-2A Firebee under the starboard wing, and a 'second-generation' Q-2C/BQM-34A Firebee under the port pylon.

machine-guns located in the nose of the A-26B, while a twin-pack of similar weapons was designed that could be carried under the wings. These units were replaced on later production aircraft by three machine-guns in each wing, making it possible for a total of 18 weapons to be carried (including those in the turrets). In practice, especially post-World War II, many Invaders lost their turrets.

Invaders arrived in the European Theater of Operations in September 1944, the 553rd Bombardment Squadron, 386th Bomb Group (BG) based at Great Dunmow, Essex, sending 18 A-26Bs on the type's first combat mission – an attack on Brest in occupied France – on the 6th of the month. By the time of its final mission in Europe on 3 May 1945, the 386th, 391st, 409th and 418th BGs were operational on the type in the theatre with the 410th converting, while the 12th Air Force's 47th BG flew them alongside A-20 Havocs in Italy. The aircraft completed 11,567 sorties, dropping 18,054 tons of bombs and shooting down seven aircraft for the loss of 67 examples.

In the Pacific two Bomb Groups were flying Invaders by August 1945. The 5th Air Force controlled the 3rd BG (and was busy converting the 41st BG onto the type), while the 7th Air Force had the 319th BG. With the end of World War II production of the A-26B and C ceased, while plans for production of the follow-on A-26D and E were ended. The last standard Invader to be delivered (44-34753) was handed over to the USAAF on 31 August 1945. A total of 2,529 Invaders had been built, including four prototypes.

Post-war service

It was the Invader's basic characteristics as one of the fastest and most versatile light bombers at the end of World War II that made sure it did not disappear from the inventory of the US Army Air Forces. The post-war draw-down allowed the Douglas A-20 Havoc, North American B-25 Mitchell and Martin B-26 Marauder to be retired completely from their primary role, allowing the relatively small number of Invaders built to equip the small number of light bomb wings with both the regular Air Force and Air National Guard.

When North Korea's forces invaded its southern neighbour on 25 June 1950, Invaders were again called back to arms, although it was more a case of being in the right place at the right time

rather than the qualities of the aircraft itself that guaranteed its involvement. The only light bomber unit available within the Far East Air Force was the 3rd Bombardment Group (8th, 13th and 90th Bombardment Squadrons) at Johnson Air Base in Japan, equipped mainly with the B-26B. It was thrown into action on the Korean peninsula, first attacking railroads on 28 June 1950 and striking the main airfield at Pyongyang the next day. Invaders continued to perform the light night intruder role throughout the conflict.

The 452nd Bombardment Group (Light), an Air Force Reserve unit based at Long Beach,

B-26s from Groupe de Bombardement 1/25 'Tunisie' are loaded with bombs for another mission against the Viet-Minh forces surrounding the French position at Dien Bien Phu in 1954. Sharing the ramp at Cat Bi are Fairchild C-119s of Détachement C-119.

California, was called to active duty to supplement the 3rd. Its pilots underwent a refresher course before three of the squadrons (728th, 719th, and 730th) were sent to Miho in Japan before going on to Pusan, South Korea. The 452nd's fourth squadron, the 731st, was attached to the 3rd Bomb Group, bringing it up to four squadrons. Later the two units were redesignated respectively

B-26s were used for a number of clandestine operations in the 1950s and early 1960s. The aircraft above, in full Cuban air force markings, was operated by the CIA during the disastrous Bay of Pigs operation, while this Nationalist Chinese aircraft (right) may also have been used on covert missions.

Brazil was one of the last air arms to use its Invaders in a front-line role, flying 16 reworked A-26Bs and Cs in the attack role until 1976. Here one releases bombs during an exercise. Replacement in the 1º Esquadrão, 10º Grupo came in the form of the AT-26 Xavante.

as the 3rd and 17th Bombardment Wings; the 17th renumbered its squadrons as the 34th, 37th and 95th Bomb Squadrons.

The two wings flew a total of 55,000 interdiction sorties, the majority at night, during which they were credited with the destruction of 38,500 vehicles, 406 locomotives and 3,700 rail cars. At the outbreak of the conflict the US Air Force had 1,054 Invaders on strength, many in storage. During the three years of the conflict 226 Invaders were lost to all causes, 56 to enemy action.

In addition to the light night intruder role, Invaders were also used for day and night photo-reconnaissance during the Korean War, flown by the 162nd Tactical Reconnaissance Squadron (12th TRS after February 1951), part of the 67th Tactical Reconnaissance Wing. The conflict spawned several special versions of the Invader, including aircraft with searchlights, radar-finding equipment and infrared detectors.

Second-line duties

Large numbers of Invaders were converted for second line duties with the US Army Air Forces and – from 1947 – the US Air Force. Many became TA-26Bs and TA-26Cs, the majority with dual controls but some just because they were used in second-line roles. Some were used as target-tugs, while the US Navy became an Invader operator when it acquired A-26Cs as the JD-1 for the same role. On 1 July 1948 the US Air Force did away with the attack category and the Invader became the B-26, the Martin B-26 Marauder having largely disappeared from airworthy service by that point. TA-26 variants became TB-26s and photo-reconnaissance FA-26Cs became RB-26Cs.

With the end of the Korean War the need to send replacement aircraft to the Far East ceased and further aircraft were available for second-line duties. Staff transports were designated as CB-26s or VB-26s, depending on the level of amenities inside the aircraft cabins, and others were converted to launch drones as DB-26s or were bailed out for research purposes as EB-26s.

The replacement for the Invader in the light bomber role with the US Air Force was the Martin B-57, a licence-built version of the English Electric Canberra, while the Douglas RB-66 Destroyer replaced many of the Invaders tasked with various forms of reconnaissance. Many of the aircraft made surplus were supplied to allied forces across the globe.

Foreign users

The first foreign service to receive Invaders was the Royal Air Force, although only a handful were received during the last years of World War II after Lend-Lease arrangements for the acquisition of the type were scaled back and eventually cancelled with the end of hostilities in Europe. Post-war, the Invader was supplied under US military aid to many countries, including Turkey (delivered from 16 March 1948), France, Peru, Chile, Saudi Arabia, Cuba and Brazil.

Of these nations it was France that became the largest operator, flying Invaders between 1951 and 1968. It received its aircraft in two separate batches, with the first being supplied to bolster French forces fighting in Indo-China in the early 1950s from surplus US Air Force stocks, retaining their USAF serials as identities. They were a mix of B-26B, B-26C and RB-26C variants. With the end of French involvement in Indo-China the majority of the aircraft were returned to the control of the United States in the Pacific, many at Clark Field in the Philippines.

For service in Europe and its North African colonies France purchased Invaders from several sources. Initial deliveries were civilian aircraft for use as test vehicles by the CEV, CEAM and a target-towing unit based at Cazaux. Further examples were acquired from July 1956 to form two bomber squadrons in the conflict in Algeria pending the arrival of the Sud Ouest Vautour bomber, while RB-26Cs in store in the UK were also supplied to form a reconnaissance squadron. Several versions unique to the French air force were developed in the early 1960s, including the RB-26P reconnaissance aircraft with French camera equipment, the B-26APQ13 radar operator trainer and the B-26N night-fighter. Other French Invaders were modified as unique testbeds.

France was not alone in developing local versions of the Invader. Chile created TB/B-26Ds with new nose gun configurations while Brazil modified its aircraft for utility transport roles. Dominica, Indonesia, Guatemala, Nicaragua, Portugal, Biafra, El Salvador and Honduras also all became Invader operators, acquiring their aircraft on the civil market or – in the case of Guatemala and Nicaragua – via CIA channels. The last operational military Invader was the sole Honduran B-26B (276, later 510) that was sold on 7 December 1982.

Second chance

The Central Intelligence Agency (CIA) made much use of the Invader during the 1950s and early 1960s for missions from covert insertion missions to overt warfare. The Agency became a major user of the type, relying on surplus US Air Force examples that had been sanitised to various degrees. CIA Invaders fought in Indonesia in 1958, Cuba (on both sides) during the abortive 1961 Bay of Pigs invasion, Laos in 1961, South Vietnam from 1961 to 1964, and in the Congo between 1964 and 1967. The Invader had commenced its clandestine operations at the end of World War II, dropping agents in southern Europe, and examples were used for similar missions against Albania and China in the 1950s.

The rise of low-intensity warfare or counter-insurgency operations in the 1960s resulted in military interest being rekindled in the Invader. Some problems were encountered with the airframe in its new role as a 'bush war fighter', such as wing structure failure and the age of some

Indonesia was among the many air arms that used their Invaders in anger, mainly on internal operations such as those accompanying the annexation of Irian Jaya. Later, the Indonesian air force used a few for target-towing, and the Invader survived in TNI-AU service into the 1970s.

In 1961 the Farm Gate detachment was established at Bien Hoa to fly missions in support of the war against Viet Cong guerrillas – it was the first USAF unit to join the escalating conflict. To circumvent political and diplomatic restrictions, the aircraft were designated 'RB-26' and their mission was cited as training for Vietnamese air crews. In fact they were a combat outfit, although a South Vietnamese observer had to be carried to authorise attacks. At left is a B-26B armed with napalm and rockets. Below a Farm Gate B-26B provides close escort for USS Providence as it sails up the Saigon River in January 1964. In the following month the Invaders were grounded after one lost a wing during a combat sortie.

of the avionics. On Mark Engineering was contracted to upgrade a motley collection of US Air Force Invaders as B-26K Counter Invaders (redesignated as A-26As in May 1966) to overcome these problems and provide a cache of aircraft to be used in operations in the growing conflict in southeast Asia. However, the adoption of the Douglas A-1 Skyraider for use by the US Air Force in Vietnam reduced the need for the Counter Invader and the aircraft were made available for other operations, although a single squadron was sent to the combat zone to operate clandestinely over Laos from Thailand between 1966 and 1969.

Five A-26As were supplied to South Vietnam in November 1969, but served only as instructional airframes. B-26Ks were also used with a glass reconnaissance nose on survey work from Panama, while CIA pilots in the Congo flew gunnosed examples between August 1964 and early 1967.

Executive transport

The plentiful availability of cheap aircraft after World War II allowed larger companies to acquire them for their own use, usually as corporate shuttles or VIP transportation. The most favoured types used included variants of the Douglas Skytrain and Lockheed's twin-engined Model 18 family, but for companies after something a bit faster many twin-engined bombers were also available and converted, including examples of the Martin Marauder, North American Mitchell and Douglas Invader.

While the Invader had a smaller main cabin than the other types, its spars delineating the forward extent of any cabin, with the addition of some windows and internal styling the area that once comprised the bomb bay and gunner's compartment could be turned into a relatively spacious cabin. These 'minimal' executive conversions allowed time-constrained businessmen to

travel in style and at speed. Many were 'one-offs', although companies like Grand Central Air Terminal, On Mark Engineering, R.G. LeTourneau and Rock Island were able to trade in executive conversions of 'stock' Invaders in the 1950s and early 1960s.

With the experience gained, some of the companies later tackled some of the basic problems inherent in the Invader airframe that limited its appeal as a corporate transport, including modifying the wing spars to increase the room available in the cabin. This trend continued with some companies such as the L.B. Smith Aircraft Corporation, On Mark Engineering and Rhodes-Berry mating what amounted to new fuselages to Invader wings. The ultimate expression of this trend was the On Mark Model 450, a new-build aircraft with some Invader components. However, by the time the upgraders had come around to this way of thinking, the days of the Invader as a topof-the-line executive aircraft had slipped away. The

early 1960s saw the appearance of the first corporate jets and piston-engined competitors were looking passé.

Other civil roles

Executive transport was not the only civilian role found for the Invader. Rock Island developed the Consort 26 especially for trials work. Other firms were involved in the conversion of airframes to act as fire-bombers to serve with companies such as Conair and Lynch Air Tankers, while some were modified for use as camera-ships for aerial filming or survey work. With the rise in interest in operating warbirds in the late 1960s, many of the exotic conversions were repainted in military markings by new owners after being acquired from the corporate transport fleet. However, a large number retain the altered profile or windows that betrayed a more sedate recent corporate history than the warpaint suggests.

David Willis

Right: Following the grounding of the Farm Gate aircraft in February 1964, the Invader returned to southeast Asia in June 1966 in the form of the On Mark A-26A. Using the callsign NIMROD, the principal role was night interdiction along the Ho Chi Minh Trail, which was undertaken until the type was withdrawn from use in November 1969.

Below: On Mark Engineering also produced numerous civilian conversions for executive use. This is a pressurised Marksman.

Invader variants

USAAF/USAF variants

The XA-26 is seen with an air data instrumentation boom projecting from the nose glazing. Dummy turrets were fitted from the outset.

XA-26-DE

The initial contract (AC17946) for the Invader was issued on 2 June 1941 and covered the production of two prototypes, one as a light bomber with a crew of three and the other as a two-seat night-fighter. The light-bomber prototype was assigned the designation XA-26-DE and, like the other prototypes, was built at the Navy-controlled El Segundo factory in California. The first flight of the XA-26-DE was undertaken on 10 July 1942 at Mines Field, California, with Douglas test pilot Ben O. Howard at the controls.

Power was provided by a pair of 2,000-hp (1492-kW) Pratt & Whitney R-2800-27 radial engines driving three-bladed propellers with large spinners. Performance and handling of the aircraft was judged to be excellent, although it did suffer from insufficient engine

cooling that resulted in some small changes being made to the cowlings and the deletion of the spinners on production aircraft.

It had a transparent nose cone that housed the navigator/bomb-aimer's position, although he was also provided with a jump seat to the right of the pilot. Compared to later Invader variants, the XA-26-DE had a flatter canopy profile. The third member of the crew operated remotely-controlled dorsal and ventral turrets from a position to the rear of the bomb bay.

As a prototype aircraft the XA-26-DE was not equipped with defensive armament at the time of its first flight, but mock-up turrets were installed to prove the aerodynamic characteristics of the units. As testing proceeded these were replaced by the real thing. These dorsal turret could be locked in a forward position, fixing the gun barrels

parallel with the centreline of the fuselage so that they could be aimed and fired by the pilot. Each turret was designed to carry a pair of 0.5-in (12.7-mm) machine-guns, with sighting for the lower unit through a periscope gunsight located immediately behind the rear of the bomb bay, a similar unit being located on the gunner's glazed canopy cover for the ventral unit. Two similar machine-guns were mounted on the starboard side of the nose. Two bomb bays could accommodate a total of 3,000 lb (1361 kg), while two racks

under each outer wing panel could carry up to 500 lb (227 kg) each for a total of 5,000 lb (2268 kg) of munitions. The aircraft was designed to be able to carry a pair of semi-recessed torpedoes in the bomb bay.

A typical range with a 3,000-lb (1361-kg) bomb load was 1,800 miles (2897 km), but a maximum 'clean' ferry range of 2,500 miles (4023 km) was possible. Internal fuel capacity was 1,050 US gal (3975 litre). The XA-26-DE was one of the fastest Invaders produced, with a maximum speed of 370 mph (595 km/h) at 17,000 ft (5182 m) and a service ceiling of 31,300 ft (9540 m).

The XA-26-DE had a wing spanning 70 ft (21.34 m) with an area of 540 sq ft (50.17 m²), and stood 18 ft 6 in (5.64 m) high on the ground, dimensions that remained constant for all of the Douglas versions of the Invader that followed. However, the length of the fuselage (51 ft 2 in/15.60 m) was to vary in subsequent aircraft with the adoption of different nose sections and armament configurations.

Serial: 41-19504 (c/n 1004), although it carried '219504' on the tail at one point

Wearing spurious serial '219504', the XA-26 is seen during early flight trials. The considerable wing dihedral was to become a trademark of all Invader variants.

XA-26A-DE

The contract for the initial two prototype Invaders stipulated that one aircraft would be completed as a night-fighter, designated as the XA-26A-DE. This designation was unusual because, as a new-built fighter type, it deserved a P for Pursuit prefix – night-fighter versions of the Havoc had become P-70s. The XA-26A-DE had a crew of two consisting of a pilot and a radar operator/gunner who controlled and monitored the MIT AI-4 centimetric radar located in the redesigned nose.

Fuselage length was shortened by 1 ft 3 in (38 cm) to 50 ft 5 in (15.37 m). A small radar scope was also provided on the instrument panel for the pilot.

The XA-26A-DE was armed with four forward-firing 20-mm (0.787-in) cannon housed in a ventral tray located underneath the forward bomb bay, in which the ammunition boxes were located. The remotely-controlled dorsal turret from the XA-26-DE was retained but was equipped with four 0.50-in (12.7-mm) machine-guns instead of two. The ventral turret was omitted. Hemispherical sighting blisters were

provided above and below the location of the gunner's compartment, but as the dedicated gunner had been dropped from the crew the dorsal turret could be locked in the forward position and fired by the pilot, doubling the forward firepower. Although designed as a night-fighter, a secondary role of night interdiction was envisaged and for this role the aircraft could carry up to 2,000 lb (907 kg) in the rear bomb bay.

The XA-26A-DE was powered by two 2,000-hp (1492-kW) Pratt & Whitney R-2800-27s with slightly larger spinners than the XA-26-DE, producing a maximum speed of 365 mph (587 km/h) at 17,000 ft (5182 m) and a service ceiling of 25,900 ft (7894 m). A typical range of 700 miles (1127 km) was demonstrated, while the maximum range was 1,420 miles (2285 km). The night-fighter Invader had an empty weight of 20,794 lb (9432 kg), a loaded

The XA-26A prototype shows the redesigned nose of the night-fighter version. The baseplate for the dorsal turret is just visible, although the turret is not fitted. Note the observation blisters above and below the rear fuselage.

weight of 25,300 lb (10660 kg) and a maximum weight of 28,893 lb (13106 kg).

The Invader proved to be well suited to the night-fighter role. However, by the time the first production Invader had flown in September 1943 the Northrop P-61 Black Widow had advanced to a point where a night-fighting version of the Invader was seen as surplus to requirements by the USAAF, allowing Douglas to concentrate on producing Invaders for the light bomber role.

Serial: 41-19505 (c/n 1005), although at one point the XA-26A had the false identity of '219505' on its tail

A-26A

The A-26A was to have been the production version of the XA-26A-DE night-fighter and night-intruder aircraft. Production was not undertaken as the US Army Air Force's requirement for a night-fighter had already been fulfilled by the Northrop P-61 Black Widow. The designation was later re-used for the B-26K-OM (see below), which became the A-26A-OM.

XA-26B-DE

Another prototype of a third variant was added to the Invader programme in late June 1941 as the XA-26B-DE. The XA-26B-DE was the prototype of a solid nose tank-destroyer version carrying a

forward-firing 75-mm (2.95-in) M4 cannon. The M4 had been developed by the US Army Ordnance Department by Captain Horace Dunn and had been produced for the North American B-25G Mitchell. It weighed 1,800 lb (818 kg) and required 14 ft (4.33 m) of space for

its installation and to accommodate the recoil when fired. In the B-25G a total of 21 rounds of the 15-lb (6.8-kg) ammunition was carried on a rack above the breech of the gun.

The need to carry and operate the 75-mm (2.95-in) weapon was one of the

reasons put forward to justify a new light attack bomber design, as the existing A-20 Havoc was deemed unsuitable to the operation of such a weapon. It was believed that it could not be accommodated in its nose. On the XA-26B-DE the weapon was located

The XA-26B was built to test the solid-nose version, originally armed with a 75-mm cannon in the starboard side of the nose. Here the aircraft wears the spurious serial '219588'.

on the starboard side of the nose so that its loading and operation would not interfere with the pilot. An aerodynamic cap was designed to fit over the end of the cannon to reduce drag, being blown off when the weapon was fired.

In the XA-26B-DE the crew returned to three, comprising a pilot and gun-loader/navigator seated in the cockpit side-by-side, and the gunner located behind the wing trailing edge to man the ventral and dorsal turrets, each with two 0.50-in (12.7-mm) machine-guns. Two viewing windows – one dorsal and one ventral – were provided for the gunner to allow him to view threats approaching from the rear hemisphere of the aircraft.

Tests proved that the Invader was

able to use the M4 successfully and plans were made for a production version to be built as the A-26B. Having shown that the Invader could use the 75-mm weapon, the XA-26B-DE was later used to test a number of different nose armament configurations.

Like the XA-26-DE and XA-26A-DE, the XA-26B-DE was completed with large spinners that were not a feature of production examples. The sole XA-26B-DE was the last Invader built at the El Segundo plant as newly opened Douglas plants at Tulsa, Oklahoma, and

at Long Beach, California, were tooled up to build A-26B (and later C) production aircraft.

Serial: 41-19588 (c/n 1006) although it carried the false serial '219588' on its tail at one point

A-26B-DL/DT

The A-26B-DL and -DT versions were the solid nose Invaders built at the Douglas facilities at Long Beach, California, and Tulsa, Oklahoma, respectively. The designation A-26B was originally assigned to a production version of the XA-26B-DE equipped with a single 75-mm (2.95-in) M4 cannon in the nose. The original US Army Air Forces plans for the Invader called for all 500 aircraft on order to carry the large cannon in the nose. However, these plans were subject to much amendment while the service decided on the exact role that it wanted its Invaders to undertake, and the M4 cannon-armed A-26B was largely overtaken by events. The attack bomber role gained more prominence within the service and, as it transpired, only the A-26B-1-DL and -5-DLs retained the weapon, along with a pair of 0.50-in (12.7-mm) machine-guns on the port side of the nose. The first A-26B-1-DLs emerged from Long Beach in September 1943. Compared to the XA-26B-DE they had revised nacelles for the 2,000-hp (1492-kW) Pratt & Whitney R-2800-27s driving 12-ft 6-in (3.81-m) three-bladed propellers without the large spinner used on the prototype. The internal fuel capacity was increased to 1,600 US gal (6057 litres) while the bombload was also raised to 6,000 lb (2722 kg), of which all but 2,000 lb (907 kg) was carried internally. The following 15 A-26B-5-DLs had slight modifications. They were the first to abandon camouflage for the natural metal scheme that became characteristic of Invaders in World War II.

The choice of nose armament for the A-26B Invader was frequently revised, but from the A-26B-10-DL a new all-purpose nose section was installed that could accommodate a wide variety of weapons without major redesign. Various armament trials were undertaken to decide the configuration of the armament for the Invader, with several early A-26Bs being used to test them. They included one 75-mm (2.95-in) cannon (starboard) and one 37-mm (1.46-in) cannon (port); a 37-mm cannon on each side; one T-13E1 37-mm cannon (starboard) installed so that it could be depressed downward 15° and two 0.5-in (12.7-mm) AN-M2 machine-guns (port); four 0.5-in machine-guns (starboard) and one 37-mm cannon (port); or six 0.5-in machine-guns, comprising two on the port side and four on the starboard. The US Army Air Force selected the last option as its standard nose armament

41-39100 was the first production A-26B. It lacked the large spinners of the prototype, but was otherwise similar.

towards the end of 1944, with each gun having 400 rounds available, compared to the 500 round for each of the weapons in the dorsal and ventral turrets.

Early combat experience demonstrated several deficiencies in the Invader, the most important being insufficient forward-firing armament and poor cockpit visibility. From the A-26B-15-DL the Invader's already considerable punch could be increased by adding four twin-gun external wing fixed gun installations of 0.5-in (12.7-mm) M2 machine-guns underneath the outer wing panels.

The canopy underwent a major modification both to improve visibility – especially to the sides – and to make it easier to escape from the aircraft in a crisis. A slightly bulged clear unit that opened outwards on both sides of the cockpit replaced the original heavily framed unit that opened upwards on the starboard side. It enabled the pilot to view his engine nacelles more easily, see that the main undercarriage was down and to look to the rear, greatly increasing his situational awareness. The new 'clam shell' canopy assembly

These two aircraft are from the early production batches, completed with the original low-profile cockpit glazing. The aircraft below carries a 75-mm M4 cannon in the nose for tests.

was initially built as a separate item in small batches to be retrofitted to existing aircraft, before becoming a standard production item from the A-26B-30-DL onwards.

Other modifications were made as production proceeded. The inlets on the wing leading edge that provided cooling for oil were redesigned, improving lift. The A-26B-45-DL introduced Pratt & Whitney R-2800-79s built by Ford, featuring a water injection system that

raised the emergency war rating to 2,350-hp (1753-kW). While combat had highlighted the problem of visibility from the cockpit that had been improved by the redesign of the component, Invader crews also wanted to increase the punch of the aircraft without cluttering the under wing surfaces with the add-on gun packs, as they had a negative impact upon performance. A mock-up of a nose section containing no fewer than 14 0.5-in (12.7-mm) machine-guns was

The A-26B-45-DL was the last block built with the asymmetric six-gun nose armament, subsequent sub-variants having an eight-gun nose.

built but rejected for production. The solution was a redesign of the gun layout in the nose to accommodate eight 0.5-in (12.7-mm) machine-guns, symmetrically mounted, and to accommodate three 0.5-in (12.7-mm) machine-guns inside each wing, allowing bombs or up to 14 5-in (12.7-cm) rockets to be carried on hardpoints or rocket cradles under the wing where the gun packs would have been.

Moving the guns inside the wing reduced drag and the stalling step, and decreased the landing run of the aircraft, while improving low-speed handling. The build-up in the number of forward-firing weapons allowed the Invader to become a more effective strafing attack bomber.

The new armament configurations first appeared with the A-26B-50-DL. From this block onwards Invaders also had an increase in internal fuel to 1,910 US gal (7230 litres). Aircraft destined for the Pacific theatre also benefited from a 125-US gal auxiliary fuel tank installed in place of the ventral turret. These changes highlighted the long-distances flown by the Invader in the Pacific theatre and were possible because the lack of fighter opposition reduced the demand for the defensive ventral gun turret. The auxiliary fuel tank was installed new in A-26B-51-DL, -56-DL, -61-DL and -66-DL block Invaders.

Later production Invaders had a refined dorsal turret to eliminate empennage buffeting when the guns were elevated forward, while engine cooling was improved with redesigned cowlings and cowl flaps.

The A-26B-60-DL had an empty weight of 22,362 lb (10143 kg), a loaded weight of around 26,000 lb (11794 kg) and maximum take-off weight of 41,800 lb (18960 kg). Surprisingly, the -60-DL was slightly lighter (by 8 lb/ 3.6 kg) empty than the A-26B-15-DL, but it was 1,600 lb (726 kg) heavier when loaded and could take-off carrying 6,800 lb (3084 kg) more than the earlier block aircraft. The length of the eight-gun nose A-26B-60-DL was 8 in (20.32 cm) more than the 50 ft (15.24 m) of the six-gun -15-DL. Both blocks had a service ceiling of 22,100 ft (6736 m), a typical range of 1,400 miles

Serial	Model	c/n	f/n
41-39100/39104	A-26B-1-DL	6813/6817	1/5
41-39105/39119	A-26B-5-DL	6818/6832	6/20
41-39120/39139	A-26B-10-DL	6833/6852	21/40
41-39140/39151	A-26B-15-DL	6853/6864	41/52
41-39153/39192	A-26B-15-DL	6866/6905	54/93
41-39194	A-26B-15-DL	6907	95
41-39196/39198	A-26B-15-DL	6909/6911	97/99
41-39201/39299	A-26B-20-DL	6914/7012	102/200
41-39300/39349	A-26B-25-DL	7013/7062	201/250
41-39350/39424	A-26B-30-DL	7063/7137	251/325
41-39425/39499	A-26B-35-DL	7138/7212	326/400
41-39500/39599	A-26B-40-DL	7213/7312	401/500
43-22252/22266	A-26B-5-DT	18399/18413	1/5
43-22267/22301	A-26B-10-DT	18414/18448	16/50
43-22302/22303	A-26B-15-DT	18449/18450	51/52
43-22305/22307	A-26B-15-DT	18452/18454	54/56
43-22313/22345	A-26B-15-DT	18460/18492	62/94
43-22350/22351	A-26B-15-DT	18497/18498	99/100
43-22352/22354	A-26B-15-DT	18500/18502	102/104
43-22355/22357	A-26B-15-DT	18504/18506	106/108
43-22358/22360	A-26B-15-DT	18508/18510	110/112
43-22361/22362	A-26B-15-DT	18512/18513	114/115
43-22363/22364	A-26B-15-DT	18515/18516	117/118
43-22365/22366	A-26B-15-DT	18518/18519	120/121
43-22367	A-26B-15-DT	18521	123
43-22368/22369	A-26B-15-DT	18523/18524	125/126
43-22370/22371	A-26B-15-DT	18526/18527	128/129
43-22372	A-26B-15-DT	18529	131
43-22373/22374	A-26B-15-DT	18531/18532	133/134
43-22375/22376	A-26B-15-DT	18534/18535	136/137
43-22377	A-26B-15-DT	18537	139
43-22378/22379	A-26B-15-DT	18539/18540	141/142
43-22380/22381	A-26B-15-DT	18542/18543	144/145
43-22382	A-26B-15-DT	18545	147
43-22383/22384	A-26B-15-DT	18547/18548	149/150
43-22385/22386	A-26B-15-DT	18550/18551	152/153
43-22387	A-26B-15-DT	18553	155
43-22388/22389	A-26B-15-DT	18555/18556	157/158
43-22390/22391	A-26B-15-DT	18558/18559	160/161
43-22392	A-26B-15-DT	18561	163
43-22393/22394	A-26B-15-DT	18563/18564	165/166
43-22395/22396	A-26B-15-DT	18566/18567	168/169
43-22397	A-26B-15-DT	18569	171
43-22398	A-26B-15-DT	18571	173
43-22399	A-26B-15-DT	18573	175
43-22400	A-26B-20-DT	18575	177
43-22401	A-26B-20-DT	18577	179
43-22402	A-26B-20-DT	18579	181
43-22403	A-26B-20-DT	18581	183
43-22404	A-26B-20-DT	18583	185
43-22405	A-26B-20-DT	18585	187
43-22406	A-26B-20-DT	18587	189
43-22407	A-26B-20-DT	18589	191
43-22408	A-26B-20-DT	18591	193
43-22409	A-26B-20-DT	18593	195
43-22410	A-26B-20-DT	18595	197
43-22411	A-26B-20-DT	18597	199
43-22412	A-26B-20-DT	18599	201
43-22413	A-26B-20-DT	18601	203
43-22414	A-26B-20-DT	18603	205
43-22415	A-26B-20-DT	18605	207
43-22416	A-26B-20-DT	18607	209
43-22417	A-26B-20-DT	18609	211
43-22418	A-26B-20-DT	18611	213
43-22419	A-26B-20-DT	18613	215
43-22420	A-26B-20-DT	18615	217
43-22421	A-26B-20-DT	18617	219
43-22422	A-26B-20-DT	18619	221
43-22423	A-26B-20-DT	18621	223
43-22424	A-26B-20-DT	18623	225
43-22425	A-26B-20-DT	18625	227
43-22426	A-26B-20-DT	18627	229
43-22427	A-26B-20-DT	18629	231
43-22428	A-26B-20-DT	18631	233
43-22429	A-26B-20-DT	18633	235
43-22430	A-26B-20-DT	18635	237
43-22431	A-26B-20-DT	18637	239
43-22432	A-26B-20-DT	18639	241
43-22433	A-26B-20-DT	18641	243
43-22434	A-26B-20-DT	18643	245
43-22435	A-26B-20-DT	18645	247
43-22436	A-26B-20-DT	18647	249
43-22437	A-26B-20-DT	18649	251
43-22438	A-26B-20-DT	18652	254
43-22439	A-26B-20-DT	18655	257
43-22440	A-26B-20-DT	18658	260
43-22441	A-26B-20-DT	18661	263
43-22442	A-26B-20-DT	18664	266
43-22443	A-26B-20-DT	18667	269
43-22444	A-26B-20-DT	18670	272
43-22445	A-26B-20-DT	18673	275
43-22446	A-26B-20-DT	18676	278
43-22447	A-26B-20-DT	18679	281
43-22448	A-26B-20-DT	18682	284
43-22449	A-26B-20-DT	18685	287
43-22450	A-26B-20-DT	18688	290
43-22451	A-26B-20-DT	18691	293
43-22452	A-26B-20-DT	18694	296
43-22453	A-26B-20-DT	18697	299
43-22454	A-26B-25-DT	18700	302
43-22455	A-26B-25-DT	18703	305
43-22456	A-26B-25-DT	18706	308
43-22457	A-26B-25-DT	18709	311
43-22458	A-26B-25-DT	18712	314
43-22459	A-26B-25-DT	18715	317
43-22460	A-26B-25-DT	18718	320
43-22461	A-26B-25-DT	18721	323
43-22462	A-26B-25-DT	18724	326
43-22463	A-26B-25-DT	18727	329
43-22464	A-26B-25-DT	18730	332
43-22465	A-26B-25-DT	18733	335
43-22466	A-26B-25-DT	18736	338
44-34098/34217	A-26B-45-DL	27377/27496	501/520
44-34218/34286	A-26B-50-DL	27497/27565	621/689
44-34287	A-26B-51-DL	27566	690
44-34288/34296	A-26B-50-DL	27567/27575	691/699
44-34297/34298	A-26B-51-DL	27576/27577	700/701
44-34299/34322	A-26B-50-DL	27578/27601	702/725
44-34323	A-26B-51-DL	27602	726
44-34324/34326	A-26B-50-DL	27603/27605	727/729
44-34327	A-26B-51-DL	27606	730
44-34328/34330	A-26B-50-DL	27607/27609	731
44-34331	A-26B-51-DL	27610	734
44-34332	A-26B-50-DL	27611	735
44-34333/34334	A-26B-55-DL	27612/27613	736/737
44-34335	A-26B-56-DL	27614	738
44-34336/34338	A-26B-55-DL	27615/27617	739/741
44-34339	A-26B-56-DL	27618	742
44-34340/34342	A-26B-55-DL	27619/27621	743/745
44-34343	A-26B-56-DL	27622	746
44-34344/34346	A-26B-55-DL	27623/27625	747/749
44-34347	A-26B-56-DL	27626	750
44-34348/34350	A-26B-55-DL	27627/27629	751/753
44-34351	A-26B-56-DL	27630	754
44-34352/34363	A-26B-55-DL	27631/27642	755/766
44-34364	A-26B-56-DL	27643	767
44-34365/34367	A-26B-55-DL	27644/27646	768/770
44-34368	A-26B-56-DL	27647	771
44-34369/34371	A-26B-55-DL	27648/27650	772/774
44-34372	A-26B-56-DL	27651	775
44-34373/34376	A-26B-55-DL	27652/27655	776/779
44-34377	A-26B-56-DL	27656	780
44-34378/34381	A-26B-55-DL	27657/27660	781/784
44-34382	A-26B-56-DL	27661	785
44-34383/34386	A-26B-55-DL	27662/27665	786/789
44-34387	A-26B-56-DL	27666	790
44-34388/34392	A-26B-55-DL	27667/27671	791/795
44-34393	A-26B-56-DL	27672	796
44-34394/34398	A-26B-55-DL	27673/27677	797/801
44-34399	A-26B-56-DL	27678	802
44-34400/34404	A-26B-55-DL	27679/27683	803/807
44-34405	A-26B-56-DL	27684	808
44-34406/34408	A-26B-55-DL	27685/27687	809/811
44-34409	A-26B-56-DL	27688	812
44-34410/34412	A-26B-55-DL	27689/27691	813/815
44-34413	A-26B-56-DL	27692	816
44-34414/34416	A-26B-55-DL	27693/27695	817/819
44-34417	A-26B-56-DL	27696	820
44-34418/34419	A-26B-55-DL	27697/27698	821/822
44-34420	A-26B-56-DL	27699	823
44-34421/34422	A-26B-55-DL	27700/27701	824/825
44-34423	A-26B-56-DL	27702	826
44-34424/34425	A-26B-55-DL	27703/27704	827/828
44-34426	A-26B-DL	27705	829
44-34427/34472	A-26B-55-DL	27706/27751	830/875
44-34473/34477	A-26B-60-DL	27752/27756	876/880
44-34478	A-26B-61-DL	27757	881
44-34479/34480	A-26B-60-DL	27758/27759	882/883
44-34481	A-26B-61-DL	27760	884
44-34482/34483	A-26B-60-DL	27761/27762	885/886
44-34484	A-26B-61-DL	27763	887
44-34485/34486	A-26B-60-DL	27764/27765	888/889
44-34487	A-26B-61-DL	27766	890
44-34488/34489	A-26B-60-DL	27767/27768	891/892
44-34490	A-26B-61-DL	27769	893
44-34491/34492	A-26B-60-DL	27770/27771	894/895
44-34493	A-26B-61-DL	27772	896
44-34494/34495	A-26B-60-DL	27773/27774	897/898
44-34496	A-26B-61-DL	27775	899
44-34497/34498	A-26B-60-DL	27776/27777	900/901
44-34499	A-26B-61-DL	27778	902
44-34500/34501	A-26B-60-DL	27779/27780	903/904
44-34502	A-26B-61-DL	27781	905
44-34503/34504	A-26B-60-DL	27782/27783	906/907
44-34505	A-26B-61-DL	27784	908
44-34506/34507	A-26B-60-DL	27785/27786	909/910
44-34508	A-26B-61-DL	27787	911
44-34509/34510	A-26B-60-DL	27788/27789	912/913
44-34511	A-26B-61-DL	27790	914
44-34512/34513	A-26B-60-DL	27791/27792	915/916
44-34514	A-26B-61-DL	27793	917
44-34515/34516	A-26B-60-DL	27794/27795	918/919
44-34517	A-26B-61-DL	27796	920
44-34518/34519	A-26B-60-DL	27797/27798	921/922
44-34520	A-26B-61-DL	27799	923
44-34521	A-26B-60-DL	27800	924
44-34522/34585	A-26B-61-DL	27801/27864	925/988
44-34587/34617	A-26B-61-DL	27866/27896	990/1020
44-34618/34753	A-26B-66-DL	27897/28032	1021/1156
44-34754/34759	A-26B-DL	28033/28038	1157/1162*
44-34760	A-26B-DL	28039	1163**
44-34761/34774	A-26B-DL	28040/28053	1164/1177*
44-34775	A-26B-DL	28054	1178**
44-34776	A-26B-71-DL	28055	1179
44-34777/34778	A-26B-DL	28056/28057	1180/1181*
44-34779	A-26B-DL	28058	1182
44-34780/35197	A-26B-DL	28059/28476	1183/1600***

** not taken on charge ** not delivered *** cancelled*

The definitive gun arrangement for the A-26B is demonstrated by this 47th BG aircraft in 1947. Eight guns were mounted in the nose and a further six inside the wings. The latter removed the need for the scabbed-on two-gun packs carried by earlier versions, and allowed a variety of underwing stores to be carried, like this load of 5-in (12.7-cm) rockets.

(2253 km) and a maximum range of 3,200 miles (5150 km).

A total of 1,356 A-26Bs (including the XA-26B-DE) was taken on charge by the US Army Air Force from the 1,381 examples built. Of these, 205 were built at Tulsa before the decision was taken at the end of 1944 for that factory to concentrate on producing the A-26C.

The A-26B-5-DT, -10-DT, -15-DT, -20-DT and -25-DTs built at Tulsa reflected the improvements incorporated in the Long Beach-built aircraft. At the end of production in September 1945, the 25 A-26Bs that remained at the Long Beach factory were scrapped, while outstanding aircraft on order (a total of 418) were cancelled. It is difficult to give an accurate figure for the total number of

A-26Bs because many A-26Cs had the solid noses of the B model fitted while in service with the US forces. The opposite was also true, many Bs receiving the glass bombardier nose of the A-26C. As the only difference between the A-26B and A-26C was the nose, an A-26B with a bombardier nose was to all purposes an A-26C, and the reverse is true for the A-26C with the solid nose. Nose changes were not limited to between the B and C models, but also among the A-26Bs themselves. Thus, many aircraft built with the six machine-gun noses received the eight-gun unit and vice-versa.

Some early A-26Bs also received the wing-mounted machine-guns standard on late production aircraft. These practices produced much confusion as to the exact designations of individual Invaders, as in US service individual Invaders retained the block numbers they were created with, even though they could be very different from when they were rolled out of the factory.

In July 1948 the A-26B became the B-26B, the former holder of the designation – the Martin B-26B Marauder – having been long since retired by then.

Above: Not all Invaders are exactly what they seem – this 'hack' transport was built as a solid-nose A-26B but has the glazed nose of an A-26C. There were many instances where noses were swapped between versions.

Below: The A/B-26B had a long career in USAF hands. This aircraft was one of those serving with the 4400th Combat Crew Training Squadron Det 2 at Bien Hoa (Project Farm Gate) in Vietnam from November 1961.

B-26B

On 1 July 1948 the Douglas A-26B was redesignated the B-26B to reflect its primary bombing role. While the B-26Ks became A-26As during 1966, the few existing B-26B variants still on US Air Force charge retained 'B for bomber' designations.

CB-26B

Following the end of World War II, and again after the Korean War, the USAF modified surplus B-26Bs for second-line tasks. Several used for transport or 'hack' duties were designated as CB-26Bs. Although each CB-26B differed in detail, most had all armament removed and seating in the cabin.

DB-26B

Several B-26Cs were converted to carry and launch remotely piloted vehicles (RPV) as DB-26Cs, and at least one B-26B was similarly modified. They had all of their armament removed and pylons mounted under the wings to carry RPVs, the Ryan Q-2A Firebee being the standard drone employed. The RPVs were used to provide live targets for the development of new air-to-air missiles or support combat training by providing targets for

The sole DB-26B carries a Ryan Aeronautical Q-2A Firebee target drone. This RPV was first glide-tested in March 1951 and made its first powered flights in the following summer. A total of 1,280 was built (including the similar KDA for the US Navy). The similar conversion of the B-26C was more numerous.

interceptor pilots. The DB-26B (44-34652, converted from an A-26B-66-DL) served from the late 1950s until the mid-1960s.

EB-26B

One A-26B-45-DL (44-34137) became an EB-26B (E for exempt, the term used for military aircraft on a bailment contract to a non-military agency). It was used as a testbed for the development of drag and braking chutes, the unusual aspect of the tests being that the aircraft was not required to fly. It was extensively modified for the test programme to make it as light as possible, allowing the aircraft to accelerate quickly to the speed required for the test chute to be deployed on a shorter length of runway. Modifications included removing the wings outside of the engines, resulting in the EB-26B becoming known as the 'wingless wonder'. All armament, the bomb bay and landing gear doors were removed while various test apparatus was

attached to the tail section. If the chute tests were for a specific aircraft type a mock-up of the drag chute compartment would be built and attached to the EB-26B. Other trials involved testing

chutes of different size for stability and effectiveness. The EB-26B was used in this unusual role during the early 1950s before being returned to the US Air Force for disposal.

The EB-26B 'Wingless Wonder' is seen testing a vortex ring parachute designed by David T. Barish and intended for production by the Pioneer Parachute Co.

GB-26B

In November 1973 TB-26B 44-35232 (coded G-5), maintained in flyable condition at the Inter-American Air Forces Academy at Albrook AFB in the Canal Zone, Panama, was redesignated as a GB-26B. The aircraft had not flown for years, having been used as a ground instructional airframe, G indicating a permanently grounded airframe. It was the last Invader on charge with the US Air Force, surviving into the mid-1970s.

RB-26B

Few reconnaissance Invaders had the solid nose of the B-26B. For night-time photography the Invader used photo flashes released from the bomb bay by the navigator in the bombardier's position in the B-26C airframe. Thus conversions of B-26Bs for the photographic reconnaissance role tended to have the Plexiglas nose of the B-26C and thus became RB-26Cs.

Although all reconnaissance Invaders were known under the generic RB-26 designation, the vast majority were C-model conversions. However, some solid-nose reconnaissance aircraft were produced by conversion as RB-26Bs.

The 42nd TRS, 10th TRW, based at Spangdahlem, West Germany, from December 1954 had two flights equipped with RB-26s, one to undertake instrumented weather reconnaissance and the other to locate and categorise

electronic emissions. One of its aircraft was a solid-nose RB-26 (44-34186) with an unusual radome located under the tip of the nose, which was devoid of armament. While both turrets remained in place, by 1955 they were also devoid of weapons. A direction-finding 'egg' was located behind the glazing on the upper rear fuselage.

The 42nd TRS flew several regular routes: Alpha over the North Sea; Bravo from Spangdahlem to Marseilles,

France; Coco Special around the Bay of Biscay; and Echo over the Mediterranean. Aircraft were also deployed to Pisa in Italy so they could fly missions along the borders with Albania and Yugoslavia to locate radar systems. The unit replaced its RB-26s with Douglas RB-66C Destroyers from 1 November 1956, the last example leaving in May 1957.

Another example of a RB-26B was 44-34159, although its exact configuration is unknown.

TB-26B (TA-26B)

From the end of World War II large numbers of Invaders were designated as trainers, initially as TA-26Bs. All A-26Bs were built for a single pilot and required a duplicate set of instruments and controls to be installed on the starboard side of the cockpit to allow dual instruction. While the vast majority

of TB-26Bs were converted to have dual control, some examples did not, as the designation reflected the relegation of the airframe to second-line duties rather than use strictly as a training aircraft. The reverse situation was also true, with second-line aircraft with the dual control modifications remaining as A-26B/B-26Bs.

Known examples: 41-39105 (French AF), 41-39182, 41-39278, 41-39316 (Colombia AF), 41-39387, 41-39423, 41-39491, 41-39499, 41-39571, 43-22476 (Peru AF), 43-22537 (to TB-26C), 44-34108, 44-34136, 44-34156, 44-34165, 44-34184, 44-34186 (French AF), 44-34401, 44-34411, 44-34450, 44-34592, 44-34593, 44-34597, 44-34602, 44-34607, 44-34608 (French

AF), 44-34633 (VA ANG), 44-34642, 44-34650 (MATS Continental Division), 44-34659, 44-34665 (to VB-26B), 44-34671 (Saudi Arabia), 44-34722 (El Salvador), 44-34735 (Peru AF), 44-34739 (Peru AF), 44-34741 (Peru AF), 44-35232 (to GB-26B), 44-35401 (French AF), 44-35466, 44-35780, 44-35975 (French AF), 44-35955

VB-26B

A number of B-26Bs were used as staff transports as VB-26Bs, some with plush interiors while others were more basic. The Air National Guard Headquarters at Andrews AFB, Maryland, used at least three examples, including 44-34610, the last airworthy Invader in the US Air Force. It was struck off charge on 12 October 1972, while 44-34665 (ex TB-26B) and 44-34360 had been retired by the same unit on 19 November 1969 and in December 1970.

Other VB-26Bs identified include 44-34160, 44-34602 (right), 44-34612 and 44-34616.

XA-26C

The XA-26C was to have been a prototype of a version of the Invader with four 20-mm (0.787-in) cannon in the nose. It was not built.

A-26C

The first use of the A-26C designation referred to versions of the Invader with four 37-mm (1.46-in) or 20-mm (0.787-in) guns in a solid nose. On 22 February 1943 the A-26C designation was re-issued to the aircraft as built.

A-26C-DL/DT

The second Invader version to use the A-26C designation – the one actually built – was equipped with a Plexiglas nose section for a bomb-aimer. A pair of forward-firing 0.5-in (12.7-mm) machine-guns was incorporated in the nose section on the starboard side. Apart from the nose section, the aircraft had essentially the same airframe, powerplants, defensive armament and other systems as the A-26B. It carried the same 6,000 lb (2722 kg) bomb load as the earlier aircraft. Empty weight for the A-26C-30-DT was 22,850 lb (10365 kg) and 27,600 lb (12519 kg)

Glass-nosed A-26Cs served in mixed units with A-26Bs, often working as bomb-leaders for the gun-nosed aircraft.

Ground crew watch carefully as an A-26C-30-DT starts engines. The view shows the flat-pane window for the bomb-aimer, and the drop-down access hatch. The Block 30 was the first A-26C built with the clamshell canopy.

loaded while retaining the 35,000 lb (15876 kg) maximum take-off weight of the early production versions of the A-26B. The length of the A-26C was 51 ft 3 in (15.62 m).

The similarities between the two variants allowed the A-26C to enter production alongside the B model with minimum interruption to the line at both the Long Beach and Tulsa plants. However, after a total of five A-26C-DLs had been built at the Long Beach plant, production of the C model ended there so that the plant could concentrate

Serial	Model	c/n	f/n	Serial	Model	c/n	f/n	Serial	Model	c/n	f/n
41-39152	A-26C-1-DL	6865	53	43-22502	A-26C-20-DT	18590	192	43-22558/22559	A-26C-20-DT	18689/18690	291/292
41-39193	A-26C-2-DL	6906	94	43-22503	A-26C-20-DT	18592	194	43-22560/22561	A-26C-20-DT	18692/18693	294/295
41-39195	A-26C-2-DL	6908	96	43-22504	A-26C-20-DT	18594	196	43-22562/22563	A-26C-20-DT	18695/18696	297/298
41-39199/39200	A-26C-2-DL	6912/6913	100/101	43-22505	A-26C-20-DT	18596	198	43-22564	A-26C-20-DT	18698	300
43-22304	A-26C-16-DT	18451	53	43-22506	A-26C-20-DT	18598	200	43-22565	A-26C-25-DT	18699	301
43-22308/22312	A-26C-16-DT	18355/18459	57/61	43-22507	A-26C-20-DT	18600	202	43-22566/22567	A-26C-25-DT	18701/18702	303/304
43-22346/22349	A-26C-16-DT	18493/18496	95/98	43-22508	A-26C-20-DT	18602	204	43-22568/22569	A-26C-25-DT	18704/18705	306/307
43-22467	A-26C-15-DT	18499	101	43-22509	A-26C-20-DT	18604	206	43-22570/22571	A-26C-25-DT	18707/18708	309/310
43-22468	A-26C-15-DT	18503	105	43-22510	A-26C-20-DT	18606	208	43-22572/22573	A-26C-25-DT	18710/18711	312/313
43-22469	A-26C-15-DT	18507	109	43-22511	A-26C-20-DT	18608	210	43-22574/22575	A-26C-25-DT	18713/18714	315/316
43-22470	A-26C-15-DT	18511	113	43-22512	A-26C-20-DT	18610	212	43-22576/22577	A-26C-25-DT	18716/18717	318/319
43-22471	A-26C-15-DT	18514	116	43-22513	A-26C-20-DT	18612	214	43-22578/22579	A-26C-25-DT	18719/18720	321/322
43-22472	A-26C-15-DT	18517	119	43-22514	A-26C-20-DT	18614	216	43-22580/22581	A-26C-25-DT	18722/18723	324/325
43-22473	A-26C-15-DT	18520	122	43-22515	A-26C-20-DT	18616	218	43-22582/22583	A-26C-25-DT	18725/18726	327/328
43-22474	A-26C-15-DT	18522	124	43-22516	A-26C-20-DT	18618	220	43-22584/22585	A-26C-25-DT	18728/18729	330/331
43-22475	A-26C-15-DT	18525	127	43-22517	A-26C-20-DT	18620	222	43-22586/22587	A-26C-25-DT	18731/18732	333/334
43-22476	A-26C-15-DT	18528	130	43-22518	A-26C-20-DT	18622	224	43-22588/22589	A-26C-25-DT	18734/18735	336/337
43-22477	A-26C-15-DT	18530	132	43-22519	A-26C-20-DT	18624	226	43-22590/22751	A-26C-25-DT	18737/18898	339/500
43-22478	A-26C-15-DT	18533	135	43-22520	A-26C-20-DT	18626	228	44-35198/35357	A-26C-30-DT	28477/28636	501/660
43-22479	A-26C-15-DT	18536	138	43-22521	A-26C-20-DT	18628	230	44-35358/35557	A-26C-35-DT	28637/28836	661/860
43-22480	A-26C-15-DT	18538	140	43-22522	A-26C-20-DT	18630	232	44-35558/35562	A-26C-40-DT	28837/28841	861/865
43-22481	A-26C-15-DT	18541	143	43-22523	A-26C-20-DT	18632	234	44-35563	A-26C-40-DT	28842	866+
43-22482	A-26C-15-DT	18544	146	43-22524	A-26C-20-DT	18634	236	44-35564/35655	A-26C-40-DT	28843/28934	867/958
43-22483	A-26C-15-DT	18546	148	43-22525	A-26C-20-DT	18636	238	44-35656/35782	A-26C-45-DT	28935/29061	959/1085
43-22484	A-26C-15-DT	18549	151	43-22526	A-26C-20-DT	18638	240	44-35783/35937	A-26C-50-DT	29062/29216	1086/1240
43-22485	A-26C-15-DT	18552	154	43-22527	A-26C-20-DT	18640	242	44-35938/35947	A-26C-55-DT	29217/29226	1241/1250
43-22486	A-26C-15-DT	18554	156	43-22528	A-26C-20-DT	18642	244	44-35948/35952	A-26C-55-DT	29227/29231	1251/1255*
43-22487	A-26C-15-DT	18557	159	43-22529	A-26C-20-DT	18644	246	44-35953	A-26C-55-DT	29232	1256
43-22488	A-26C-15-DT	18560	162	43-22530	A-26C-20-DT	18646	248	44-35954	A-26C-55-DT	29233	1257***
43-22489	A-26C-15-DT	18562	164	43-22531	A-26C-20-DT	18648	250	44-35955	A-26C-55-DT	29234	1258
43-22490	A-26C-15-DT	18565	167	43-22532/22533	A-26C-20-DT	18650/18651	252/253	44-35956	A-26C-55-DT	29235	1259*
43-22491	A-26C-15-DT	18568	170	43-22534/22535	A-26C-20-DT	18653/18654	255/256	44-35957/35996	A-26C-55-DT	29236/29275	1260/1299
43-22492	A-26C-15-DT	18570	172	43-22536/22537	A-26C-20-DT	18656/18657	258/259	44-35997/36005	A-26C-55-DT	29276/29284	1300/1308**
43-22493	A-26C-15-DT	18572	174	43-22538/22539	A-26C-20-DT	18659/18660	261/262	44-36006/36009	A-26C-55-DT	29285/29288	1309/1312***
43-22494	A-26C-20-DT	18574	176	43-22540/22541	A-26C-20-DT	18662/18663	264/265	44-36010/36011	A-26C-55-DT	29289/29290	1313/1314*
43-22495	A-26C-20-DT	18576	178	43-22542/22543	A-26C-20-DT	18665/18666	267/268	44-36012/36025	A-26C-55-DT	29291/29304	1315/1328***
43-22496	A-26C-20-DT	18578	180	43-22544/22545	A-26C-20-DT	18668/18669	270/271	44-36026/36047	A-26C-55-DT	29305/29326	1329/1350**
43-22497	A-26C-20-DT	18580	182	43-22546/22547	A-26C-20-DT	18671/18672	273/274	44-36048/36060	A-26C-60-DT	29327/29339	1351/1363**
43-22498	A-26C-20-DT	18582	184	43-22548/22549	A-26C-20-DT	18674/18675	276/277	44-36061/36062	A-26C-60-DT	29340/29341	1364/1365***
43-22499	A-26C-20-DT	18584	186	43-22550/22551	A-26C-20-DT	18677/18678	279/280	44-36063	A-26C-60-DT	29342	1366**
43-22500	A-26C-20-DT	18586	188	43-22552/22553	A-26C-20-DT	18680/18681	282/283	44-36064/36097	A-26C-60-DT	29343/29376	1367/1400***
43-22501	A-26C-20-DT	18588	190	43-22554/22555	A-26C-20-DT	18683/18684	285/286	44-36098/36797	A-26C-DT	29378/30076	1401/2100***
				43-22556/22557	A-26C-20-DT	18686/18687	288/289				

+ to XA-26E * not taken on charge ** not delivered *** cancelled

solely on the A-26B. Production of the A-26B at Tulsa ended after 205 had been completed so that it could became the centre for A-26C production.

Improvements in the A-26B production run were mirrored in that of the A-26C. The new improved visibility canopy became standard from the A-26C-30-DT, while the C-45-DT introduced the Ford-built R-2800-79 powerplants with water injection, increased fuel tankage, internally mounted machine-guns in the wings, and the provision to carry and fire rocket projectiles that had been incorporated from the A-26B-45-DL.

A total of 1,144 A-26Cs had been built by the time production ended in August 1945, of which the US Army Air Force took 1,091 on charge. The unaccounted aircraft were either delivered but not taken on charge with the US Army Air Forces or not delivered from Douglas and scrapped *in situ* at the factory. Most of those built but not taken on charge with the USAAF were offered on the civil market. A further 756 examples on order were cancelled.

The paragraph under the A-26B-DL/DT heading relating to the transfer of noses between A-26Bs and Cs and among the different blocks of each also relates to the A-26C. Thus the exact number of A-26Cs that existed at one time or another is difficult to state with any form of authority.

When the US Air Force was formed and the attack category was deemed redundant in July 1948, the A-26C became the B-26C, the B-26C Marauder having been withdrawn from US service by then. Some B-26Cs were later equipped with an AN/APS-15 X-band bombing and navigation radar (equivalent to the British H$_2$S) in a radome between the nose undercarriage well and the bomb bay.

Above: The C model was more widely adapted to other roles than the A-26B. This A-26C has a target-towing fitment.

Below: A few B-26Cs, like 44-35992 here, were fitted with APS-15 bombing radar. It also has SHORAN equipment.

A-26C Speedee carrier

At least one A-26C was modified to test the Highball anti-shipping bomb, known as Speedee to the USAAF. A spherical weapon weighing 1,260 lb (572 kg) – including 600-700 lb (272-318 kg) of Torpex – with flattened sides, the Speedee weapon was rotated to a speed of 800-900 rpm in the bomb bay of a modified A-26C before release, allowing it to 'bounce' across the water until it hit a solid object (such as the sides of a warship) and sank to explode below the waterline. The A-26C was modified with holes in the bomb bay doors through which the weapon could be dropped, with a ram air turbine to provide power to 'spin up' the weapon.

Between 4 March and 8 May 1945 tests were undertaken at Eglin AFB, Florida. It was determined that a gun-nosed A-26B would be more suited to carrying the weapon on operations as it would be able to provide suppressive fire during the bomb run. An A-26C was selected for the tests as it allowed an observer to be carried. Tragedy struck the Speedee tests on 28 April when the A-26C released a weapon at around 10 ft (3 m) above the sea during rough water conditions. The bomb ricocheted into the aircraft aft of the ventral turret and went through the rear fuselage, severing the tailplanes. The aircraft crashed into the sea and disintegrated. Prior to this accident the undersides of the aircraft had been damaged by a column of water from a dropped weapon, highlighting the need to maintain a safe minimum height above the water. Given the lack of suitable targets for the weapon in the Pacific, the fact that modified Invaders would not be readily convertible back to the standard bomber configuration, and the availability of other methods of attacking warships (such as torpedoes), the tests were abandoned.

A-26C ramjet testbed

A-26C-40-DT 44-35572 was used by the US Navy as a testbed for a ramjet in 1946. Operated by the Pilotless Aircraft Unit based at NAS Mojave, California, with the code '880', the A-26C carried the ramjet on struts below the bomb bay. In appearance the ramjet looked similar to the unit that powered the German Fieseler Fi 103 (V-1) that was adopted by Ford for the JB-2 flying bomb. Operation of the ramjet was controlled by test personnel located in the gunner's compartment. At least the ventral and probably the dorsal turret were removed before the test programme began.

A-26 container dropper

Towards the end of World War II at least two Invaders were modified to carry an air-droppable container. They were A-26B-61-DL 44-34606 and A-26C-45-DT 44-35678, equipped with a suspension apparatus on their starboard wings. A remotely-operated release mechanism was built into the apparatus so that the pilot could drop the container where needed. The container underwent several redesigns but remained a bulky but streamlined box with doors on each side to allow it to be loaded or unloaded. A tricycle undercarriage with a nose wheel and skids was fitted, no parachute or other device being added to slow its descent. One version was equipped with a vertical tail.

During the tests, at least one of the Invaders (44-35678) was painted with calibration lines down the fuselage and nacelles on the starboard side parallel to that on the container itself so that accurate measurements of the container's movement in the air and when being dropped could be taken.

Tests were marginally successful. While the ability to air-deliver material to precise locations was of great value in several theatres, and the Invader proved to be capable of carrying the cargo, the task was deemed to be better suited to transport or light aircraft that were able to land in confined areas. The idea was abandoned before it saw use.

FA-26C

Only a small number of A-26Cs (including 44-35691) were modified as FA-26C night-photographic reconnaissance aircraft. The armament was usually deleted and a camera suite was installed in the bomb bay and – sometimes – in the nose. For the night photography role photo-bomb flashes were carried in the bomb bay. In 1948 the US Air Force replaced the F for photographic prefix with R, while the A for attack category was abandoned and the Invader became the B-26. Thus in 1948 the small number of FA-26Cs became RB-26Cs.

B-26C

The A-26C was redesignated as the B-26C on 1 July 1948, retaining the designation until it was retired from US Air Force service.

B-26B/C and RB-26C Korean War modifications

While the US Air Force had already started the process of finding a jet successor for the Invader when the forces of North Korea invaded its southern neighbour in June 1950, Invaders had a large role to play in the Korean conflict. Indeed, Invaders flew the first (on 28 June 1950) and last (on 27 July 1953) bombing missions of the conflict. During the Korean War combat (as opposed to reconnaissance) Invaders were assigned the Light Night Intruder (LNI) role, interdicting the movement of Communist supplies at night. The targets the crews were after were summed up by the phrase 'trains, trucks and tracks'. Several modifications were made to the Invaders involved in the conflict to help them undertake this role more effectively, while others were more extensively modified for niche roles.

During the war Navy Mk IV flares were modified so that they could be dropped from B-26s, with both the 3rd and 452nd Bombardment Wings assigning the flare-dropping task to specific aircraft. These aircraft had a slightly modified bomb bay able to carry just over 50 flares. Deployment of the flares was not exactly high-tech; a crew member would crawl into the bomb bay space and throw them out as requested.

An idea put forward by the personnel of the 3rd Bombardment Wing during the Korea War to help identify targets at night was equipping the Invader with a powerful searchlight. Operations with C-47 Firefly flare-ships had proved to be disappointing and, after the AN/AVQ-2 70-million candlepower carbon-arc searchlight – developed for the US Navy's anti-submarine warfare blimps – was demonstrated to the wing's personnel at Langley AFB, Virginia, a handful of Invaders were modified in 1951 to carry the unit under the starboard wing between the engine nacelle and hardpoints. The searchlight had a range of around half a mile (0.8 km), at which range it produced a 6-ft (1.8-m) wide beam. In the Invader the navigator directed the beam using a small joystick controller. Each operation of the searchlight was limited to less than a minute's duration before it required a cooling-down period of around five minutes. As it weighed 154 lb (70 kg) and created a lot of drag it had a negative effect on the aircraft's range.

When the searchlights began arriving in Korea during July 1951, the commander of the 3rd Bombardment Wing (Colonel Nils O. Ohman) limited each squadron in his command to two modified aircraft each, as the idea of flying over hostile territory in an aircraft with a bright light on it was not altogether popular with the crews. Also, supplies of the unit were limited. Initial problems were encountered in retaining the AN/AVQ-2 on the airframe, as the brackets holding it on the wing were not strong enough. Others caught fire and had to be dropped.

Tactics with the searchlight were developed by both the 3rd and 452nd Bombardment Wings (BW). Truck convoys would be found by observing the vehicles' headlights, with the first pass releasing fire-bombs to mark their position. The searchlight would then be used to attack the convoy. Some successes were recorded by searchlight-equipped Invaders and it was in one such configured machine (B-26B 43-49770) that Captain John S. Wolmsley of the 8th BS, 3rd BW, won a posthumous Congressional Medal of Honor while illuminating targets for his wingman on the night of 14 September 1951. However, flying the searchlight

This line-up of B-26Bs and Cs in Korea in June 1953 was from the 17th Bomb Wing. The aircraft nearest the camera – originally built as an A-26B – is a B-26C fitted with SHORAN equipment in place of the dorsal turret.

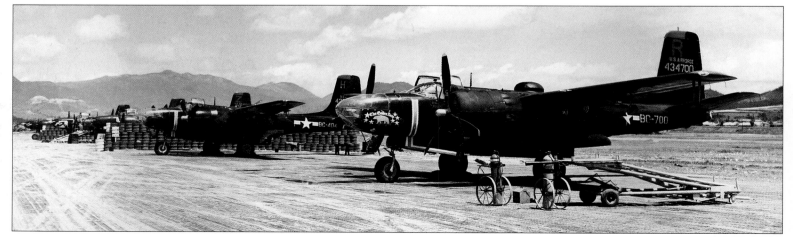

41-39401 was one of three aircraft converted with a Bell infra-red sensor in the nose for night interdiction duties, primarily against the North Korean rail network. This aircraft served with the 3rd BW's 13th Bomb Squadron and is seen at K-8 Kunsan in January 1953.

equipped aircraft was not popular. When switched on the searchlight destroyed the pilot's night vision and made the aircraft visible to any local anti-aircraft gun battery. When turned on the narrow beam had to be moved to the area of interest manually as its start position was not fixed, eating into the unit's limited operating time. Some pilots suggested that it would have been better synchronised in line with the guns. Operations with searchlight-equipped aircraft were abandoned in October 1951.

Another modification that was undertaken to locate targets at night over Korea involved mounting a Bell Telephone Laboratories infra-red detector in the nose of the Invader. The gear was known as Mac, after the project engineer for the system, Demetro Mac Cavitch. Mounted in the nose of B-26Cs, the unit covered the top half of the Plexiglas, jutting above the wrap-around lower half of the original transparency. Behind the detector the Plexiglas was reinforced so that it looked like a solid-nosed aircraft that had six windows panes on each side of the nose.

The modification was undertaken to help the aircraft find locomotive engines used by the North Koreans and Chinese to move supplies to the front line. An infra-red source found by the unit would spin the hands of a dial inside the cockpit. This would alert the crew to drop a flare, allowing a standard B-26 following the modified aircraft to attack any target illuminated.

Three aircraft were modified at Wright Field and all carried the non-standard code 'Mc' on the tail. One built as an A-26B-30-DL, 41-39401, named *Mac's Gadget*, had the standard gloss black finish with the red tail tip and triangle design on the front of the nacelles and around the wing oil cooler inlets used by 13th BS, 3rd BW. It was noted in Korea and at Iwakuni AB, southern Japan, at the end of the Korean conflict, still retaining both the dorsal and ventral turrets. Another was built as an A-26C-50-DT 44-35867. It had a yellow tail tip when noted at Iwakuni in January 1953, indicating use by the 8th BS, 3rd BW. The final Mac-equipped Invader carried the nose art *We go Pogo* and was in storage at Iwakuni in 1954, minus its engines.

In service from 1953, the Mac aircraft were flown by select crews from squadrons of the 3rd and 17th BWs, the first with the 13th BS under the code name Red Bird. While the system worked – allowing a pair of aircraft to destroy five locomotives on the Wonsan to Vladivostok line during one night – it was not adopted for widespread use.

Some Invaders were equipped with an AN/APA-64 radar signal analyser during the conflict under Project Buster. The AN/APA-64 had been developed as part of the electronics suite of the Lockheed P2V-4 Neptune. It was used to locate North Korean air defence radars and signals from the Chinese air defence system located in denied territory during the conflict. Invaders equipped with the system had a large aerial on the nose with a cross in the horizontal plane on top and small vertical dipoles at the ends of the longitudinal arms of the cross.

The 12th TRS, 67th TRW used some of these modified aircraft to locate radar stations installed by the North Koreans to provide warning for newly constructed airfields. By December 1951 the unit had one aircraft modified, but it took until November 1952 before the first of a batch of four more arrived. By February 1953 a total of 112 radar search missions had been flown by the 12th TRS, increasing to 187 by June of that year. It is possible that the radar-hunting aircraft were also flown by other Invader units active in Korea. An example of an Invader equipped with the system was a B-26B that carried the nose art *KTTV Channel 11*.

In July 1953, soon after the Korean war ended, the 12th TRS was involved in Project Bird Dog, a continuation of the radar location work. RB-26 45-35231 was equipped with radar homing equipment, AN/APN-3 and AN/APA-54 for the SHORAN navigational system to increase the accuracy of locating radar sites. However, the sole operator in the rear compartment of the Invader could not operate all of the electronic gear effectively by himself, resulting in the unit seeking permission to modify an aircraft to accommodate a second mission specialist. This was granted in August 1953, resulting in 44-35825 and 44-35909 being modified. The success of the project resulted in the 67th TRW gaining a fourth squadron, the 11th TRS for both radar location and weather reconnaissance roles.

DB-26C

In a similar manner to the DB-26B, several Plexiglas-nosed Invader were also modified as drone-carriers for use in missile development and interception training roles. Three DB-26C-DTs have been identified, consisting of 43-22494 (ex A-20C-20-DT), 44-35350 (A-26C-30-DT) and one with the last three of '666 (either A-26C-45-DT 44-35666 or A-26C-25-DT 44-22666). All three were based at Holloman AFB, New Mexico, presumably with the 3225th Drone Squadron of the US Air Force Missile Development Center. Like the DB-26B, the standard types of remotely piloted vehicles (RPV) carried were the Ryan Q-2A Firebee and Radioplane OQ-19. Each Invader had been stripped of armament and had reinforced mount points added to the wings to enable the special pylons for the Firebees to be carried. The bombardier nose allowed camera gear to be carried to monitor the launch of the RPV.

Operating mostly from Holloman, the USAF's DB-26C fleet was used to launch launched Q-2/BGM-34 Firebee drones over the White Sands Missile Range.

JDB-26C

DB-26C 44-22494 was used as a temporary testbed as a JDB-26C towards the end of its US Air Force career. It had been relegated to gate-guard duties at Davis-Monthan AFB, Arizona, by 1960, joining the Pima County Air Museum, Tucson, circa 1969.

EB-26C

Several B-26Cs became EB-26Cs, usually for short periods. As with the EB-26B, the E stood for exempt (rather than electronics) meaning the aircraft was bailed out to non-military agencies or companies, usually for trials. Two aircraft have been identified that flew as EB-26Cs. They were former A-26B 44-35300, converted as a A-26C with a Plexiglas nose, and A-26C-50-DT 44-35839. '300 was used by the Wright Air Development Center, Air Research and Development Command, as an EB-26C for trials related to missile guidance development in the mid-1950s. At some point it had a small radome at the tip of the nose attached to the glass nose cone. The other EB-26C was also used for missile guidance development, before being retired to the Military Aircraft Storage and Disposal Centre (MASDC) at Davis-Monthan AFB, Arizona, by April 1960.

EB-26C 44-35300 is seen during trials with a small bullet fairing in the nose. It wears ARDC markings.

GB-26C

Several B-26C Invaders were relegated as permanent grounded airframes as GB-26Cs. 44-35265 was at MASDC, Davis-Monthan AFB, Arizona, by April 1960. 44-22606 was donated to the City of Boise, Idaho, on 1 February 1961 for display, and was last noted there in 1972. 44-35986 (ex TB-26C) owned by the Benjamin Davis Vocational School, Detroit, Michigan, was used as an instructional airframe at Detroit-Metro Airport, Michigan, between 1961 and 1986. It later went to Selfridge ANGB, Michigan, as a display airframe.

JB-26C

JB-26C 44-35627 was flown by Bendix Research at Detroit, Michigan, for trials work concerning the development of avionics and electronics until July 1965. It was noted at Davis-Monthan AFB, Arizona, in November 1965, but by May 1971 it was on display at Dodge City Airport, Kansas. At least one other JB-26C existed as it formed the basis of one of the 40 B-26K Counter-Invader upgrades produced in the early 1960s. It is believed that this aircraft was used in a test programme somewhere in Utah.

RB-26C

The RB-26C designation covers a number of different configurations, including the original FA-26Cs redesignated as such in 1948. These initial aircraft were joined by many other conversions, although the generic designation RB-26 was widely used as some of the aircraft started out as solid-nose B-26Bs before being converted with a C-style nose and cameras.

The shortage of tactical reconnaissance assets resulted in the US Air Force ordering more Invaders to be converted as RB-26Cs for the conflict in Korea, with many aircraft specialising in night reconnaissance. The configuration of the camera suite used by RB-26s was never standardised and varied from aircraft to aircraft, as well as in time. One camera package was a pair of 12-in (30.5-cm) K-24 split-plane vertical cameras behind the rear bulkhead, although most RB-26s had cameras in the nose as well as in the bomb bay. Cameras in the nose were positioned to view through the original bombardier's position as well as from a dedicated camera window positioned on the lower left-hand side.

Navigating at night over Korea was made easier by the adoption of LORAN (LOng-RAnge Navigation) and SHORAN (SHOrt RAnge Navigation) by many of the aircraft. The systems worked by timing transmissions from several ground stations, allowing the crews to make sure they were over the right area to photograph at night. The SHORAN transmitter was installed in the Invader in a novel way. Most of the reconnaissance aircraft (and many of those assigned to the night intruder role) had the ventral and dorsal turrets removed, allowing the T-11/APN-3 transmitter for the system to be installed in the upper turret housing,

One of a number of Invaders used for shadowy reconnaissance operations in the 1950s, this RB-26C served with the 42nd TRS at Spangdahlem in West Germany. The three non-standard underfuselage radomes and rear-fuselage antennas indicate a Sigint-gathering role.

which was raised about 8 in (20.32 cm) to house the electronics. The AT-14/APN-3 receiving antenna was located on the tip of the tail, while the AT-176/AP transmitting antenna was located under the fuselage, usually in line with the leading edge of the national insignia.

SHORAN-equipped aircraft usually carried a navigator/radar operator in the rear compartment with a high-frequency liaison radio transmitter, displays for the navigational equipment, a seat and a small plotting table. On many aircraft the original gunner's escape hatch on the starboard rear fuselage was replaced by a window. The pilot had a direction indicator for SHORAN located on the left side of the cockpit panel.

LORAN was installed on many of the Invaders that operated over Korea, not just the RB-26s. The R-65/APN-9 receiver-indicator was installed in the nose of B-26Cs, visible as a large black box inside the starboard side of the nose Plexiglas.

One modification that originated from experience in Korean was the conversion of the front and rear bomb bay to allow S1 bomb racks to be installed – plus extra racks on the wings – to increase the photoflash and parachute flare load of the RB-26. Photo flares were released by the navigator who sat in the nose against the bulkhead in front of the cockpit section. Modified aircraft were available in theatre from February 1953 under project Highlight, six having been upgraded by 6 June 1953. The usefulness of the Highlight modification resulted in it also being adopted for reconnaissance Invaders outside the Far East Air Force.

Several programmes evaluated systems for wider use on RB-26s. For example, a specially modified M9 bombsight to aid more accurate photoflash drops was evaluated by the 67th TRW in Korea under the code-name Night Owl and, as part of Project Vector, the AN/APN-60 S-band navigation beacon was evaluated by the RB-26 (and Lockheed RF-80 Shooting Star) as part of a plan to extend the range of the AN/MPQ-2 portable ground radar control stations.

The end of the Korean War saw the creation of an air buffer zone between the Communist and United Nations forces. In order to monitor the opposing side the 67th TRW equipped a RB-26 with a K-38 camera mounted obliquely. The aircraft flew along the edge of the demilitarised zone every 15 days taking photographs.

In addition to the aircraft equipped for photographic reconnaissance a small number of US Air Force RB-26C Invaders were equipped with search

The 'RC' buzz-codes immediately distinguishes this RB-26C from a standard bomber, although there is little other evidence of its reconnaissance role. Note the D/F teardrop above the rear fuselage, which became a common fit on RB-26s.

radars. Radars were attached to the wingtips and under the fuselage to allow horizon-to-horizon search.

Another version of the RB-26C (of which 44-35762 was an example) exhibited several radomes containing radar or electronic gear. One was located under the nose just behind the lower glass transparency; a similar example sat behind the bomb bay doors, while a larger radome was located at the front of the bay doors. The dorsal turret was replaced by a SHORAN dome on top of the fuselage and a direction-finding 'egg' was located behind it. Mounted under the rear fuselage were a number of blade-type antenna. The aircraft is believed to have been operated by the 42nd TRS, 10th TRW from Spangdahlem, West Germany, in the mid-1950s on electronic intelligence gathering flights (see RB-26B).

RB-26Cs were also used by several foreign operators, sometimes with locally-manufactured camera gear. The Armée de l'Air operated around 19 RB-26Cs in Indo-China, Africa and Europe. These aircraft were at times equipped with French systems and cameras, and eight were later upgraded as RB-26Ps.

Three RB-26Cs were flown by two survey companies in West Germany. Although civil aircraft, they were hired at times by the West German Ministry of Defence to check navigation aids and undertake photographic survey work. Government-owned military cameras and other equipment was fitted for these missions. It is rumoured that this work included clandestine reconnaissance work in Africa and the Middle East, although this cannot be confirmed. The first RB-26C was acquired in May 1962 by Prakla-Seismos AG of Hannover as D-BELE (44-35622), quickly being re-registered as D-CELE. It was equipped with a PM-22 proton magnetometer, an LK-20 flight path

RB-26Cs served alongside the B-26B/Cs on the Farm Gate detachment in Vietnam in the early 1960s. The aircraft carried KA-1 cameras in the nose compartment, able to peer forwards theough the nose glazing or to the side through windows fitted with airflow deflectors.

camera, Radan-Doppler 502 navigation system and RCA APN-1 radar altimeter. The aircraft was lost on 14 November 1963 in a crash at Cotonou, Dahomey. In early 1963 a second RB-26C was acquired (44-35638) as D-BELI (later D-CELI) for Photogrammetrie GmbH of Munich and used in a similar role to the original Prakla-Seismos aircraft over the Middle East instead of Africa. It was retired in June 1966, spending its last years based at Erding. Prakla-Seismos acquired a second aircraft (D-CADU, 44-35682) that was predominantly based in Nigeria. On 13 May 1967 it was sold to Trans-Peruana of Lima, Peru.

Known serials: 43-22502, 44-34718, 44-35207, 44-35216, 44-35223, 44-35245, 44-35250, 44-35256, 44-35257, 44-35262, 44-35271, 44-35307, 44-35322, 44-35358,

44-35359, 44-35363, 44-35375, 44-35385, 44-35444, 44-35456, 44-35457, 44-35490, 44-35493, 44-35497, 44-35500, 44-35508, 44-35512, 44-35559, 44-35581, 44-35582, 44-35583, 44-35585,

44-35590, 44-35596, 44-35599, 44-35606, 44-35607, 44-35617, 44-35621, 44-35622, 44-35626, 44-35631, 44-35643, 44-35644, 44-35654, 44-35660, 44-35678, 44-35679, 44-35682, 44-35705,

44-35725, 44-35726, 44-35741, 44-35762, 44-35763, 44-35779, 44-35782, 44-35785, 44-35804, 44-35808, 44-35813, 44-35819, 44-35822, 44-35858, 44-35889, 44-35938

TB-26C (TA-26C)

The TB-26C was a Plexiglas-nosed Invader used for second-line training duties, including target-towing. Like the TB-26B, redesignation as a TB-26C did not imply that the aircraft had been modified for two-pilot operations.

N8018E (ex 43-22652) of Aerojet General Corp., later Aerojet Electrosystems of Azusa, California, was used to test various military tracking sensors between 1961 and 1972 during its time with the company.
Known serials are 43-22537 (ex TB-26B), 43-22546, 43-22624, 43-22652,

43-22657, 43-22673, 43-22710, 43-22724, 44-22717, 44-34136 (ex TB-26B), 44-34622 (MATS Continental Division, ex A-26B), 44-35258 (SAC), 44-35371, 44-35385 (ex RB-26C), 44-35752, 44-35807, 44-35901, 44-35923, 44-35975, 44-35986 (to GB-26C), and 44-35994.

JTB-26C

One TB-26C (44-35729) was assigned to unspecified test duties as a JTB-26C. The J prefix indicated a modification of a temporary nature and that the aircraft could economically be modified back to its original configuration.

WB-26C

During the Korean War a small number of RB-26C Invaders were used for weather reconnaissance missions attached to the 6166th Air Weather Flight, its aircraft and activities coming under the control of the 12th TRS

(Night-Photo) of the 67th TRW. Painted black and devoid of all armament, the aircraft carried meteorological equipment to map out weather fronts that would affect operations over the Korean peninsula. Within the unit the aircraft were referred to as WB-26Cs, although this designation is believed to

have been a local, unofficial one.
Missions were flown to the west of Korea over the Yellow Sea close to Shanghai, China, with the dual purpose of counting shipping and observing their deck cargo as well as gathering meteorological data. These weather reconnaissance flights were usually

flown at around 1,000 ft (305 m) above the sea, but the aircraft descended much lower when entering storm fronts. Day and night missions were also flown into North Korea itself, allowing the crews to gather information of use to personnel planning the following day's strikes.

XA-26D

A-26B-71-DL 44-34776 is believed to have become the sole XA-26D-DL, the prototype of the A-26D that was envisaged as a replacement for the solid-nose A-26B on the Long Beach

production line. However, there is no official record of the aircraft having been upgraded as the XA-26D-DL. It was available for the US Army Air Force from the Long Beach factory on 14 August 1945 but was not accepted into service until 15 March 1946, although no

mention of conversion was entered on the aircraft's Data Card. By then the war was over and interest in new-built Invaders had gone. It is possible that the variant was of interest as a conversion programme for existing A-26Bs. Power for the XA-26D-DL was

to have been two Chevrolet-built R-2800-83 radial engines, giving it a top speed of slightly over 400 mph (643 km/h), while the fixed forward armament would have consisted of eight 0.50-in (12.7-mm) machine-guns in the nose section and six more in the wings.

A-26D-DL

A total of 750 production examples of the XA-26D-DL was ordered from the Long Beach production line as A-26D-DLs, but were cancelled soon after VJ-Day. Some sources state that

A-26B-40-DL 41-39543 and A-26B-45-DL 44-34100 were converted as A-26Ds before interest in the variant disappeared. 44-34100 was accepted by the US Army Air Force on 31 January 1945, but was not available until 31 October 1945, having been retained

by Douglas at Long Beach. It was first officially listed as an A-26D-DL on 11 November 1945, although photographs of the aircraft reportedly as an A-26D dated 31 July 1945 exist. It is possible that 41-39543 was used to test some of the features of the A-26D.

Serials	c/n	f/n
45-22393/22792	36929/37328	1601/2000
45-54825/55174	40579/40928	2001/2350
both batches cancelled		
Plus: 41-395543?, 44-34100?		

XA-26E

While the XA-26D-DL was a re-engined version of the solid-nose A-26B, the

XA-26E-DT was a similar upgrade for the Plexiglas-nosed A-26C. A-26C-40-DT 44-35563 was modified as the prototype, probably while still on the

Tulsa production line. As with the XA-26D-DL, two 2,100-hp (1567-kW) Chevrolet-built R-2800-83 engines were installed. The XA-26E-DT reportedly flew

for the first time in 1945 before interest in the A-26E-DT production version disappeared among the post-war cutbacks.

A-26E-DT

The A-26E-DT was due to replace the A-26C-DT on the production line at

Tulsa, Oklahoma, as a production version of the XA-26E-DT. A total of 3,350 A-26Es was on order when the Japanese surrender signalled the end of

World War II, resulting in all production examples of the variant being cancelled.

Serials	c/n	f/n
45-17343/19342	37329/39328	2101/4100?
45-53575/54824	41604/42853	4101/5350?
both batches cancelled		

XA-26F (XB-26F)

In 1945 A-26B-61-DL 44-34586 (c/n 27865) was converted as the sole XA-26F-DL, powered by a pair of 2,100-hp (1567-kW) R-2800-83 piston engines and a single-stage centrifugal General Electric J31-GE turbojet rated at 1,600 lb st (7.12 kN) located in the rear fuselage. The R-2800s initially had the

standard three-bladed propellers but they were replaced by four-bladed units by mid-November 1945. Later, spinners similar to those used by the first prototype Invaders were added. The intake for the 850-lb (386-kg) turbojet was located behind the cockpit in a neat installation that looked similar to that used by the Bell P-39 Airacobra. The exhaust for the engine vented out under

the tail of the redesigned rear fuselage. Fuel for the jet engine was carried in the area used by the bomb bay on earlier Invader variants, limiting any

production version of the XA-26F-DL to the strafing role. Initially this was not of concern because the aircraft had been designed to test the J31 engine (also

The XB-26F is seen in its original guise with three-bladed propellers and no gun armament. The three-engined 'hot-rod' Invader did not offer a sufficiently large increase in speed to warrant further investigation.

used in Bell P-59A/B Airacomets), but as testing proceeded the aircraft was equipped with eight 0.5-in (12.7-mm) machine-guns in the nose and six more in the wings.

The increase in speed to a maximum of 435 mph (700 km/h) at 15,000 ft (6804 m) with all three engines running

did not warrant the variant entering production as a combat aircraft, as it was not seen as a sufficient leap over the 405 mph (652 km/h) maximum speed of the R-2800-powered A-26D and E versions, while being more expensive to buy and operate. However, the aircraft was retained as a testbed long enough for it to be redesignated as the XB-26F-DL in July 1948. Between 1962 and 1972 the aircraft was registered as N66368 with the Lindsay Hopkins Vocational School, Miami Airport, Florida. It is presumed to have been scrapped since.

'A-26Z' (A-26G/A-26H)

Post-war development of the Invader was continued by Douglas under the unofficial 'A-26Z' designation. It was envisaged that the new variant would have more powerful versions of the Pratt & Whitney R-2800, a cockpit with

an improved layout and a raised canopy to further improve the pilot's view. Fuel capacity would be increased by adding jettisonable wingtip tanks, with the wing structure beefed up to cope with the strain of the tanks when manoeuvring.

Crew entry was envisaged through the nosewheel well.

Like previous production versions of the Invader, it was planned to offer the 'A-26Z' with both solid and Plexiglas noses, with the two production versions due to become the A-26G and A-26H. This plan did not come to fruition

because work on the 'A-26Z' family ended after the US Army Air Force's Air Technical Service Command concluded in October 1945 that it had enough existing Invaders to fulfil its immediate post-war needs.

On Mark YB-26K-OM Counter Invader

American use of Invaders in clandestine campaigns during the 1950s and 1960s was extensive. Interest in upgrading the basic aircraft to overcome some of its shortfalls resulted in the development of the B-26K-OM Counter Invader. The YB-26K-OM prototype was 44-35634. It had originally been built as an A-26C-40-DT before becoming an RB-26C. In

August 1961 it was converted with a solid-nose as a B-26B and dual controls, and was used as a crew trainer and hack until being transferred to the On Mark Engineering Company of Van Nuys, California, in October 1962 for conversion as the YB-26K-OM.

The upgrade was extensive, incorporating many of the

improvements made to the company's Marketeer and Marksman executive conversions (see below). The fuselage was remanufactured, as was the tail assembly with a slightly larger rudder fitted, and the dorsal and ventral turrets were omitted. The aircraft had a solid nose with eight 0.50-in (12.7-mm) machine-guns. Experience with earlier versions of the Invader had demonstrated a tendency for the aircraft to shed its wings during high-*g* pull-outs and the upgrade aimed to overcome this problem. Only the spars from the original wings, strengthened with steel straps above and under the wing, were used in the conversion, in conjunction with a totally redesigned wing with 165-US gal (625-litre) tip-tanks, similar to those used by the Lockheed T-33A Shooting Star. De-icer boots and anti-icing equipment was fitted.

Two views show the YB-26K prototype. At left the aircraft carries bombs, napalm tanks and drop tanks. The first few production aircraft were delivered in this green/white scheme.

A total of eight wing pylons, built and designed by Baldwin Locomotive, were located under each wing, while six machine-guns were installed internally, three in each wing. In place of the original 2,000-hp (1492-kW) R-2800-79s, 2,500-hp (1865-kW) R-2800-103W engines with water injection were installed, propelling square-tipped fully reversible Hamilton-Standard propellers with automatic feathering. The YB-26K-OM had elegant propeller hubs, a feature that was not included on the B-26K-OM production upgrades. The wheels had heavy-duty brakes derived from Boeing KC-135 Stratotanker components, an anti-skid system and nose-wheel steering. Internally the aircraft was refurbished, the cockpit having new avionics and dual controls as standard. The communications system included FM, HF, UHF and VHF radios while navigation aids included an ILS, LF ADF TACAN and VOR.

The YB-26K-OM featured an increase in speed from 240 to 265 knots. Range was more than doubled from 210 to 500 nautical miles, including a 90-minute loiter over the target. Rate of climb, service ceiling and other performance criteria were also improved. Armament weight climbed from 7,500 lb (3402 kg) to 12,000 lb (5443 kg). On the negative side the aircraft could no longer be trimmed for hands-off flight as it was less stable than the original versions.

The YB-26K-OM was painted with white undersides and dark upper surfaces by the time it made its first post-conversion flight on 28 January 1963. While some sources state that it was allocated a new serial this is incorrect, the YB-26K-OM retaining 44-35634 as its identification. A successful evaluation of the YB-26K-OM at Hurlburt Field, Florida – the home of the USAF Special Air Warfare Center – in June 1963 resulted in the order for 40 production-standard B-26K-OMs. One of the 40 upgrades included the YB-26K.

On Mark B-26K-OM (A-26A) Counter Invader

A contract worth $12.6 million for the upgrade of 40 Invaders to B-26K-OM standard was signed by the US Air Force with the On Mark Engineering Co. in November 1963. The B-26K-OMs differed from the YB-26K-OM in several respects. The spinners were deleted from the propellers while the R-2800-52W was installed in place of the R-2800-103Ws used on the prototype, the rating of 2,500-hp

(1865-kW) remaining constant. The six machine-guns in the wings of the YB-26K-OM were deleted in the production aircraft. B-26K-OMs could be configured to have either an eight-gun solid nose or a glazed 'bombardier nose', and although the latter was seldom used it was noted on the first production aircraft and 64-17643. Nose sections could be changed in four man-hours.

The airframes destined to be converted as B-26Ks comprised B-26Bs, TB-26Bs, two B-26Cs and a JB-26C, as well as the original YB-26K-OM prototype. Twenty-nine came from the Military Aircraft Storage and Disposal Center (MASDC) at Davis-Monthan AFB, Arizona, 10 from Hurlburt Field, Florida, and one from a test programme in Utah (presumably the JB-26C). The average number of hours flown by the aircraft

selected for upgrade was around 800. The first B-26K-OM conversion made its maiden flight on 26 May 1964.

B-26K-OMs were ordered to replace the existing B-26s and RB-26s in use in Vietnam. However, wing spar failure resulted in those Invaders being retired earlier than planned and they were replaced by Douglas A-1 Skyraiders, resulting in the B-26K-OMs becoming surplus to requirements. The first B-26K-OM was handed over to the Air Force Test Center at Edwards AFB,

California, on 15 June 1964, the subsequent two aircraft going to Hulburt Field, Florida. The next three were delivered to the CIA for use in the Congo, where they were later joined by another pair. The last of the 40 aircraft was handed over to the US Air Force on 14 April 1965, by which time 26 had gone to the 6th Fighter Squadron (Composite) – originally the 602nd FS(C), another seven to the 605th Air Commando Squadron at Howard in the Panama Canal Zone, and two had crashed.

In May 1966 the B-26K-OM was redesignated as the A-26A-OM. The redesignation was in response to the need to base the aircraft in Thailand for operations in southeast Asia. The Thai government in the mid-1960s was opposed to the basing of US bomber aircraft in its country, but the expedience of changing the Counter Invader from the bomber to the attack category overcame this barrier. The A-26A-OMs were used to attack the multitude of trails and roads in Laos over which supplies from North Vietnam to the South were carried. Attached to the 606th Air Commando Squadron based at Nakhon Phanom Air Base, the A-26A-OM deployment was known as Big Eagle with the aircraft officially forming Detachment 1 of the 603rd Air Commando Squadron (the renamed 6th FS(C)). As the US Air Force did not officially operate over Laos the aircraft had their national insignia painted out. To aid the night-time interdiction role some were equipped with the AN/APS-2 Starlight scope to enhance the chances of finding targets.

By the end of December 1966 the A-26A-OMs were reassigned to the 634th Combat Support Group at Nakhon Phanom, being transferred again in April 1967 to the 609th Air Commando Squadron, 56th Air Commando Wing. In 1968 it became the 609th Special Operations Squadron, 56th Special Operations Wing.

A typical underwing load was a pair of SUU-25 flare dispensers, two LAU-3A

B-26K/A-26As flew on night interdiction sorties against the Viet Cong supply lines, operating from Nakhon Phanom in Thailand. Others undertook less publicised work in the Congo and Central America.

rocket pods and four CBU-14 cluster bomb units. Later the rockets and flares were often replaced by 500-lb (227-kg) BLU-23 or 750-lb (340-kg) BLU-37 finned napalm bombs. M31 and M32 incendiary clusters could also be carried, as well as M34 and M35 incendiary bombs, M1A4 fragmentation clusters, M47 white phosphorus bombs and CBU-24, -25, -29 and -49 cluster bomb units. General-purpose bombs such as the 250-lb (113-kg) Mk 81, the 500-lb (227-kg) Mk 82 and 750-lb (340-kg) M117 could also be carried.

The A-26A-OMs were phased out of service in southeast Asia in November 1969, flying their last combat mission on the night of the 9th. Of the 30 examples that had served in Thailand, 12 had been lost. Five were transferred to the South Vietnamese Air Force but were only used for ground instruction duties, being blown up before falling into North Vietnamese hands. The last US Air Force examples were retired to Davis-Monthan AFB, Arizona, in February 1973.

B-26K serial	former serial
64-17640	44-35896
64-17641	44-35322
64-17642	44-35435
64-17643	44-35392
64-17644	44-35451
64-17645	44-35546
64-17646	44-35375
64-17647	44-35904
64-17648	43-22732
64-17649	43-22720
64-17650	44-35766
64-17651	44-34119
64-17652	44-34361
64-17653	41-39378
64-17654	41-39491 (ex TB-26B)
64-17655	44-34184
64-17656	44-35847
64-17657	43-22649
64-17658	44-35865
64-17659	41-39564
64-17660	44-35608
64-17661	44-35433
64-17662	41-39462 or 44-35458
64-17663	41-39462 or 44-35458
64-17664	43-22665
64-17665	44-34145
64-17666	44-35483
64-17667	44-35468
64-17668	44-34652
64-17669	44-34606
64-17670	44-35634 (ex YB-26K)
64-17671	44-35820
64-17672	44-35251
64-17673	44-34135
64-17674	41-39573
64-17675	44-34173
64-17676	41-39596
64-17677	44-34108 or 44-35205
64-17678	44-35205 or 44-34108
64-17679	44-34198

On Mark RB-26K-OM Counter Invader

The B-26K-OM was built with the provision for a complete reconnaissance package based on that used by the RB-26L. Ten F-477 sensor suites were

This view (above) shows the bomb bay camera pallet fitted to the RB-26K. in the foreground is a KA-56A panoramic camera, behind which are two split-vertical cameras and a K-38 high-altitude camera. The sensors peered through windows in the bomb bay doors, as seen on this RB-26K (right), which also has a camera in the nose.

purchased by the US Air Force for the B-26K-OMs, the aircraft being known as RB-26K-OMs when installed. However the system was seldom used, the one exception being in Central America where B-26Ks based at Howard AFB, Panama, were used for aerial surveys.

The conversion from B-26K to RB-26K included replacing the nose with a glazed example, adding a removable bomb bay suite consisting of five cameras and flash ejectors, and new bomb bay doors. M123A1 flash ejectors, consisting of two blocks of 20 cartridges located at the rear of the bomb bay, were carried with role-specific bomb bay doors that had apertures in them so that when they

closed the blocks were flush to the undersides of the fuselage. A K-47 split vertical photo-electric cell camera was used in conjunction with the photo flashes for night-time photographic missions.

For daytime missions the suite could include the K-47, LC-1b or T-11 mapping camera, KA-56 panoramic camera system, and 12-in (30.48-cm) K-17 or 6-in (15.24-cm) KA-2 vertical cameras. Three circular glass doors, one on the starboard bomb bay door and two on the port door, and a viewing blister on the port bomb bay door for a panoramic camera, allowed the camera equipment to be carried and operated without having to open the bomb bay doors. A smaller window was located on the port bomb bay towards the front of the bay

for a tracking camera. The camera suite in the bomb bay had an inflight processing capability.

The nose transparency of the RB-26K-OM had an optically flat panel so that a forward oblique reconnaissance camera (either a KA-1 or K-38 camera with 36-in/91.22-cm lens that would be operated by the navigator) could be installed, while a vertical camera was also located in the rear fuselage area under the tail. The aircraft was equipped with a P-2 strike camera to record machine-gun or bomb attacks. Further P-2 cameras could also be carried in a pod on a wing pylon.

Serials: 64-17643, 64-17648, 64-17669 are known to have carried the camera suite

RB-26L

While the RB-26L was outwardly similar to the RB-26C, its reconnaissance capabilities were a generation ahead of the earlier version. The RB-26L was conceived under project Sweet Sue as a state-of-the-art night photography aircraft using equipment integrated by General Dynamics at Fort Worth, Texas, with E-Systems of Greenville providing some of the electronics. Among the systems installed in the aircraft was the Reconofax VI infra-red night-time aerial mapping system, reportedly developed for the Convair B-58 Hustler. In addition, the RB-26Ls could carry a similar camera suite to that used by the RB-26K.

Many sources state that only two RB-26Ls were completed but this is not the case. One source states that a total of four RB-26Ls was produced, two of which were able to undertake night reconnaissance missions. At least three were produced and have been identified. Both 44-34718 and 44-35782 were both originally built as A-26Bs before being converted as RB-26Cs and then RB-26Ls. They were produced to replace the RB-26Cs used in the conflict in southeast Asia and departed for the Farm Gate deployment at Bien Hoa on 3 March 1963. The two aircraft carried a strange mixture of US Air Force and South Vietnamese Air Force markings, later joining Detachment 1 of the 33rd Tactical Group at Tan Son Nhut. The

infra-red mapping system did not function well in the humidity of Vietnam and the utility of the RB-26Ls was thus lower than the RB-26Cs it was due to replace.

44-35782 was lost on 6 December 1963 during a reconnaissance mission south of Saigon to photograph territory inland from the coast where the tributaries of the Mekong River emptied into the 'mouths of the Mekong' area of the South China Sea. The aircraft crashed onto mudflats approximately 3,280 ft (1000 m) offshore and in 5 ft (1.52 m) of water, about 3 miles (4.8 km) southeast of Xem Cuo Tieu village. While it was empty when inspected, the crew – consisting of pilot Captain Gary W. Bitton, instructor

navigator Captain Thomas F. Gorton, navigator Captain Norman R. Davison, photographer Airman 2nd Class Richard D. Hill and an unidentified South Vietnamese Air Force crewman – all perished. The RB-26L had flown over a Viet Cong machine-gun nest that had hit one of the engines. Sister-ship 44-34718 survived its time in southeast Asia and is believed to have been scrapped at Hill AFB, Utah, around mid-1964.

The third example, 44-35779, remained in the United States for training and development tasks, spending most of its time at Eglin AFB, Florida. One unusual task it was used on was a project to count moose in the Isle Royale National Park, Michigan.

US Navy variants

XJD-1

US Navy interest in the Invader stemmed from a need for a high-performance target-tug to replace its Martin JM-1/2 Marauders. The first two Invaders for the US Navy were

designated as XJD-1s. They were an A-26B-45-DL (ex 44-34217) and an A-26C-40-DT (ex 44-35467) that had the Bureau Numbers (BuNos) 57990 and 57991 applied after transfer from the US Army Air Force in 1945. They were both later redesignated as JD-1s.

JD-1 (UB-26J)

The Invader joined the US Navy in the final days of World War II when surplus A-26Cs were passed to the service as target-tugs and general utility hacks. Designated as JD-1s but known as the 'Jig Dog' or 'Julie Delta' to those that flew or serviced them, the US Navy's Invaders had a crew of four consisting of a pilot, plane captain, towman and safety towman.

The first batch of 86 JD-1s were A-26C-DTs that were assigned BuNos 77139 to 77224. Some (if not all) of the 33 aircraft allocated to the RAF as Invader B.Mk 1s in store at Sacramento were assigned to the Navy in August or September 1945. The only confirmed tie-up with an RAF serial is BuNo. 77154, which was KL702. The two Invader B.Mk 1s evaluated in the United Kingdom as KL690 and KL691 aircraft were boxed up and shipped in crates across the Atlantic in early 1947. It is

almost certain that these two aircraft became JD-1s BuNos 80621 and 80622. The next 10 JD-1s were all ex-US Army Air Force A-26C-35-DTs, becoming BuNos 89072 to 89081. Their previous identities were 44-35370, 44-35379, 44-35381, 44-35390, 44-35397, 44-35399, 44-35406, 44-35418, 44-35424 and 44-35526. The final 52 JD-1s were surplus US Air Force B-26Cs transferred to the US Navy as BuNos 140326 to 140377, making a total of 150 aircraft. To these must be added the two XJD-1s that were redesignated as JD-1s following initial trials.

The JD-1 had a new nose configuration that set it apart from other Invaders. It featured a new Plexiglas transparency, similar looking to that used by late models of the Lockheed Neptune. Many JD-1s (such as BuNo. 77192 of VU-5, and 'JE'-coded examples 77173, 77211 and 140344) had a centimetric radar scanner installed

This JD-1 was used for ejection seat trials from Mustin Field, Philadelphia, in late 1948.

behind the Plexiglas, while on other JD-1s the Plexiglas was painted over. All armament, including the dorsal and ventral turrets, was removed from the A-26Cs before they entered service as JD-1s. A single hardpoint was retained under each wing.

Target winches with up to 7,000 ft (2134 m) of braided cable (or armoured wire from 1957) were located in the bomb bay, attached to a target that usually consisted of an 18 ft (5.49 m) fabric tube about 18 in (45.72 cm) wide at its throat and approximately 30 in (76.2 cm) at the closed end. The target

had a metallic strip inside it to allow it to be used as a radar target. The JD-1 would unwind the targets from its bomb bay at around 120 knots, before increasing to 240 knots to fly within range of the guns of the warships or aircraft it was providing facilities for.

In Navy service the JD-1s served in squadron strength with VU-2, VU-3, VU-4, VU-5, VU-7 and VU-10 among others, painted in bright colours denoting their target-towing role. The 'Jig Dog' survived in Navy service long enough to be redesignated as the UB-26J when the Tri-Service Designation System was introduced in 1962. UB-26Js were replaced by target-towing variants of the US-2 Tracker.

One JD-1 was modified as an ejection seat testbed, with an open cockpit for test seats located in the position occupied by the dorsal turret on Air Force Invaders. A glass windshield was provided for the test cockpit while its headrest formed a streamlined bulge on the top of the fuselage. The testbed was in active use in late 1948. BuNo. 77149 was also used for an unidentified test programme, carrying a nose boom in March 1950 when flying from China Lake, California.

The naval Invader had a revised nose compartment, and was used almost exclusively for target facilities. This UB-26J served with VU-5.

JD-1D (DB-26J)

Several examples of the JD-1 were converted to carry and launch Ryan

remotely piloted vehicles as JD-1Ds. Launch rails were located under the outer wings. Several photographs of JD-1Ds show a RPV under the starboard

side and what appears to be a fuel tank under the port to balance out the weight of the drone. Known examples included BuNos 57990, 71142, 77156, 77183,

89075 (of VU-3) and 140356. The JD-1D was redesignated as the DB-26J in 1962.

Foreign operator variants

Brazil

The Força Aérea Brasileira operated around 30 Invaders. In the late 1960s 16 were reworked in the US to a standard that equated roughly to the B-26K.

CB-26/C-26B

On 21 June 1966 a highly modified Invader was seized by the authorities at Brasilia, Brazil, after being implicated in illegal activity, probably drug-smuggling. It remained semi-derelict at Brasilia until 1970, when it was impressed into the Força Aérea Brasileira (FAB, Brazilian Air Force). The aircraft, with the solid nose of the B-26B, was a composite of several aircraft rebuilt by R.G. LeTourneau Inc. of Longview, Texas, in 1954, and was equipped with a number of unusual and expensive modifications, including a large window on both sides of the fuselage, airstairs, metal fuel tanks in the wings, installation of a floor instead of the original bomb bay doors and two large passenger doors. The

aircraft later also gained seating for up to nine passengers plus the pilot. The avionics suite was extensive and included four radio transceivers/receivers and associated aerials.

Produced as N115RG, it became CB-26 5176 (later C-26B 5176) in FAB service with the Parque de Material Aeronáutico de Recife based at Recife (Pernambuco) as a cargo-hauler. The aircraft was offered for sale by the FAB in January 1975, later going to the Museu de Armas e Veículos Motorizadas Antigos at Bebedouro, São Paulo.

A second Brazilian Invader, B-26B 5156 (ex 44-35586), was also reported to have been operated as a CB-26/C-26B. After being retired from flight duties it was displayed at the FAB Academy at Pirassununga, São Paulo, in 1º/10º GAv colours as 'C-26B 5156' until 1987, when it left for Parnamirim Air Base. It 1994 it moved to Agusto Severo Air Base, Natal, marked as 'B-26B 5156'.

Above is a standard B-26B, serving with 5º Grupo de Aviação. Below is a B-26C after rework in the US. The Brazilian Invaders were retired in 1976.

Chile

B-26D

During the 1960s at least three B-26Cs and a B-26B of the Fuerza Aérea de Chile (FACh) Grupo 8 were converted with new locally-produced 'semi-hard' noses with a six and even eight machine-gun configuration. The six-gun conversion involved mounting four on the starboard side of the nose with another two on the port. Known conversions were B-26Cs FACh 840 (41-39537), FACh 842 (43-22728), FACh 845 (44-35908) and B-26B FACh 846 (44-35937, converted in 1966).

In addition to the B-26D conversions, two Chilean Invaders were modified as high-speed couriers with Grupo 10 based at Santiago by July 1963. Although the exact nature of the

conversion is unknown, the two aircraft had the black paint that was the standard finish of Chilean Invaders removed. Some sources state that the two were B-26Bs, which should make them FACh 824 (44-35753) and FACh 833 (44-35919).

TB-26D

Fuerza Aérea de Chile (FACh) TB-26B FACh 848 (44-34741) was converted as the sole TB-26D, presumably with the same 'semi-hard' machine-gun nose fitted to the air force's B-26Ds.

With a row of F-80Cs behind, this is a regular Chilean B-26C, displayed with an array of weapons. Chile received 34 B-26Cs between December 1954 and March 1958, plus four B-26Bs.

France

B-26N

In 1961 the Armée de l'Air (French Air Force) elected to modify eight B-26Cs as night-fighters to intercept Armée de Libération Nationale (ALN, the Algerian national liberation army) light aircraft in north Africa. Conversions were undertaken by the Union Aéromaritime de Transport (UAT), the first (41-39579) being completed in January 1961. The B-26C's glass nose was replaced by a radome containing an AI.Mk X radar removed from Meteor NF.Mk 11s, while armament consisting of four 0.5-in (12.7-mm) machine-guns was mounted in packs under the wings, along with two Matra 122 rocket packs with 19 SNEB rockets each. The first B-26N was test flown from Bône in Algeria from the last day of February to mid-March 1961.

Deliveries were undertaken to Escadron de Chasse de Nuit (ECN) 1/71 from 9 May 1961, joining the unit's radar-equipped MD.315R Flamants. The first interception was undertaken on 18 August 1961, but the helicopter being hunted escaped. Chances of an interception became increasingly rare as the number of targets diminished, so

the B-26Ns were assigned flare-dropping and even daytime strafing missions by late 1961. The signing of a ceasefire on 18 March 1962 resulted in ECN 1/71 moving to Reims, France, on 31 August 1962, by which time it only had a single B-26N on strength. The seven survivors were passed onto ECN 1/30 Loire at Creil, 2/30 Normandie-Niemen at Orange and 3/30 at Reims to act as trainers for the units' Vautour IINs. The last B-26N was retired in May 1965 by ECN 3/30.

Serials: 41-39358, 41-39482, 41-39523, 41-39579, 43-22609, 44-34210, 44-34213, 44-35926

RB-26P

The RB-26P was a French conversion of B-26Cs and RC-26Cs to carry French reconnaissance systems, including one Omera 30, two Omera 31 and three Sephot-Omera 11 cameras. RB-26Ps could be identified because they had a redesigned, more rectangular, camera window on the port side of the nose. By the time that the RB-26P appeared, French Invaders carried a single letter at the base of the fin denoting the version, RB-26Ps logically being identified by 'P'.

The prototype RB-26P upgrade was completed in August 1960, being delivered to the French air arm as such. It was followed by the conversion of six

unarmed B-26Cs from French stocks in early 1961, the work being undertaken by French Invader specialist Union Aéromaritime de Transport (UAT). Eight RB-26C were also upgraded as RB-26Ps, the last being redelivered to the Armée de l'Air by mid-1962. While the majority remained unarmed, the four RB-26Ps of ERP 1/32 deployed to Fort Lamy, Chad, in mid-December 1964, were equipped with underwing gun pods. It was an RB-26P (44-35512) that undertook the last flight of an Armée de l'Air Invader on 2 July 1968, landing at Cazaux.

Although no RB-26Ps were exported to other military services as such, one

From the starboard side the RB-26P looked like a B-26C. However, the 'P' on the fin denoted the variant, which had a large camera window on the port side.

The large radome under this Invader's belly identifies it as a B-26APQ13 radar trainer, of which six were produced.

example (44-34312) belonging to Société Carta was smuggled in August 1967 to the Biafrans fighting to carve out an independent homeland from Nigerian territory. Ferried via Lisbon, Portugal, the aircraft was abandoned after a landing accident at Port Harcourt on 2 or 3 December 1967.

Serials: 44-34729 delivered as RB-26P
ex B-26C: 41-39511, 41-39546, 43-22615, 44-34308, 44-34312, 44-34493
ex RB-26C: 44-35216, 44-35223, 44-35257, 44-35457, 44-35512, 44-35583, 44-35599, 44-35607

B-26R

Two French Centre d'Essais en Vol (CEV) Invaders used for radar testing were referred to as B-26Rs, 41-39531 and 44-34401. The CEV is the French aviation research establishment. Both aircraft commenced their French military service as B-26Bs before gaining a re-profiled nose containing an Antilope radar. 44-34401 also tested the Rafal radar, being withdrawn circa 1968.

On 11 July 1966 '531 was put up for sale by the CEV, going to the Pan Eurasian Trading Company of Luxembourg on 2 August, before it was resold on 22 September to Mr Earnest A. Koenig as N12756, although it may not have carried the identity. The destination for the Invader was Biafra, the aircraft arriving at Enugu Airport on 29 June 1967 in an unpainted silver scheme. Painted black all over after its arrival, the aircraft was flown in combat by mercenary pilots including Jean Zumbach and Jacques Lestrade. It became known as *The Shark* after a set of teeth was painted on the nose. A single machine-gun was installed, the barrel slightly protruding from the empty nose radar cone. *The Shark* had to be abandoned at Enugu Airport after it was damaged in an attack by a Nigerian Air Force Hunting Percival Jet Provost T.Mk 51 on 6 September 1967. It is believed that it flew its last mission for the Biafrans on 21 August and had been grounded by a lack of spares before it

was attacked. It fell into the hands of the advancing Nigerian troops in October 1967.

B-26TMR

Two French B-26Cs (44-34756/157 and 44-34773/052) were used by the CEV as testbeds for laser range-finders as B-26TMRs. At least one ('773) subsequently reverted to B-26C standard in time for its delivery flight to the Musée de l'Air at le Bourget in 1970, making what is believed to have been the last flight of a French military Invader.

B-26Z

Some confusion exists as to the B-26Z designation issued to a possible nine French Invaders that carried the serials Z001 to Z009. It may apply to the nine aircraft acquired with 'Z'-prefixed serials, to the seven aircraft originally marked Z003 to Z009 or only to aircraft used by the Centre d'Experiences Aériennes Militaires (CEAM), based at Mont-de-Marsan.

In reality, the aircraft with Z serials consisted of two batches of both B-26Bs and B-26Cs acquired from two sources. B-26Bs Z001 and Z002 were acquired for the CEV from private owners in Mexico in the spring of 1953. They carried the names *El Fantasma* and *El Indio*, but did not survive long in service.

The second batch of seven included at least two B-26Cs, and they were bought for the Armée de l'Air. They were acquired in March 1953 from the Eastern Aircraft Sales Corporation of

Z006 was one of the B-26Z batch acquired commercially in the US. It was used for target-towing at Cazaux.

New York, most retaining their civilian schemes during their early French service before adopting a natural metal finish. Four examples – Z004/41-39505, Z006/41-39538 (although it may have been 44-39539 it was listed as '538 in French records), Z008/41-39154 and Z009/41-39223 – were equipped with target-towing equipment for Base Ecole 706 at Cazaux. They were used over the nearby ranges between March 1954 and February 1964, with all but Z004 (destroyed on 21 June 1956) reverting to their US military identities when the French standardised their Invader serial system in March 1959.

Later in their service careers French Invaders wore variant codes on the base of their tails. During its last 11 months in service 41-39223 (ex-Z009) was stripped of its target-towing equipment and was flown by CEAM, carrying the variant letter 'Z' on its tail. CEAM had been the original recipient of B-26B Z003/41-39531, and B-26Cs Z005/41-39512 and Z007/41-39162. They were non-standard aircraft – some had sealed bomb bays – and were definitely referred to as B-26Zs.

B-26APQ13

The B-26APQ13 was a radar operator trainer version of the Invader developed for the Armée de l'Air. It carried the Western Electric AN/APQ-13 X-band bombing radar mounted on the front of the bomb bay, the radome protruding below the fuselage. The aircraft could still carry a bomb load, although it was reduced from the standard load. The B-26APQ13s carried a pair of radar trainees, one in the nose and another seated on the starboard side of the cockpit in a rearward-facing seat next to the pilot. A radio operator could also be accommodated in the former gunner's position, the B-26APQ13 carrying no guns.

B-26Z Z007 was reportedly used for development of the B-26APQ13 version. Two B-26Cs (44-35859 and 44-35957) were acquired and converted by Union Aéromaritime de Transport (UAT) to B-26APQ13 standard before delivery to the Armée de l'Air, the first being handed over to CEAM in 1960. They were passed onto CIB 328 in November 1960, and were followed by four further aircraft converted from the Armée de l'Air's existing B-26C stock during 1961 and 1962.

Serials: ex B-26C: '44-34274'/ 44-34214, 44-35317, 44-35504, 44-35565
Delivered as B-26APQ13; 44-35859, 44-35957

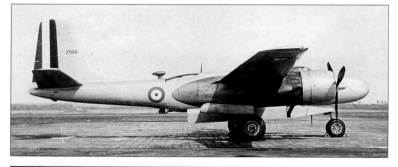

Invader Mk I

The first Invader to be transferred to a foreign air arm was A-26B 41-39158. It left the United States on 28 June 1944 and arrived in Great Britain two days later, initially being lent to the Aeroplane and Armament Experimental Establishment at Boscombe Down, Wiltshire, on 11 July, retaining its US serial and stars and bars even after it was officially transferred to the RAF in August. Via No. 12 MU at RAF Kirkbride, the aircraft entered service with No. 2 Group, part of the 2nd Tactical Air Force. Although it crashed, experience derived from operating it prompted 480 Invaders to be included among the 'wish list' of aircraft to be supplied under Lend-Lease to the RAF during 1945. However, this was cut back until on 10 November 1944 General Arnold of the USAAF agreed to release 140 A-26Cs to Great Britain in the first half of 1945.

The aircraft were designated Invader B.Mk 1s by the Air Ministry, the military markings KL690 to KL829 being allocated to the aircraft. Two A-26Cs were allocated to Boscombe Down on 18 December 1944 for armament and handling trials, although one went initially to Cunliffe-Owen for unspecified trials. The serials TW222 and TW224 were allocated but, instead, the aircraft adopted the first two serials of the 140 allocated aircraft. The two aircraft were officially taken on charge in early February 1945.

For RAF use it was thought essential that the A-26Cs was equipped with Gee and GH radio navigation aids; the bomb switch panel required moving from the cockpit to the bomb aimer's position; provision was needed to carry two more 500-lb (227-kg) bombs and the engine exhausts required flame-damping equipment. The two Invaders under test at Boscombe were capable of only

using the internal bomb bay, as no evidence of the underwing bomb-arming circuits could be found on the aircraft. This limited the aircraft to carrying only six 500-lb (227-kg) bombs – two less than the Douglas Boston – resulting in the requirement for an increase in the bomb-carrying capability of RAF Invaders to reach parity with the older design.

Aircraft for the RAF were assembled and stored at Tulsa pending service modifications. However, by April 1945 the need for the aircraft had diminished and, on 7 April, it was decided that only 12 were required, but this soon dropped to two aircraft – the trials aircraft already in use. By the time Douglas heard about the cancellation 33 A-26C-DTs (43-22604, 44-35283/35284, 44-35290/35291, 44-35297/35298, 44-35304/35305, 44-35311/35312, 44-35318/35319, 44-35325/35326, 44-35332/35333, 44-35374, 44-35398,

44-35414, 44-35422, 44-35438, 44-35446, 44-35454, 44-35462, 44-35470, 44-35478, 44-35486, 44-35494, 44-35502) in storage at Tulsa had been allocated to the RAF. They were flown out to Sacramento for further storage before they were handed over for the US Navy in August and September 1945, becoming JD-1s (which see). KL690 and KL691 were stored at No. 12 MU before being returned to the US in February 1947. They joined the other prospective Invader B.Mk 1s towing targets for the US Navy.

Serials

KL690	ex 43-22479	to JD-1 BuNo. 80621?
KL691	ex 43-22482	to JD-1 BuNo. 80622?
KL692	ex 43-22604	wore RAF serial but not delivered, to JD-1
KL693 - KL829		not delivered

Civilian conversions

Several Invaders were converted in the post-war era for civilian use, covering executive transports and fire-fighters. A number of these returned to military or government service as covert operations platforms, VIP transports or for various trials.

Grand Central executive conversions

Several Invaders were modified for executive use at the Grand Central Air Terminal, Glendale. Unlike some of the modifications that followed, the Grand Central executive conversions were limited to adapting the existing airframe to its new role, without greatly modifying the external lines of the aircraft.

A-26B 44-34758 was one of the aircraft not delivered to the US Army Air Force, going straight to the Reconstruction Finance Corp for disposal after World War II before being sold at Kingman Field, Arizona, in February 1946 to Charles H. Babb Co. of Glendale, California. It was registered

as NL67908, dropping the L in the registration on sale to the Ford Motor Co. on 6 February 1948. After its sale the aircraft was converted to an executive configuration with underwing fuel tanks at Grand Central Air Terminal. In February 1952 the aircraft was sold to the Centre d'Essais en Vol (CEV) at Brétigny-sur-Orge, France, becoming 908. It regained its former US military serial in line with the standardisation of French military Invader serials around March 1959 and later flew with CEAM and CIEES 343. It may have gained the nose of a B-26C at some time during its French service.

B-26B 44-34768 was similarly

delivered to the Reconstruction Finance Corp., becoming N67162 and then N4852Y before being converted as a pressurised executive aircraft for the Brown Paper Mill Co. of Monroe, Louisiana. It was later used as a mineral geo-survey aircraft.

A further possible Grand Central conversion was B-26B N67161 (ex 44-34767) that was, along with 908, one

The CEV's B-26B 908 was a Grand Central executive conversion that returned to military use in the trials role.

of five Invaders sold to the French government for the CEV, becoming 161. Suspicion falls on it being a Grand Central conversion is raised because it had a sealed bomb bay and passenger cabin when delivered.

Lockheed Air Services Super 26

Lockheed Air Service of Ontario, California, was an offshoot of the airframe manufacturer. The Super 26 was a pressurised executive transport modification of the A-26 Invader with a new, redesigned fuselage cabin capable of accommodating between five and nine passengers in comfort. Internally the new cabin was 22 ft (6.71 m) long, 6 ft (1.83 m) high and 4 ft 8 in (1.42 m) wide, the length being restricted as only the rear spar of the Invaders wing was replaced by a steel ringspar. The new

fuselage was mated to existing wings, engines, undercarriage and tail. Lockheed Constellation windows and cockpit transparencies were used, while an airstair door was located on the rear starboard side. A new fibreglass nose housed electronics and a baggage compartment. The first – and only – Super 26 (N5052N) first flew following conversion at Ontario, California, in the late summer of 1960. It was later delivered to the Mesta Machine Co. of Pittsburgh, Pennsylvania.

Lynch Air Tankers B-26STOL

The Invader was a popular choice as a fire-fighting tanker with several companies modifying aircraft for the

role. At least five Invaders modified as fire-fighters to carry a 1,200-US gal (4542-litre) tank by Lynch Air Tankers of

Billings, Montana, also had modified wings to make them more manoeuvrable at low level. The converted aircraft were known as B-26STOLs. They were N3426G (ex

RB-25C 44-35497), N4818E (ex TB-26C 44-35371), N4805E (ex Consort 26 and A-26B 44-34121), N4060A (ex B-26C 44-34102) and N9425Z (ex A-26C 44-35721).

On Mark Marketeer

The Marketeer was an Invader conversion by On Mark Engineering, the company to which Douglas had passed all rights to produce spare parts for the Invader in the mid-1950s. The Marketeer was produced as an aircraft certified on a Supplemental Type Certificate, limiting its weight to the 35,000 lb (15876 kg) gross weight of the original Invader. By removing all armour, weapons and the bomb bay structure the Marketeer was able to carry 11 passengers in an unpressurised cabin with two pilots. Early conversions retained the twin wing spars through the fuselage that limited the space available for the cabin, but this was replaced by a new steel ring structure to eliminate the rear spar in later aircraft.

A new lengthened nose section with either a large baggage component with its own access ladder, a fuel tank or weather radar was attached to the existing fuselage. The cockpit had a new solid roof and double glass windows installed. An autopilot was a standard item, while those equipped with a weather radar had a 6-in (15.24-cm) screen to display its returns on the instrument panel. The rear fuselage had considerable re-skinning to increase the size of the cabin, which

This Marketeer has a baggage hold (with integral loading ladder) in the nose and an airstair for the well appointed passenger cabin.

effectively extended back to 2 ft (0.61 m) in front of the fin.

The Marketeer could either be equipped with an airstair under the rear fuselage or a door with integral steps on the starboard side, the latter being the more popular choice. A large panoramic window was arranged behind each wing along with a smaller window, while two small windows were added in front of the wing. Other windows were arranged above and below the wing to create a light cabin. Several different cabin configurations were offered, including the Administrator, Director or Secretarial layouts. The wings were equipped with wingtip fuel tanks, while

the original 2,000-hp (1492-kW) Pratt & Whitney R-2800-75/79 powerplants could be replaced by 2,500-hp (1865-kW) R-2800-CB-16s, requiring an increase in rudder area but raising speed from 315 mph (507 km/h) to 365 mph (587 km/h).

The first Marketeer flew in 1955. At least 29 Marketeer conversions of B-26B, B-26C and RB-26Cs have been identified, although some sources state as many as 48 had been completed by May 1961. None appears to have entered direct military service, the majority finding employment within the flight departments of corporations. However, at least two were used by the Central Intelligence Agency (CIA).

N800V (ex 'N5001X', 44-35698) and

N900V (ex N5002X, 44-34415) were redelivered to the CIA front Intermountain of Marana, Arizona, after modification by On Mark in 1964. They were then further modified for development and testing of covert operations systems. LTV's Temco Aerosystems at Greenville, Texas, installed an extensive navigation and communications suite in the aircraft. Provision was also made (at least in N900V) for a Texas Instruments AN/APQ-99 J-band forward-looking multi-purpose radar and AN/APN-125 radar set allowing the aircraft to operate down to around 200 ft (61 m) altitude in zero visibility by coupling the aircraft's autopilot with the radar. The CIA's Marketeers were also reportedly capable of carrying electronic countermeasures.

The cabin floors had rollers so that around 500 lb (227 kg) of cargo on a pallet could be air-dropped out of a cargo drop ramp. To accomplish this the crew included an Air Freight Specialist (known as a 'kicker') in addition to the pilot and co-pilot. If required, a navigator could also be carried but the aircraft were equipped with a military TACAN system, an ADF, VOR navigation system and an instrument landing system. A comprehensive communications suite included UHF, VHF and HF radios and an IFF encoder.

N800V was used from January 1964 and may have survived until 1971 at Norton AFB before being scrapped. It was reportedly falsely painted as

N46358 in 1966. One of the two aircraft was used to train pilots on methods of penetrating Chinese airspace from Taiwan in 1965 and 1966 before plans for such activity were cancelled. Those flights were undertaken from Brownsville, Texas.

N900V was flown from 25 February 1964 until April 1967 on trials work before it was transferred from Intermountain to Air America via the Pan Aero Investment Group, leaving for southeast Asia as N46598. Known as the 'Blue Goose' because of its dark blue scheme with white trim and the 'Blivit' because of the large amount of gear carried inside, the Marketeer was used to drop supplies to road surveillance teams in Laos, operating from Udorn, Thailand. As a drop aircraft the Marketeer was not ideal as its minimum speed was 140 kt (163 mph; 262 km/h) and the view from the cockpit was somewhat limited, both factors reducing the time the crew had to line up on drop zones. The supply drop project was cancelled in October 1967. In late 1968 N46598 was reported as having been transferred to Overseas Aeromarine Inc. and scrapped, although it almost certainly remained at Udorn beyond this date.

In addition to the two CIA Marketeers, the Department of Commerce's Weather Bureau flew one with a radar in the nose wearing dual civilian/military identities as N800W/35725. Modified from RB-26C 44-35725, it was retired to Davis-Monthan AFB, Arizona, in 1965.

On Mark Marksman

The Marksman was a pressurised development of the Marketeer, On Mark Engineering receiving a new Supplemental Type Certificate for the conversion in January 1961. Only 15 percent of the original Invader structure was retained in the Marksman. The headroom of the fuselage cabin and cockpit was increased by increasing the depth of the fuselage, two windows being incorporated each side above the wings while three larger windows were added aft of the wing's trailing edge. The pressurisation system provided an equivalent pressure of 7,800 ft (2377 m) at 20,000 ft (6096 m). Wing and tail de-icing was offered as standard, while the broader rudder was also fitted.

Three versions of the Marksman were offered, differing in the choice of powerplant. The Marksman A was powered by the 2,000-hp (1492-kW) Pratt & Whitney R-2800-75/79 engines, as used on 'stock' Invaders. The Marksman B was similar to the Marksman A except it was powered by the 2,100-hp (1567-kW) Pratt & Whitney R-2800-83AM-4A engines and had an increased fuel capacity. Pratt & Whitney R-2800-CB16/17 engines powered the Marksman C, the only difference between the variant and the Marksman B. Reversible-pitch Hamilton Standard 43E60 propellers were used as standard by all three versions. The vast majority of the 12 identified Marksman conversions were of the C model, although the exact total of conversions undertaken is open to debate.

While no air arms are known to have used Marksmen, several were involved in military development programmes. In 1969 N161H (ex A-26C-55-DT) owned by the Government Electronics Division of Motorola Inc. of Scottsdale, Arizona, was used as a testbed for the Motorola AN/APS-94 side-looking airborne surveillance and mapping radar for the Grumman OV-1D Mohawk. On 5 August 1970 it was purchased by Grumman Ecosystems, complete with the radar, and further modified with a Wild RC-8 camera (a multi-spectral four-lens sensor), Doppler navigation system and a Collins HF-618T transceiver and antenna to undertake surveys.

Calspan Corp. of Buffalo, New York, based Marksman N237Y (ex A-26B 41-39516) at Edward AFB, California, between 1977 and 1992. Between 1972 and 1986 Calspan also operated TB-26B N9146H (ex 44-34165), while a further company Invader, B-26B N9417H (ex 44-34653), was written-off on 3 March 1981 at Edwards after its left wing structure failed. Calspan's Invaders were modified to demonstrate variable stability, being used for both Navy and Air Force test pilot training among other tasks.

On Mark Model 450

While some executive conversions of the Invader changed the aircraft's external shape little, making use of the existing attributes and components of the attack bomber, others were more radical, using little of the original airframe. The On Mark Model 450 belongs to the later category. The Model 450 was planned as a 14-passenger executive aircraft development of the Marketeer powered by two Allison 501D turboprops, with a new wider, deeper fuselage with an airstair opening forward, located adjacent to the wing root. The cabin was extended forward as far as the cockpit, with a line of windows along its length. The vast majority of the airframe would have been new, although some Invader components were incorporated.

On Mark envisaged a market for 120 aircraft, with possible sales to the US Air Force, although the launch customer was Maytag Aircraft Corporation. However, the launch of the more capable Grumman G-159 Gulfstream that made its first flight in August 1958 resulted in On Mark deciding to shelve Model 450 development, and what would have been one of the ultimate executive Invader versions remained a paper project only

R.G. LeTourneau executive conversions

At least five Invaders (both B-26Bs and C variants) were upgraded by the R.G. LeTourneau company of Longview, Texas. The conversions were limited in comparison to some others, having the upper observation windows faired over and a six-passenger cabin layout installed with a single window on each side. All military equipment was removed to give the aircraft an empty weight of 22,000 lb (9979 kg), allowing a total load of 13,000 lb (5897 kg) to be carried under the Supplemental Type Certificate. The conversions remained unpressurised but oxygen masks were provided for each seat in the cabin.

After modifying three Invaders for its own use, R.G. LeTourneau offered similar conversions on the open market in 1959. Exactly how many were modified is unknown. One of the aircraft (A-26C N4974N, ex 44-34134, later to N115RG) was later impounded at Brasilia, eventually being impressed into the Força Aérea Brasileira (Brazilian Air Force) as a CB-26.

Rhodes-Berry Silver 60

The Silver 60 was an executive conversion of the A-26B Invader undertaken by the Rhodes-Berry Company of West Los Angeles. It was probably one of the ugliest executive Invader variants produced, its appearance being determined by the need to increase the number of passengers that could be carried while using the existing wing spars. The upper fuselage was reskinned and mated to a new lower fuselage to create an enlarged central fuselage with room for 14 passengers with enough head room (6 ft 6 in/1.98 m) to stand up in. The conversion retained the Invader's existing spar, wings, cockpit and rear fuselage, giving the Silver 60 a distinctively 'pregnant' look. Twelve windows were provided for the passenger cabin, while a main entry door for the cabin and a crew/baggage door were located ahead of the wing on the rear of the port side. The nose undercarriage retraction sequence was modified so that it retracted straight into the bay instead of being rotated 90° before retraction, decreasing drag during the retraction process.

It was planned that a kit would be supplied to existing executive Invader operators wanting to upgrade their aircraft to Silver 60 standard. The prototype (N5510V, ex 41-39262) was converted by Volitan Aircraft Inc. of Pasadena, California. It first flew following upgrade on 25 June 1960, later going to the Whiteman Manufacturing Co. One further conversion was undertaken, flying on 11 February 1961, before work on the Silver 60 was abandoned.

At least two other versions of the Silver 60 were planned but never flew. They were a cargo-hauler with a rear ventral loading ramp and a troop-carrier variant aimed at the military services of South America countries and capable of carrying 20 soldiers.

The deep fuselage of the Silver 60 provided standing room in the cabin.

Rock Island Monarch 26

The Rock Island Oil & Refining Co of Hutchinson, Kansas, developed an executive conversion of the Invader from 1958 as the Rock Island Monarch 26. Monarch 26s were stock Invaders with a re-contoured and re-skinned fuselage to provide a cabin for six passengers. A laminated rear ring spar replaced the original example to increase the space in the cabin, while the rear fuselage was redesigned to include an airstair on the starboard side and panoramic windows were added on both sides. The cabin, outfitted by Horton & Horton Custom Works of Fort Worth, Texas, accommodated both a lavatory and a galley. The nose of the Monarch 26 was increased by 30 in (76 cm) when compared to a stock Invader, and customers could decide to use the extra space for either a weather radar, radios or a baggage compartment. New internal wing fuel tanks added an additional 87 US gal (329 litres) to the

Executive Invader conversions have provided good sources for warbirds. Although fitted with an eight-gun nose and other 'A-26B' features, this aircraft retains the long rear cabin windows from its days as a Monarch 26.

fuel capacity. Internal tanks were selected after wingtip tanks were rejected because of the strain they would have placed on the wing structure.

A new dual-control cockpit layout was installed with the overhead transparencies replaced by a metal canopy that housed overhead control panels, while a Sperry SP-20 autopilot, longitudinal flight control system, safe flight control system, Sperry engine analyser, fire detection and extinguishing systems were also added to the cockpit.

The first Monarch 26 (N6840D) flew in September 1959 on a Supplemental Type Certificate (STC). It proved to be fast (400 mph/644 km/h at only 2,500 ft/ 762 m), had the feel and stability of an aircraft much larger than its dimensions and could carry a full load of passengers, baggage and fuel and still be well below the 35,000 lb (15876 kg) maximum take-off weight stipulated by the STC.

Three conversions were produced by Rock Island itself. They were all ex-Armée de l'Air aircraft consisting of B-26Cs 44-34390 (originally built as an A-26B-55-DL, to N6836D) and 44-35911 (A-26C-50-DT, to N6840D) and RB-26C 44-35643 (originally A-26C-40-DT, to N6841D). A fourth conversion was also produced using VB-26B N8392H (ex TB-26B and A-26B-61-DL 44-34602) with the nose modified by Rock Island, the avionics in Houston, Texas, and the rear fuselage converted by Hamilton Aircraft of Tucson, Arizona. Like many of the executive Invader conversions, the Monarch 26 appeared at the start of the boom in executive designs and its prospects suffered accordingly.

Rock Island Consort 26

The Consort 26 was the name used for Invaders converted by Rock Island for research and development duties, using the expertise gathered producing the Monarch 26s. Three were produced, all to differing standards according to the tasks they were used for. They all had their military equipment removed, bomb bays sealed and a reinforced floor added, and were sold or leased to aerospace companies for development roles.

Consort 26 N4805E (ex A-26B-45-DL 44-34121) was retained by Rock Island and Koch Industries (as the company became in 1972) from 1966 until it was sold to Lynch Air Tankers Inc. in 1975, being modified as a STOL 26. N6838D

(ex A-26C-55-DT 44-35440) served with Rock Island between 1966 and 1971 before going to Conair Aviation Ltd as a fire-bomber. It was used for a number of Rock Island projects before being leased to Douglas and equipped with a number of sensors for use in the US Army's SAM-D programme. (SAM-D eventually became the Pershing ballistic missile.) During the spring of 1968 it was used by Learjet as an icing tanker, equipped with a large boom under the fuselage, a water tank and AiResearch starter engine to provide power to pump the water out the boom. Flying in front of the Learjet 24 prototype, the Consort 26 released water that would instantly turn to ice, allowing the effects

of ice build-up on the jet's airframe to be catalogued. N6838D was also used for similar tests with the Swearingen Merlin.

The Hughes Aircraft Company purchased N6839D (ex A-26B-61-DL 44-34538) in 1966 as an airborne radar systems and guided missile sensor testbed. Initially used in simultaneously transmit and receive (STAR) and low probability of intercept (LPI) radar work based at Culver City, California, on moving to Van Nuys, California, it became a testbed for missile seeker heads, primarily for the AGM-65 Maverick air-to-surface missile, with an extended nose cone. During such test flights the aircraft would simulate the missile's flight-path, including diving on the target and pulling up at the last

A Hughes technician examines the seeker head of a Maverick missile in the company's Consort 26 testbed.

moment. Hughes used the aircraft until the company sold it 1987. It later became N34538 and gained the nose of an A-26C.

Smith Biscayne 26

The L.B. Smith Aircraft Corporation recognised the good qualities of the Invader that made it a popular corporate aircraft and also some of its limitations, including the intrusion of the wingspars into the available cabin space. The company decided to replace the wingspars on its executive conversions with a large aluminium wing spar. One of the consequences of this was that the centre section of the wing was increased, allowing the engines to be positioned around 20 in (50.8 cm) further away from the fuselage, reducing the noise levels in the cabin. New wheels and brakes, tip tanks, a lengthened nose with a baggage compartment and radar, a redesigned cockpit, nine cabin windows and an airstair door on the port side were added to produce the Biscayne 26. However, the Biscayne 26 retained the original 2,000-hp (1492-kW) Pratt & Whitney R-2800-75/79 powerplants. Only the prototype Biscayne 26 was completed before the company decided to use the experience gained to develop the more radically altered Tempo series.

Smith Tempo I

While the Biscayne 26 involved modifying the existing Invader fuselage, the Tempo mated the wings, tailplanes engines and undercarriage to a new cabin. The Tempo fuselage was 10 ft (3.05 m) longer than the standard Invader and had a 28-ft (8.53-m) walk-through cabin with room for up to 13 passengers. The depth of the fuselage was increased by 15 in (38.1 cm) and was wider than that of the standard Invader. Wingtip tanks, updated hydraulic, pneumatic and electrical systems, and new wheels and brakes were incorporated. Two versions of the Tempo were envisaged, the Tempo I being an unpressurised version. Tempo Is were to be built so that they could be upgraded as pressurised Tempo IIs at a later date. However no Invaders were converted as Tempo Is, work proceeding on the Tempo II.

Smith Tempo II

The Tempo II differed from the Tempo I in that it had a pressurised cabin. The prototype conversion was N4204A (ex 44-35640) that made its first flight from Miami in October 1959. Testing continued until 1962. One of the more unusual features of the aircraft was that JATO bottles could be used for take-offs from high-altitude airfields.

Unfortunately for the L.B. Smith Aircraft Corporation, the Tempo II came up against the first generation of purpose-built corporate aircraft and the prototype remained the sole example converted. It ended its days flying as a weather research aircraft for the University of Nevada, being written-off in an accident on 2 March 1980.

The Tempo II was perhaps the most radically modified Invader derivative, with lengthened nose and cabin. It is seen here carrying meteorological research equipment.

Westland B-26

Westland & Son Inc. of Los Angeles, California, modified a single B-26B Invader as a three-passenger VIP transport aircraft at the behest of an agent of the President of Mexico in 1949. It had curtained passenger windows and a direction-finding 'egg' above the rear fuselage and cabin. By the mid-1950s the aircraft was registered as XB-PEK, named *Sierra Hermosa*, and based at Mexico City in the presidential fleet. It was replaced during December 1962 by an executive Martin B-26C Marauder (XB-PEX), passing to the Commander of the Fuerza Aérea Mexicana (Mexican Air Force), later becoming FAM 1300. It was last seen operational in late 1971 at Mexico City, being noted there again but in poor condition during late 1975. Its use with the Fuerza Aérea Mexicana was limited because of safety concerns.

Wold B-26 Invader Executive

A-26B N4000 (ex 44-34762) was modified as a Wold B-26 Invader Executive in the mid-1950s for the Swiflite Aircraft Corp. of New York, NY. In 1956 it went to Slick Airways. 44-34762 was one of the batch of A-26Bs not delivered to the Air Force but handed over to the Reconstruction Finance Corp. for disposal. The exact configuration of the Wold executive Invader is unknown, although it was one of the less extensive modifications.

XD813

Vickers Valiant
V-bomber pioneer

Designed with just one purpose in mind, the history of the Valiant is inextricably linked to that of the early days of the United Kingdom's airborne nuclear deterrent. The delivery means by which Britain assumed its seat at the 'nuclear top table', the Valiant went on to pioneer every role to which the V-bombers were put. Most importantly, it was the first – and only – British aircraft to ever drop an atomic weapon.

WB210, the prototype B.9/48 (yet to be named Valiant), sits on the ground at Wisley in May 1951 at around the time of its first flight. Noteworthy are the large flaps, including scalloped sections that deployed from under the engine jetpipes.

On 12 May 1945, just four days after the German surrender in Europe, British Prime Minister Winston Churchill sent a telegram to President Truman in which he asserted 'an iron curtain is drawn upon their [Russian] front. We do not know what is going on behind…' This now famous document spelled out a vision of post-war Europe in which the continent was split in two along ideological lines. Two months later Churchill was ousted from power by Clement Attlee's Labour government, and the US exploded its first atomic device at the Trinity site in New Mexico. On 6 and 9 August atomic bombs were dropped on Hiroshima and Nagasaki, bringing the war with Japan to an end.

British scientists had played a vital part in the development of an atomic weapon. It had been the 1941 Maud Report that had been the spur to the US decision to proceed with the Manhattan project, in which British scientists had been heavily involved – notably Dr William G. Penney.

In the aftermath of the war the Attlee government and its military advisors discussed the creation of an independent British bomb. The climate was generally favourable to such development as most politicians and military staff shared Churchill's bleak view of the European situation. Furthermore, the McMahon Act – passed in the US on 1 August 1946 – effectively prevented any collaboration between the US and UK. Days later, on 9 August 1946, Air Staff Operational Requirement (OR) 1001 was issued that outlined the requirements for an air-dropped atomic bomb, and on 8 January 1947 the government authorised research and development work. Britain's bomb was under way.

Where the Valiant led, the other V-bombers followed. As the first of the trio to enter service, the Valiant was entrusted with pioneering all the roles with which the V-force was eventually tasked. A role which assumed increasing importance was inflight refuelling. Here a No. 90 Squadron BK.Mk 1 refuels a Vulcan B.Mk 1A of the Waddington Wing in the early 1960s.

Under the guidance of Penney, an RAF team at Fort Halstead, Kent, began the task of designing the OR1001 bomb. Thanks to Penney's close involvement in the Manhattan project, it was essentially a copy of the plutonium weapon dropped on Nagasaki. As development and the production of materials continued, the project was given new impetus by eventsas the Soviet Union tightened its grip in eastern Europe. On 29 August 1949 the first Russian atomic bomb was exploded, signalling an end to the US monopoly.

After a significant national effort involving all three services and a sizeable portion of the scientific community, at 0930 on 3 October 1952 – under Operation Hurricane – Britain's first atomic device was successfully detonated in the hull of HMS *Plym*, anchored near the Monte Bello islands off the northwest coast of Australia. Britain, at last, had a bomb – but as yet no means to deliver it.

Development of the Valiant

From the outset of the bomb project it was envisaged that the only practical means of delivery were high-speed, high-

Today it is staggering that a country the size of Britain could have built three nuclear bombers, but such were the demands of the Cold War. Of the three the Valiant was the least capable, but it was the first in service. Below the prototype flies low over Farnborough during the 1951 show.

WB210 – the B.9/48 prototype

Built in the experimental workshop at Foxwarren (close to Brooklands), the B.9/48 was constructed quickly and without any complications. Its first flight was originally scheduled for January 1952 but construction proceeded so smoothly that this was brought forward to May 1951 (it flew on the 18th). The aircraft was of conventional construction, but the attention to streamlining – producing an almost perfect cigar-shaped fuselage – and its polished aluminium alloy finish gave it a very futuristic look when it appeared. With the exception of the air intakes, the shape of the Valiant remained essentially unchanged throughout its life.

Above: This view of WB210 at Wisley highlights the original 'letter-box' intakes, fitted with guide vanes.

Right: By contrast, the second prototype had the much larger 'spectacle' intakes. Initially the outer intakes had vanes fitted, but they were later removed. Here the aircraft is seen on the Farnborough runway during the September 1952 SBAC show. The first prototype had been lost in January.

altitude jet-powered bombers. The bomb itself was to be 10,000 lb (4536 kg) in weight and 5 ft (1.52 m) in diameter. These dimensions provided designers with a cornerstone around which to craft an aircraft. An initial requirement (OR230) was drafted in early 1946, specifying an aircraft that could carry the bomb to a target 2,000 nm (3704 km) from base. This was felt to be too ambitious and was dropped. On 7 November OR229 was issued in draft form with the range reduced to 1,500 nm (2778 km). As well as its nuclear requirements, the new bomber had to be capable of carrying a variety of conventional weapons – it would still have a use if the atomic bomb project was terminated for any reason.

OR229 was issued in final form on 7 January 1947, the day before atomic bomb development was authorised. Four companies were invited to tender for Specification B.35/46 – Avro, Armstrong Whitworth, English Electric and Handley Page. Two others – Shorts and Vickers – prepared technical documents. While OR229 was being discussed, it was also felt prudent to order an entirely conventional design as an insurance policy. Accordingly, OR239 was issued in January 1947, leading to Specification B.14/46 of August 1947. By the end of the year three companies were invited to proceed with their designs – Avro (Vulcan) and Handley Page (Victor) for B.35/46, and Shorts (Sperrin) for B.14/46.

Late in 1947 interest in Vickers's original design began to grow and by early in the following year that interest gained serious momentum. The Air Staff realised that the Shorts design would fall below the already reduced B.14/46 specification, leaving a considerable gap between the Sperrin and the ambitious Avro and Handley Page designs. The Vulcan and Victor would not be in production for some years yet, and would not be ready until some time after the planned bomb delivery

schedule. With some changes the Vickers design could fill the 'interim' bomber requirement much better than the Sperrin, and be ready in time for the first bombs. Two prototype Sperrins were built and flown, and although they were later used for drop trials of the casing for Britian's atomic bomb, the type was relegated to become a footnote to the RAF's nuclear deterrent history.

As a result of the resurgence of interest in its work, Vickers was issued with an invitation to proceed with Specification B.9/48 in April 1948, and was given a contract for two prototypes the following February. In anticipation of this order Vickers had already begun work using its own funds.

The Valiant design

Sized around the dimensions of the OR1001 bomb, the new B.9/48 bomber featured a cigar-shaped fuselage, the lines of which were only broken by the bulges above and below the forward fuselage – above to accommodate the flight deck and below to provide a visual bomb-aiming position. The shoulder-mounted wings were of moderate sweep, with straight trailing edges but with a compound taper on the leading edge. Four

WB215, the second B.9/48, was also built at Foxwarren. It was initially rolled out in polished aluminium (left) but was subsequently given a metallic paint finish (above).

Originally assigned the in-house designation Type 667, WB215 was modified to Type 709 with long-range underwing drop tanks for participation in the London-New Zealand Air Race. Although it was not ready in time to take part, the modification work was useful to clear the tanks for subsequent service use. Each tank provided an additional 1,500 Imp gal (6819 litres) of fuel.

Valiant B.Mk 2 – the Pathfinder

The Vickers Type 673 was a one-off prototype for the Valiant B.Mk 2, a target-marking version of the B.Mk 1. The operational concept envisaged for the aircraft was to fly ahead of the Main Force, descending to low level to rove around the target area to locate and mark targets. Its main range requirements were the same as the B.Mk 1 which it would support, but it needed additional fuel for the low-level search. It was also much stronger to cope with extended high-speed operations at low level. In August 1952 the Air Ministry was forced to cut back its expenditure with the result that the B.Mk 2 order for 17 Type 718 Conway-powered aircraft was cancelled. Completion of the prototype was allowed to proceed, but only on the basis that it would be used to trial new systems for the B.Mk 1. It first flew on 4 September 1953.

Above: The 'Black Bomber' cruises over southern England, demonstrating the pointed undercarriage nacelles outboard of the jetpipes. The new nacelles allowed the wings to carry a greater amount of fuel.

Below: Another significant change in the B.Mk 2 – evident in this view – was the forward fuselage stretch of 4 ft 6 in (1.37 m).

Rolls-Royce R.A.3 Avon turbojets were mounted in the wing roots. The engine bays were sized to accommodate Sapphires as an alternative. The tailplane was mounted half-way up the fin to keep it clear of the jet efflux. The twin-wheel main undercarriage units retracted outwards to lie flat in the wings. The shoulder wing configuration required only a short nose-wheel unit, although there was limited ground clearance for bomb loading.

To alleviate this problem, the bomb bay doors retracted up inside the bay, so increasing the ground clearance to the sides of the bay. For large loads the nose of the aircraft could be jacked up to provide additional clearance. Bombs were lifted up into the bay by a hoist that could be positioned above the aircraft. A similar arrangement was used to attach the large external tanks that could be mounted at roughly mid-span.

Apart from the single 10,000-lb (4536-kg) atomic weapon, the Valiant could carry a conventional weapon of similar weight, or up to 21 1,000-lb (454-kg) bombs. A single 10,000-lb bomb or 12 1,000-lb weapons could also be carried in bomb-carrying nacelles that replaced the drop tanks under the wings. Development of these nacelles was cancelled in 1956, however.

No defensive armament was fitted, although Vickers looked at a tail barbette with 20-mm cannon, and even towed missiles, as a means of defending against fighter attack. Later a single 30-mm cannon was proposed, and then a B-36-style tail turret.

Apart from the brakes and steering, the Valiant was all-electric, a major step for such a large aircraft. The aircraft had large flaps to keep landing speeds acceptably low. The inboard flap sections included scalloped sections beneath the jetpipes. Divebrakes were fitted above the wings.

Originally the V-bombers were to be fitted with a completely jettisonable crew compartment, which would fall to earth by parachute. This proved to be too difficult from a technical standpoint, and would have imposed weight penalties and structural complexity. Instead the two pilots were provided with upward-firing ejection seats while the three other crew members baled out through the air-baffled entry hatch. Work continued on the jettisonable cabin, but this never became reality. The lack of ejection seats for the rear crew remained controversial throughout the career of all three V-bombers – which all shared the same crew configuration.

The new trailing-edge nacelles allowed the B.Mk 2 to adopt a four-wheel bogie undercarriage rather than the tandem twin-wheel arrangement of the B.Mk 1.

The navigation/bombing system (NBS) was based on the H2S Mk 9a Yellow Aster radar with a navigational ballistic computer. Development of the NBS took some time and many early Valiants were delivered without it. Other navigation equipment included the Gee-H that worked with ground stations and the Green Satin Doppler.

Flight test

Construction of the first prototype Vickers Type 660 (WB210) proceeded at the Foxwarren experimental workshop quickly and without complication. When complete it was taken by road to the grass airfield at Wisley, from where it made its first flight on 18 May 1951 (beating the Sperrin into the air by three months). At the controls for the maiden flight were 'Mutt' Summers and 'Jock' Bryce. Three more flights were made from Wisley before flight trials moved to the better-equipped airfield at Hurn (Bournemouth). In July the aircraft was named Valiant, although Vickers would have preferred to reuse the name Vimy.

Flight trials proceeded relatively smoothly at both Hurn and Wisley, punctuated by the Valiant's public debut at the 1951 Farnborough air show in September. On 12 January 1952 tragedy struck the programme: during engine relight trials the prototype caught fire and was abandoned. While the other crew ejected or baled out safely, co-pilot Squadron Leader Foster was killed when his ejection seat hit the aircraft's tail.

Designated Type 667, the second prototype (WB215) took to the air for the first time on 11 April 1952. As a result of information from the crash of WB210, the second aircraft incorporated several changes, most notably enlarged 'spectacle' intakes. It was also fitted with R.A.7 Avons of 7,500 lb (33.37 kN) thrust. The aircraft showed adequate performance, although in climbing to its intended operational altitude of 45,000 ft (13716 m) it was found necessary to level the aircraft out and accelerate part way through the climb.

Bomb trials were undertaken from 2 July, when WB215 dropped a series of 1,000-lb (454-kg) weapons. Ground tests included fitting dummy OR1001 bombs. In October 1952 the

aircraft was damaged when its port undercarriage collapsed near the end of the landing run. The wing structure was undamaged although repairs were needed to the flaps. During the lay-up the aircraft had R.A.14 Avons installed.

During 1953 WB215 was prepared for participation in the New Zealand Air Race between London and Christchurch. It had drop tanks installed under the wings and was given

Early production tests

Following the first flight of a production Valiant (WP199) on 21 December 1953, most of the early batch of aircraft were used for some sort of trials: WP199 (handling, bomb release), WP200 (radar), WP201 (Blue Danube development), WP202 (radio), WP203 (RAE armament), WP204 (Boscombe Down handling), WP205 (PR), WP208 (autopliot), WP209 (armament), WP210 (Mach number trials), WP214 (Farnborough radar trials), WP218 (performance). Of the 'missing' aircraft, WP206, 207, 211, 212, 213, 215 and 216 were used by the Bomber Command Development Unit at Wittering before forming the initial equipment of No. 138 Squadron, while WP217 was the first B(PR).Mk 1 for No. 543 Squadron. A small number of the early test aircraft found their way into operational service but most stayed within the trials fleet.

An aircraft without a mission: the Valiant B.Mk 2 was cancelled a year before the prototype flew, and the Pathfinder role for which it was intended was not adopted.

WP201 was an armament trials aircraft that was dispatched to Wittering for trials with the Blue Danube atomic bomb.

A fine view captures WP209 during operations with the RAE at Farnborough. This aircraft was mainly used for bomb ballistic trials, going to Woomera in Australia in July 1955. It ended its days at the Stansted fire school.

This line-up shows the first aircraft permanently assigned to No. 232 Operational Conversion Unit at Gaydon, where all Valiant crew trained throughout the type's career.

WP204 was assigned to the A&AEE Handling Squadron at Boscombe Down. It was later modified for trials of the Blue Steel missile.

Armstrong Siddeley Scarab RATO packs to cope with heavy take-off weights in hot conditions. By the time the race got under way on 8 October, the Valiant had not accumulated sufficient test hours in its new configuration and was not entered. Honours for the event went to a Canberra PR.Mk 3 of the RAF, which arrived at Christchurch in 23 hours 50 minutes. The intended flight plan for the Valiant was nine minutes quicker.

B.Mk 2 – the 'Black Bomber'

Long before the Valiant first flew, Vickers had proposed a number of developments, including a pathfinder aircraft that would mark targets for the main bomber force. It was to carry extra fuel and be strengthened to operate at low level. Reconnaissance was another role for which the design was proposed, perhaps in the same version as the target-marker. In July 1950 OR285 was raised to cover the pathfinder. OR279 for the PR aircraft was shelved while Vickers attempted to cover both tasks with one version, and the follow-on OR287 for a long-range PR aircraft was also subsequently dropped.

Work on the OR285 aircraft led to a number of changes, the most obvious being a move of the undercarriage to 'speed pods' under the wings and a change to four-wheel bogies. The wing structure was left completely clear for the carriage of fuel, which, with the optional bomb bay tanks, was raised to a capacity of 9,720 Imp gal (44187 litres). Moving the undercarriage aft also moved the centre of gravity. To offset this the forward fuselage was made 4 ft 6 in (1.37 m) longer. Designed to carry 6,000 lb (2722 kg) of target-markers, the OR285 aircraft (Type 673, Valiant B.Mk 2) could search at low level for its targets. It could also carry a full bombload if required. It should be noted that an earlier Mk 2 proposal had featured extended wings, bicycle main undercarriage and outriggers.

Power was, initially at least, to be provided by the R.A.14 Avon, although the promising Rolls-Royce Conway Co.3 would be installed in later production aircraft. Fitment of the Conway would have greatly enhanced the aircraft's performance, especially in terms of speed and range. Vickers geared up to produce what it termed the B.Mk 1½ and 2½ – Conway-powered versions of the Mk 1 and 2, respectively – but in the end no Conways were ever used in a Valiant.

Vickers received a contract for a single B.Mk 2 prototype to Specification B.104 on 8 November 1950, followed by a production order for 17 (two for tests and 15 for operational use). However, in August 1952 the B.Mk 2 production order was cancelled with the prototype over half complete. The 17-aircraft order was switched to further Mk 1s. As the Mk 2 prototype was so far advanced, Vickers was instructed to complete it so that it could be used to aid Mk 1 clearance trials.

On 4 September 1953 'Jock' Bryce and Brian Trubshaw took the black-painted B.Mk 2 aloft for the first time at Weybridge. Early flight trials included a dive to Mach 0.905 and a low-level run at 552 mph (888 km/h), an impressive figure compared to the B.Mk 1's 414-mph (666-km/h) low-level limit. In September 1954 it was publicly displayed at Farnborough, its sinister colour scheme earning it the nickname 'Black Bomber'. Subsequently it was used to test equipment destined for the B.Mk 1, such as the Sprite take-off rockets and refuelling equipment. In May 1958 the B.Mk 2 was dismantled at the end of a promising trials career.

There were numerous further Valiant proposals, such as the Conway-powered Mk 3 with fully swept wings, and various Mk 4 studies, most of which had Olympus engines and some had T-tails. The bomber was also studied as the basis for the

V.C.5 jet airliner, initially schemed with shoulder-mounted wings but later adopting a low-wing configuration. As the Type 1000, a prototype was ordered (XD662) and construction began. In September 1955 six were ordered for the RAF as high-speed troop transports. In late 1955, with BOAC and other airlines having withdrawn their interest in the Type 1000, the project was terminated and the RAF bought Bristol Britannias instead. The prototype Type 1000 was nearing completion, but was immediately scrapped.

Valiant into service

Even before the prototype's first flight, 25 Valiant B.Mk 1s were ordered on 8 January 1951. In the midst of a difficult time for the programme, when aileron flutter and rear fuselage acoustic cracks on WB215 were being hastily remedied, the first production aircraft (WP199) took to the air on 21 December 1953 under the power of R.A.14 Avons. It, and the 103 aircraft which followed, were all built at Weybridge and then flown from the very short strip at Brooklands. Many of the early aircraft were assigned to various trials programmes. Later the improved R.A.28 Avon became the standard production powerplant.

Meanwhile, the OR1001 bomb team had been progressing, refining the weapon and its circuitry, and developing a ballistic case to accommodate the package. The second prototype and other early Valiants were widely used during the ballistics effort, making many drops at the Orfordness range. By late 1953 the bomb was ready for use, the first examples being delivered to the Bomber Command Armament School at Wittering on 7 and 14 November 1953.

Most of the later bomb-drop trials had been carried out by a Ministry of Supply unit within the RAF, designated No. 1321 Flight. Formed in April 1954 at Wittering, the unit had flown most of its trials from Weybridge using the third production

aircraft, WP201. On 15 June 1955 this aircraft was transferred to Wittering to join the bombs: theoretically, at least, this day marked the start of a UK nuclear airborne delivery capability.

Deliveries of aircraft to the RAF began on 8 February 1955 when WP206 flew to Gaydon to join No. 138 Squadron, the first operational Valiant unit. In July the unit moved in alongside 1321 Flight at Wittering, the trials flight becoming 'C' Flight of No. 138 Squadron in early 1956. Soon after, 'C' Flight became No. 49 Squadron to prepare for the upcoming Buffalo weapons trials. No. 232 Operational Conversion Unit, the Valiant training outfit, received its first aircraft at Gaydon in July 1955.

With the training, trials and first operational bomber unit working up, it was next the turn of a photographic reconnaissance unit to stand up. No. 543 Squadron officially reformed on 29 September 1955 at Gaydon, but it had been operating two Valiants for some weeks prior. Photo-reconnaissance was seen as something of a priority in the creation of the V-force. Although a dedicated PR version to meet OR279 had been cancelled, Vickers had continued to work on a PR-dedicated version (Type 710) of the standard B.Mk 1. This was a fairly simple conversion with extra fuel in the front of the bomb bay, and a pannier with cameras in the rear. The bay doors were replaced by new units with sliding windows for the cameras held within. The standard day reconnaissance suite consisted of eight 36-in (91.4-cm) F.52 cameras, three 6-in (15.2-cm) F.49s in a trimetrogon arrangement for wide-area coverage, and a single vertical F.49 for survey. For the night role it was planned that B(PR).Mk 1s would carry photo-flash bombs in large panniers under the wing pylons, but this did not proceed despite successful wind tunnel tests.

Above: No. 138 Squadron, the first Valiant unit, put up this fine four-ship. The establishment of an operational squadron before the training unit was an unusual step for the RAF, but highlighted the perceived need to get operational nuclear bombers into service as soon as possible.

Above left: The Weybridge plant produced 104 Valiants between 1953 and 1957. The aircraft were built on two production lines to satisfy the demands of the production schedule.

These three Valiant B(PR).Mk 1s were part of the initial equipment of No. 543 Squadron, which was formed at Wyton during 1955 to undertake the photo-reconnaissance role. This was accomplished with some urgency to allow planners to plot ingress routes for the growing V-bomber fleet.

Operation Musketeer – the Valiant goes to war

Following the Egyptian nationalisation of the Suez Canal, British and French forces were sent to the Mediterranean to exert pressure on Egypt's President Nasser to return the Canal to its Anglo-French owners. The force included 24 Valiants. When a final ultimatum expired, an air campaign was unleashed on 31 October 1956. The first phase lasted for two days and involved attacks against Egyptian airfields to remove the threat of the Egyptian air force. Valiants were heavily involved in this phase. For the next four days they also turned their attentions to other targets while continuing attacks on airfields. In the face of international opposition, the campaign was halted on 5 November, despite a successful assault landing.

Above: A Valiant is parked at Luqa, Malta, during the Suez campaign. In the background are other Valiants and Canberra bombers, and a Hastings transport.

Below: The Valiant second prototype gets smartly airborne during trials of the de Havilland Super Sprite rocket-assisted take-off gear (RATOG).

Having been built as a standard B.Mk 1, WP205 tested the camera installation, and was followed by 10 production aircraft, which were given the designation B(PR).Mk 1. Additionally, five B(PR)K.M k 1s were produced later, based on tanker-capable machines. WP217 was the first B(PR).Mk 1, flying on 27 April 1955. It was delivered to the nascent No. 543 Squadron on 11 May and moved with the rest of the squadron to Wyton in November.

Operation Buffalo

Although a theoretical operating capability in the nuclear role had been achieved with the co-location of bomb and aircraft at Wittering in June 1955, the creation of a meaningful

deterrent force was far from complete. By the end of January 1956 No. 138 Squadron had eight aircraft on strength at Wittering, and a number of bombs were available. Nevertheless, without live testing, no-one could be too sure of whether the system would work. The aim of Operation Buffalo was to put theory into practice.

On 1 May 1956 No. 49 Squadron was reformed under Squadron Leader D. Roberts DFC AFC. It was, in fact, the successor to No. 1321 Flight, and its task was to prepare for and undertake live nuclear tests. The first of these was conducted at Maralinga in Australia on 11 October 1956, as part of Operation Buffalo, which also involved three other ground-based shots. For its part in Buffalo No. 49 Squadron received two specially modified Valiants: WZ366 and WZ367. They had airflow baffles for the bomb bay to ensure that the OR1001 bomb, also known as Blue Danube, separated cleanly. A Mk 10 autopilot and a T.4 visual bombsight were fitted. The aircraft were also given recording equipment.

Painted white to protect against nuclear flash, the two Buffalo Valiants deployed to RAAF Edinburgh in August, where the crews resumed training with the new equipment. By 11 October all was ready, and WZ366 lifted off from the

Left: *Views from the front and rear show the Super Sprite RATOG installation on WB215. Next to the starboard undercarriage in the rear view is the portable hydrogen peroxide and nitrogen pressurisation system used to top off the rockets prior to flight. The rockets were angled down to give an element of lift as well as thrust during the take-off run. After use they were jettisoned and returned to earth by parachute. The flight trials were not without incident and nearly caused the loss of WB215. When WB215 was retired due to a cracked wing spar, further trials were performed with XD872.*

Maralinga strip clutching a live weapon in its belly. The crew was led by Squadron Leader E.J.G. Flavell. After three fly-overs to ensure everything was in place and working, the weapon was released from 30,000 ft (9144 m) at 1527 local, after which the aircraft banked sharply away. The weapon – a Blue Danube round with its yield reduced to around 3 to 4 kilotons – performed correctly and detonated at an air burst height of 500 ft (152 m) near target 'Kite'. Both WZ366 and WZ367 – following a few minutes behind – recorded the explosion. Buffalo was deemed a complete success – Britain finally had a proven airborne deterrent.

Building the force

Prior to the Buffalo drop, the growing Valiant force was beginning to flex its muscles. In September 1956 under Operation Too Right, two aircraft from No. 138 Squadron flew to Australia and New Zealand via Iraq, Sharjah, Karachi, Ceylon and Singapore to prove the ability of the aircraft to undertake long-range deployments. As well as the trials unit, No. 49 Squadron, two more Valiant squadrons were formed in early 1956 when Nos 207 and 214 Squadrons began operations from Marham. Later in the year No. 148 Squadron also formed at Marham, while No. 7 Squadron began operations from Honington. Thus, by the end of the year, the V-force had six Valiant bomber squadrons, one photo-recce unit and the OCU. However, during the year the first Avro Vulcans were delivered to No. 230 OCU, an event that heralded the eventual demise of the Valiant as the RAF's prime deterrent platform. Two more Valiant squadrons formed in 1957, No. 90 at Honington in the bomber role and No. 199 at the same base in the radio coun-

termeasures role. The last Valiant from the production line, BK.Mk 1 XD875, was delivered in September 1957.

As the force grew, so did its range of operations. In September 1956 a No. 214 Squadron aircraft made the first

Radio countermeasures

The Valiant was a natural candidate for the escort jamming role, and in 1956 WP214 was fitted with an extensive RCM suite for trials. It remained the RCM test aircraft for much of its career. Based on its work, seven aircraft (WP211, 212, 213, 215, 216, WZ365, 372) were modified for service with No. 199/18 Squadron, distinguished by an additional air scoop on the port side of the rear fuselage.

Above: *After brief service with No. 232 OCU, WZ365 was modified with RCM gear for No. 199 Squadron.*

Below: *WP213 was one of No. 18 Squadron's seven RCM-equipped Valiants.*

non-stop transatlantic flight, flying from Loring AFB, Maine, back to Marham. Deployments were made to the Mediterranean, mostly to RAF Idris in Libya, from where Valiants tested NATO's southern defences and explored ingress routes on the southern flank of the Soviet Union.

Operation Musketeer

Fresh from the success of the Buffalo weapons drop and early deployment operations, the RAF was brought back to earth later in October 1956 during the Suez campaign. In July Egypt's President Nasser had 'nationalised' the Suez Canal Company, a joint Anglo-French concern. To protect the interests of both countries aircraft and troops were sent to the Mediterranean. The RAF dispatched 24 Valiants to Luqa, Malta, to complement 10 squadrons of Canberras. No. 138 Squadron deployed its full complement of eight aircraft, Nos 148 and 207 provided six aircraft each while No. 214 sent four aircraft. When Nasser failed to respond to a 31 October ultimatum to hand back the canal company to its owners, Operation Musketeer was initiated, with air attacks launched against Egypt. Israel also joined in the attack.

Valiant raids were delivered initially against airfields, beginning with Almaza and Abu Sueir and broadening to include Cairo West, Fayid, Kabrit, Kasfareet and Luxor. Other targets such as marshalling yards, barracks and a submarine repair depot were also hit. Sorties were generally around five hours in length. No. 148 Squadron's XD814 was the first aircraft to drop in anger, hitting Almaza as part of a six-ship attack which

was target-marked by a Canberra. In all Valiants flew 49 sorties over Egypt.

The Suez campaign, which lasted from 31 October to 5 November, highlighted some serious deficiencies in the Valiant fleet as it stood then, chiefly concerning navigation and bombing equipment. The Gee-H equipment could not be used

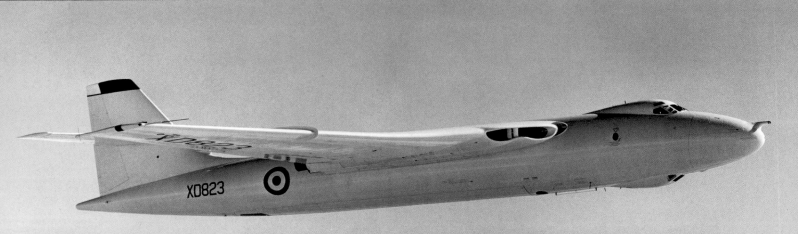

as there were no ground stations in the region. Not all aircraft had the NBS system and those that did found the equipment suffered from a high rate of unserviceability. Only the Green Satin Doppler drift-measuring equipment came out with any credit, although that required a suitably rough sea state before it could work adequately. The V-force – optimised for a war in Europe – had been found wanting in its first foray outside the Cold War scenario. Nevertheless, the Valiant had become the first V-bomber to drop bombs in anger. Until the 1982 Vulcan Black Buck raids against the Falkland Islands, it was the only V-bomber to have seen combat.

V-force operations

Following the end of Musketeer a number of Valiants deployed to Malta to deter any further aggression from Egypt under Operation Goldflake. Malta was also a regular destination for long-range navigation exercises as the Valiant force returned to the task of preparing for nuclear war. Aircraft routinely deployed to Goose Bay in Labrador, and from there occasionally tested NORAD's defences.

In October 1957 crews from Nos 138 and 214 Squadrons took a total of five Valiants to participate in Strategic Air Command's bombing competition at Pinecastle AFB, Florida. Two aircraft were flown in the 90-aircraft competition and the RAF crews acquitted themselves well, one crew finishing 11th overall. In October 1958 Valiants returned to the SAC competition, this time at March AFB, California. Again the Valiants performed well, one crew achieving 9th place out of 164.

On 29 October 1957 three Valiants of No. 214 Squadron

left Marham for a deployment that was to become an important commitment for the V-force for a number of years: support of the Far East Air Force. Under Exercise Profiteer the aircraft deployed to Changi in Singapore, establishing a regular V-bomber detachment in the Far East. The Valiants flew local training missions, and ranged further afield to Laos and Thailand. The primary aim of the Profiteer detachment was to provide training in the theatre, but the aircraft could also be called upon to act as a back-up to the ongoing operations in Malaya.

At the time the British were still engaged in Operation Firedog, fighting communist insurgents in what was known as

Above: Arguably the best-known of the Valiants, and the only one to survive today, XD818 dropped Britain's first H-bomb. It is seen here in full Grapple fit, including the unusual recording gear carried in the truncated tailcone. White paint or white plastic covering was used on every conceivable item to avoid the effects of the bomb's heat blast.

Above: One of the No. 49 Squadron Valiants returns to Wittering. For the first Grapple series seven of the squadron's eight aircraft were deployed: subsequent detachments used four aircraft with occasional rotations.

Left: This scene from Christmas Island was recorded around the time of the first Grapple test in May 1957. XD818 was the actual drop aircraft while XD824 flew the 'Grandstand' mission, providing a platform from which the explosion could be observed and recorded.

Vickers Valiant BK.Mk 1
No. 214 Squadron, RAF Marham

This Valiant BK.Mk 1 is depicted as it appeared in No. 214 Squadron colours, having earlier served with Nos 90 and 7 Squadrons. It is shown refuelling a Gloster Javelin FAW.Mk 9R of No. 23 Squadron. Valiant tankers from both Nos 214 and 90 Squadrons were instrumental in developing long-range deployment techniques for fighters, and No. 23 Squadron was a pioneer among the fighter community. In October 1960, under Operation Dyke, Valiants refuelled four Javelins out to Singapore in a rapid reinforcement exercise, the first large-scale use of this technique. Later, in the March 1964 Operation Chive, Valiants took another four Javelins to the Far East by a shorter route with several long overwater sectors, notably those between Khormaksar and Gan, and Gan and the final destination at Butterworth.

The Valiant was hardly ideal as a tanker due to its poor offload figures and lack of performance at high weights. It was better suited to the refuelling of large aircraft as it only had one drogue. Nevertheless, it was readily available thanks to the delivery of later V-bombers, and for the two years it was officially assigned to the tanker role it was used to establish tanking routines that are still in use today.

Until it was officially re-roled for tanker duties in April 1962, No. 214 Squadron maintained its Main Force bomber commitments. However, the squadron's duties had been turning increasingly to the tanker role since 1957. Its normal squadron markings of a winged nightjar were incorporated into a new design that alluded to the squadron's new-found role.

the Malayan Emergency. While the Valiants were considered far too sophisticated to be of much use in the counter-insurgency war, their presence in the Far East certainly sent out a message to a wider audience in an increasingly volatile region. No. 90 Squadron subsequently sent Valiants to Changi, and No. 148 Squadron deployed four to the RAAF base at Butterworth in Malaysia. Vulcans later took over the deployment. While mainly of training benefit, Profiteer laid solid foundations for further V-bomber detachments (Chamfrom/Matterhorn) to the Far East, which acquired a more operational aspect during the Indonesian Confrontation.

No. 543 Squadron's photo-reconnaissance specialists were also very busy, alternating their Cold War role of recce and mapping the approaches to the Soviet Union with deployments to survey various parts of the world. Mapping and survey deployments were made to British Honduras (later Belize), the south Pacific and Africa.

Super Sprite

Part of the original specification for the Valiant stated that the aircraft should be able to take off in less than 6,000 ft (1829 m) at its fully loaded weight of 160,000 lb (72576 kg). Although the Avons had water methanol boosting to provide 43 seconds of extra power for the take-off run, this was insufficient to meet the specification when operating in hot conditions. To overcome the shortfall, Vickers studied rocket-assisted take-off gear.

Initial studies focused on the Armstrong Siddeley Scarab, which comprised seven 3,000-lb (13.35-kN) thrust rockets in a

WZ376 and WZ390 were the much-photographed Valiant tanker trials aircraft, seen here during what was possibly the first hook-up between the two. Tanker crews received instruction from the makers of the refuelling equipment, Flight Refuelling Ltd at Tarrant Rushton, which had been testing inflight refuelling equipment with various aircraft since the end of World War II.

bundle. One Scarab would be mounted on either side of the rear fuselage. Burn time was six seconds and the entire installation provided a hefty 42,000 lb (186.9 kN) of thrust – more than doubling that available from the engines alone. Ground tests were conducted using a Valiant fuselage section, but the idea was dropped in favour of the de Havilland Super Sprite.

This liquid-fuelled rocket used a mix of hydrogen peroxide and kerosene, and gave a thrust of 4,200 lb (18.69 kN) for up to 40 seconds. Thrust and burn time could be altered. Weighing 1,460 lb (662 kg) when full, the Super Sprite was carried in a pod that was attached below and between the Valiant's engines, angled downwards to give a lift component. After the aircraft was airborne the rocket packs could be jettisoned, returning to earth by parachute for reuse.

The Valiant B.Mk 2 was the first aircraft to be fitted with Super Sprites, flying with them fitted from Wisley on

6 October 1954. The test was not a success: the starboard pack broke away, and the port pack had to be jettisoned. When the starboard pack had broken away it caused serious damage to the aircraft, notably to the starboard undercarriage, which would not lower for landing by normal or standard back-up means. The crew had to use the 'last-ditch' undercarriage release system – explosives in the uplock system – to free the gear and land safely.

Nevertheless, the Super Sprite showed sufficient promise to proceed, and in 1956 the second Valiant prototype, WB215, went to Hatfield for a trial installation and a series of flight trials. Despite a further pod break-away incident, jettison trials were completed and in June WB215 took off for the first time with the Super Sprites at full power. Early trials were generally successful. On 29 April 1957, however, disaster nearly struck the programme. During a simulated engine-out take-off with

WZ376 streams its drogue during an early test. The bomb bay provided more than adequate housing for the refuelling gear, aided by the large panel at the rear of the bay which could be raised into the fuselage (initially fitted to ensure clean bomb separation).

Right: Precise flying was required during the refuelling process as the aircraft could get quite close to each other. XD870 was a BK.Mk 1 delivered direct to No. 214 Squadron.

WZ390 was initially equipped only as a receiver, but later gained the modifications to accept a tanker package. It was built as a B(PR)K.Mk 1 with the ability to also carry the camera pallet, although it would appear it never did so.

the rockets running a loud bang was heard, accompanied by a violent shaking of the aircraft. Nothing appeared to be wrong and the aircraft landed safely. Ground inspection revealed that the rear wing spar had cracked. At the time the fatigue was put down to the heavy trials use to which WB215 had been subjected – much later the problem was to appear again, and ultimately spell the end of the Valiant.

WB215 did not fly again, but parts of it were used for a series of ground trials of the Super Sprite. They were concerned mainly with the ability of the rockets to operate successfully after long periods of inactivity, as might be experienced during extended alerts. Having passed these trials the Super Sprite went on to tropical tests, for which Valiant XD872 was used, and then on to the Bomber Command Development Unit at Wittering. Throughout the trials there were numerous niggling problems: packs did not fire, packs were damaged when released or the parachute would not function correctly. Although the Valiant was eventually cleared for operational use with the Super Sprite, by the time it was achieved the emphasis was on more powerful engines to overcome the take-off performance shortfall, and rocket-assisted take-offs were sidelined.

RCM aircraft

Throughout World War II the black art of radio countermeasures had become increasingly important, especially in the night bomber offensive against Germany. It came as no surprise to Vickers when the company was asked in 1952 to prepare an RCM version of the Valiant. WP214 was chosen to be the testbed, and RCM specialist aircraft would be operated by No. 199 Squadron, then flying Lincolns and Canberras.

As the trials aircraft, WP214 commenced RCM testing in December 1956. It had a complicated equipment suite fitted covering a mix of UK and US equipment. In the tail was the Orange Putter warning radar, while the rear fuselage was strengthened and redesigned to accommodate the main bulk of the RCM equipment fit. Jammers included the powerful Red Shrimp, Airborne Cigar, ALT-7 and APT-16, while US APR-4 and APR-9 receivers were installed. Window (chaff) dispensers were installed, and a large cooling intake was added to the port rear fuselage. The Air Electronics Officer's position was modified, while a new RCM operator position was installed above the visual bombing station. The weight of the RCM equipment in the tail required ballast to be added to the forward fuselage.

In May 1957 the first of seven RCM Valiants was delivered

Valiant refuelling trials encompassed tests with many other receivers. Here WZ376 passes fuel to Victor XA930. As well being able to refuel the V-bomber trio, Valiant tankers also proved their ability to refuel the much slower Argosy transport.

A typical Cold War scene shows Valiant B(PR)K.Mk 1 WZ390 of No. 214 Squadron being towed past a field of Bloodhound SAMs. The aircraft had been one of the initial trials pair, but was subsequently brought up to full service standard and issued to the RAF. It did undergo a number of other trials later, before being damaged and retired in 1962.

Above: Valiants were not only tested with RAF aircraft. As well as trials with Royal Navy types such as the Scimitar, the Valiants refuelled UK-based USAFE aircraft like this Douglas RB-66 Destroyer.

Above: Another USAFE type tested with the Valiant tanker was the McDonnell F-101A/C Voodoo. Single-seat nuclear fighter-bomber Voodoos served with the 81st Tactical Fighter Wing at RAF Bentwaters/Woodbridge.

to 'C' Flight of No. 199 Squadron at Honington to serve alongside Canberras. When that type was withdrawn the Valiants moved to Finningley in December 1958, where the unit was renumbered as No. 18 Squadron. By the very nature of its role the unit shunned publicity, and there is little detail of its operations. By the late 1950s it was realised that such a small force could not provide RCM protection for the widely dispersed V-force as a whole, and it was a better option to

provide the Main Force bombers with their own jammers. This work proceeded and in April 1963 No. 18 Squadron was disbanded.

Operation Grapple

A month after the first British atomic device had been successfully tested in October 1952, the US exploded its first hydrogen bomb. Less than a year later the Soviets followed suit. Although the Monte Bello and subsequent tests could be viewed as quite an achievement for the United Kingdom in the austere post-war environment (and without any assistance from the US), the fact remained that the UK lagged some way behind the other two nuclear powers. Britain needed the H-bomb, and Dr Penney's team at Aldermaston forged ahead with its development.

As the bomb neared fruition, thoughts turned to the testing of such a device. Clearly the expected explosive yield ruled out Monte Bello or Maralinga in Australia, so a location was sought in the southwest Pacific. The uninhabited Malden Island was chosen as a suitable drop site, while Christmas Island – around 400 miles (644 km) away – would provide a suitable location for the construction of an airstrip. Naturally, No. 49 Squadron and its Valiants were chosen to perform the air drops. The whole operation was to be known as Grapple.

Eight aircraft were selected on the Valiant production line to be modified for the Grapple tests. The modifications were extensive, from replacement of the aircraft's dielectric radome with a metal nose fairing, through shielding of cables and antennas with PVC coating, to the provision of shutters for the cabin windows. The aircraft were painted all-over anti-flash white, recording gear was installed and elements throughout the aircraft were strengthened.

With their new aircraft No. 49 Squadron began a series of visual bomb drop training sorties while a joint-service force completed the Christmas Island base. Seven of the aircraft were deployed to the Pacific. As well as the Valiants, a fleet of Canberras was also deployed – B.Mk 6s from No. 76 Squadron for air sampling and PR.Mk 7s from No. 100 Squadron for meteorological reconnaissance. Two Valiants were selected to drop the Britain's first H-bomb – XD818 and XD823. Their finish was kept scrupulously clean to provide maximum protection against nuclear flash for the crew and the aircraft's systems.

On 0900 local time on 15 May 1957 XD818 lifted off from Christmas Island, commanded by No. 49 Squadron's CO, Wing Commander Ken Hubbard OBE DFC. In its belly was the Short Granite weapon, a Green Granite Small thermonuclear warhead housed in a Blue Danube casing. Following close behind was XD824 that was to act as the 'Grandstand' aircraft.

Above: Vulcan B.Mk 1 XH478 was the first different type to be tested as a receiver. It was assigned for probe trials to the A&AEE.

Right: The tanker tanked: trials aircraft WZ376 hooks up with a Boeing KB-50J. Fuel from the Valiant's probe was ducted around the pressurised crew compartment by an external conduit.

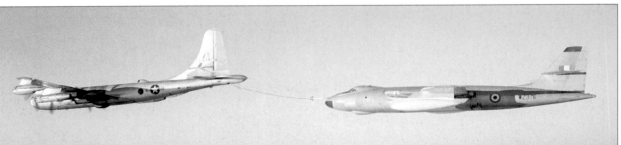

After a single dummy run, at 1036 local the weapon was released over Malden Island from an altitude of 45,000 ft (13716 m) and a speed of Mach 0.76. Immediately after release Hubbard executed a 60° bank to the left, pulling 1.7 g for 40 seconds until the aircraft rolled out to fly away from the area on a reciprocal heading. The 'Grandstand' Valiant, following some ½ mile behind, had initiated its escape manoeuvre a few seconds before. The bomb detonated at its programmed 8,000-ft (2438-m) air-burst point. After two and a half minutes XD818's crew reported feeling the air blast, although it was not violent, and after five minutes were instructed to open their cabin shutters so that they could see the mushroom cloud. Britain had achieved something of a 'first' by being the first nation to air-drop an H-bomb from a jet.

'Fizzling' bomb

As the euphoria of a successful air-drop test subsided, it became clear to the scientists that the Short Granite test had actually been something of a failure. Explosive yield was only 0.3 MT and the bomb had 'fizzled' rather than exploded correctly. In the second Grapple test, flown by Squadron Leader Dave Roberts in XD822, a device known as Orange Herald was dropped. This was a back-up fission weapon and was much more powerful than the hydrogen fusion bomb test. A third Grapple test on 19 June involved Valiant XD823 captained by Squadron Leader Arthur Steele dropping the Purple Granite fusion weapon. Like the first test this, too, was disappointing, resulting in only a 0.2-MT blast.

While publicly the government lauded the success of the Grapple trials, the reality was that the UK had not yet fully conquered the scientific challenges associated with thermonuclear warheads. Instead, and without publicity, the large fission device was chosen for production to provide an 'interim megaton weapon', eventually fielded in a Blue Danube case as the Violet Club. Valiants did not carry this weapon operationally, however. In the meantime, a series of kiloton trials was under-

taken at Maralinga as part of Operation Antler. Two of No. 543 Squadron's Valiants provided reconnaissance support for the three bursts, two of which were tower-mounted while the other was suspended from a balloon.

Undeterred by the Grapple failure, the Atomic Weapons Research Establishment team forged on with its H-bomb development. Progress was swift and in October 1957 four Valiants from No. 49 Squadron returned to Christmas Island for the Grapple X trials, which involved a single live drop. Unlike the previous trials, a new aim-point just off the southern tip of Christmas Island was chosen. Round C, a revised Green Granite Small warhead again in a Blue Danube casing, was dropped on 8 November 1957 by a crew led by Squadron

XD822 was a double-bomb veteran of the Grapple trials, having released the Grapple 2 and Flag Pole 1 weapons. Here it is seen after modification to standard BK.Mk 1 configuration, with H_2S radar inside the nose radome and No. 49 Squadron's greyhound badge emblazoned on the fin. It was with the unit when the squadron came under NATO control.

Production details

Valiant production deliveries began in December 1953 and ended in August 1957. As well as the three prototypes, 104 aircraft were ordered in batches on 20 April 1951 (25 aircraft – WP199 to WP223), 1 October 1951 (24 aircraft – WZ361 to WZ384, plus 17 switched from the B.Mk 2 order – WZ389 to WZ405), and 20 March 1953 (38 aircraft – XD812 to XD830, XD857 to XD875). Eighteen additional aircraft from the latter order were cancelled (XD876 to XD893), as was a final order for six aircraft (XE294 to XE299).

A small crowd gathers at Brooklands (above) to witness the departure on 27 August 1957 of XD875, the 104th and last Valiant from the production line. The aircraft took off (left) and then beat up the airfield before delivery to No. 207 Squadron. The photo shows to good effect the large undercarriage doors needed to cover the bays for the twin mainwheels, and also the narrow width of the wheel assemblies when viewed from the front – a necessity to fit them into the confines of the wing without having to add bulges.

WP200 was the second production Valiant, first flying on 9 March 1954. It spent much of its career with the Radar Research Establishment at Pershore, flying on trials of the H₂S Mk 9 bombing radar. It was struck off after overshooting the runway during an abandoned take-off on 14 March 1961.

Below: This view shows a scale drop-test model of the Avro Blue Steel in the bomb bay of a Valiant.

Leader Barney Miller in XD824. The results this time were more rewarding for the scientific team: a 1.8-MT burst.

Grapple Y was the codename for the Green Granite Large test conducted by Squadron Leader Bob Bates in XD825 on 28 April 1958. This bomb was spectacular: a 3-MT blast that was the largest recorded for a UK weapon. AWRE scientists were understandably pleased – Britain at last had a true megaton H-bomb. Grapple Z followed up with four more bomb tests, two dropped from Valiants (Flag Pole 1 and Halliard 1) and two suspended from balloons (Pennant 2 and Burgee 2). Flag Pole 1 was conducted by Squadron Leader Bill Bailey's crew in XD822 on 2 September 1958, while Halliard 1 was flown by Flight Lieutenant Tiff O'Connor's crew in XD827 on 28 September. Unlike previous drops, Grapple Z involved the Valiants employing radar bombing, achieving excellent accuracy. The blasts were also successful.

Sadly the successes of the Grapple X, Y and Z trials came too late for Britain's H-bomb programme. Having proven beyond all doubt that the technology had been mastered, the AWRE team was not able to productionise its design as a US-designed warhead had been chosen instead. The service UK thermonuclear warhead – the Red Snow used in the Yellow Sun Mk 2 free-fall bomb and Blue Steel missile – was based on the American Mk 28.

Grapple Z marked the last of the UK's above-ground tests – further trials were conducted at the underground test site north of Las Vegas in Nevada. Including the earlier Buffalo A-bomb test, the Valiants of No. 49 Squadron had successfully delivered eight live weapons with considerable precision, underlining the value of the V-force as a whole in its crucial role of nuclear deterrence.

Valiant tankers

At the start of the Valiant programme the priority was to get bombers into the main force as quickly as possible. However, as the Vulcan and Victor appeared in the second half of the 1950s, it was assumed that the Valiants could be released for the role of inflight refuelling. This was eventually accomplished but it was far from easy as there was considerable resistance to the idea within the upper echelons of the RAF and the government. Many felt that the diversion of Valiants to the tanker role unduly weakened the strike force.

Nevertheless, work on a Valiant tanker began at an early stage, initially using the redundant B.Mk 2 prototype. Fitted with a Flight Refuelling Ltd hose and drum in the bomb bay, and with a nose-mounted probe, the aircraft flew trials with Flight Refuelling's Canberra B.Mk 2. On 17 March 1954 the Valiant made a dry contact with the Canberra, but further trials were shelved while more pressing matters were attended to.

In 1956, however, trials began again using a pair of dedicated test aircraft. WZ376 was selected as the tanker, and was fitted with a hose-drogue unit (HDU). WZ390 became the receiver, although both aircraft were fitted with probes. Later, WZ390 also gained an HDU in the bomb bay. Trials continued through 1957 and 1958, testing a variety of drogue designs.

In late 1957, as the Vulcan force grew, the RAF's No. 214 Squadron at Marham began gearing up for a shift to the tanker role, its 'A' Flight taking part in a number of trials. Although it retained a strike role, the squadron's activities increasingly turned towards tanking, its aircraft being converted to BK.Mk 1 standard with underwing tanks, probe and bomb-bay HDU. During 1959 the squadron began a series of long-range, non-stop flights to far-flung corners of the Empire. On 2/3 March 1960 a Valiant flew for 18 hours 5 minutes, covering 8,500 miles (13679 km).

No. 214 Squadron was considered operational in the tanker role by the end of 1959. By then it had begun trials refuelling Vulcans, the first hook-up being made in October. On 20/21 June 1961 nine Valiants refuelled a Vulcan of No. 617 Squadron during a 20-hour 5-minute non-stop flight from Scampton to Sydney, impressive proof of the value of inflight refuelling. Valiants also began refuelling other aircraft types, notably the Gloster Javelin and English Electric Lightning interceptors. Indeed, it was fighter deployment support that became the main tasking for the Valiants. Operations were also conducted with the Royal Navy's Scimitars and Sea Vixens.

During 1961 conversion of a second squadron – No. 90 at Honington – to the tanker role began, overseen by No. 214. By the end of the year No. 90 was operational in its new role. On 1 April 1962 both units officially became tanker squadrons, operating 16 aircraft between them and finally relinquishing any Main Force bomber commitments. They operated successfully for the next two years until the fleet was grounded.

Valiant replacement

Valiants did more than any aircraft to firmly entrench the role of inflight refuelling in RAF doctrine. However, they were not ideal tankers and their replacement was sought at an early stage. In the FR role the Valiant had a total transferable fuel quantity of 45,000 lb (20412 kg), whereas the figure for the Handley Page Victor K.Mk 1 was 98,500 lb (44680 kg). The Valiant was restricted to Mach 0.74 compared with the Victor's Mach 0.91, and its maximum refuelling altitude was 32,000 ft (9754 m) as opposed to 40,000 ft (12192 m). Furthermore, the Valiant's single-point refuelling capability was not conducive to fighter operations, whereas the Victor offered two- and three-point capability. Besides, the Valiant's fatigue life was due to expire in 1968.

With Victor Mk 1s becoming available through changes to the V-force's posture and delivery of Mk 2 aircraft, it was inevitable that the Victor would replace the Valiant in the tanker role, even though it had a longer take-off run. Although

Blue Steel tests

Although it was never intended to carry the Blue Steel weapon operationally (that job being entrusted to Vulcan and Victor B.Mk 2s), the Valiant played an important role in its development by being the main drop-test vehicle during the early years. At least five Valiants were assigned to the programme at one point or another. The aircraft dropped small-scale models of the missile – including powered examples – and also full-size glider versions, in either controlled or uncontrolled configurations. Powered tests of full-size missiles were conducted by a Vulcan B.Mk 1.

WP204 was one of the hardest-working Blue Steel test vehicles, having been modified for the role in early 1958. It spent some time at RAAF Edinburgh for trials over the Woomera range. Here it is seen taking off with a full-scale Blue Steel nestling in its bomb bay.

Left: Technicians perform the tricky task of mating a full-scale Blue Steel into the bay of WP204. As with other large bomb loads, the missile was hoisted into the bay by means of a crane, which was lowered through a hatch in the spine to lift the load into position. For some loads it was necessary to jack up the Valiant's nose to increase ground clearance. Visible on the Blue Steel are the blanked off nozzles for the booster and sustainer motors.

Below: With the full-scale Blue Steel installed there was very little ground clearance, as demonstrated by WP204. The Valiant's bomb bay doors were removed and replaced by fairings which fitted snugly around the missile to form an aerodynamic seal.

WP210 was assigned to the A&AEE at Boscombe Down and spent a long time on the unglamorous but important role of navigation trials. It undertook the Decca/Dectra trials of Green Satin, Decca and VOR/DME navigation equipment that involved long overwater flights over the North Atlantic. It was one of the few long-term test aircraft to be upgraded to full service standard, being delivered to No. 49 Squadron in January 1959 and still active with the Marham Wing when the fleet was grounded.

development of the Victor tanker was well advanced, the sudden retirement of the Valiant fleet nevertheless left a considerable 'tanker gap' that took some time to cover.

As well as the Valiant BK.Mk 1s there were 14 aircraft converted to B(PR)K.Mk 1 standard, which could carry either the camera pallet or tanker gear. Most of them served in the PR role with No. 543 Squadron, but a few were used by the tanker squadrons, and by No. 148 Squadron.

Blue Steel

As Soviet defences grew in complexity and capability, so new methods were devised to deliver nuclear attacks. An idea that showed clear promise from the early days was the air-launched guided missile. Vickers itself schemed several weapons, but it was the Avro Blue Steel that was chosen to arm the V-force. While no Valiants ever carried it operationally, they were at the heart of the development programme.

Designed to meet OR1132, the Avro W.100 used a two-chamber rocket motor: a booster stage to climb the missile and accelerate it, and a sustainer motor for its high-altitude flight. As it approached the target it pitched over into a near-vertical dive to deliver a thermonuclear warhead (Red Snow). Avro was loaned several Valiants to assist in the development of this sophisticated weapon.

The first was WZ370, which was initially used from Avro's Woodford airfield to carry a 2/5th scale model known as the 19/15. The model was carried in the bomb bay and was used for carriage tests only. Following these tests, WP206 was modified to carry full-scale Blue Steels for unpowered drops, comprising the W.100 operational weapon in stainless steel, the W.102 alloy full-scale model with fixed surfaces and no motor, and the W.103 full-scale alloy model with controls and power.

Meanwhile, WZ370 was dispatched to Australia in August 1957 to begin a series of powered trials with the 2/5th models, and WZ375 was also modified for the same work in the UK.

For full-scale tests a two-phase approach had been formulated, the first involving 'un-navigated' trials using Valiant WP206 (this aircraft was never modified for this phase, WP204 being substituted) and a Vulcan B.Mk 1. The second phase would be 'navigated' and would involve Valiants WZ373 and WZ375. Marshalls at Cambridge undertook the conversions to the three Valiants, the first of which (WP204) was redelivered in February 1959. After a series of captive-carry flights, the first W.102 unguided, unpowered round was undertaken on 13 May 1959. Buffet problems surfaced when carrying the Blue Steel, requiring a redesign of the bomb bay outer surfaces to seal the missile aerodynamically into the Valiant. On 15 August 1960 WP204 left for RAAF Edinburgh in Australia: powered tests would be conducted at the Woomera range. WZ373 and WZ375 followed some months later. For the most part they flew captive-carry tests and trialled evasive manoeuvres.

Ejection seat tests

All three V-bombers shared the same crew arrangement: two pilots on upward-firing Martin-Baker 3A ejection seats and three rear crew who were forced to bale out through the entry hatch. The three sat side-by-side facing backwards. Throughout the history of the V-force the lack of a suitable escape system for the rear crew members remained a subject of controversy, an argument sparked when a Vulcan crashed on approach to Heathrow airport in 1956. The pilots ejected safely, but the rear crew perished. The crash highlighted the main problem: if an accident happened close to the ground, during take-off or landing, the rear crew had virtually no chance of escape.

As noted before, a complete crew escape capsule had been studied for the Valiant, but had been rejected on cost and complexity grounds. Downward-firing seats were also rejected as they would have required too much altitude and were of no use in the take-off/landing scenario. Upward-firing seats were the only answer. Martin-Baker proposed a solution using three seats mounted on lateral rails. The principle was that the central seat was fired first through a hatch. The port seat was then slid into the central position along the rails before firing, with the starboard seat following. The arrangement was elegant but the process inevitably took some time to complete. Nevertheless this would surely have been quicker than the rear crew having to unstrap and bale out, and the ejection fired them upwards away from the aircraft.

Martin-Baker's concept showed enough promise for an aircraft to be assigned for trials, the first production Valiant (WP199) being dispatched to company's test airfield at Chalgrove, Oxfordshire, in September 1959. Rather than test the whole three-seat arrangement, the company chose to initially demonstrate the ejection of a single seat. It was mounted in the centre and had blast panels fitted on either side.

On 27 June 1960 WP199 roared down the Chalgrove runway at 100 kt (185 km/h). The seat was fired and functioned perfectly, achieving sufficient altitude for a fully braked, controlled descent. Two further tests were accomplished successfully at low level before a live ejection trial was undertaken. On 1 July 1960 Martin-Baker's legendary seat trial specialist 'Doddy' Hay took his place in the seat. At a speed of 250 kt (463 km/h) and an altitude of 1,000 ft (305 m) he ejected over the airfield. The seat again functioned perfectly. It was Britain's first rear-facing ejection.

Martin-Baker had proven the rear-facing seat concept and offered to install a three-seat package in a Vulcan at its own expense. Vickers had already warned that fitting the escape system to the Valiant would entail significant structural changes, and the same was true of the Vulcan and Victor. These difficulties, combined with the cost of the alterations, were considered prohibitive by the Air Ministry, which declined Martin-Baker's offer. The V-bomber seat controversy continued until the last Victor was retired over 30 years later.

Prior to the live-human rear-facing seat trial, Martin-Baker undertook three with dummies. This is the first, conducted on the Chalgrove runway on 27 June 1960.

Live firing trials were undertaken using Vulcan XA903, but the Valiants had played a significant part in advancing the programme to this stage. Blue Steel achieved a limited operating capability in September 1962. Before long it was adapted for a low-level launch in line with RAF doctrine.

Assignment to SACEUR

From 1955 to 1960 the Valiant bomber force was assigned to national control, and was armed primarily with the Blue Danube kiloton bomb and the later tactical Red Beard. At the end of the 1950s, with nuclear relations restored with the US, a number of US Mk 5 weapons were made available under Project E to the V-force at three bases. One of them, Honington, had a squadron of Valiants (No. 90) and they were armed with US weapons for a while before their conversion to the tanker role.

However, in 1960 the Valiant adopted a new role within what was termed the Tactical Bomber Force. The TBF was assigned to NATO's SACEUR (Supreme Allied Commander Europe) and initially comprised 64 Canberras. With Valiants becoming surplus to requirements due to deliveries of Vulcans and Victors, the RAF looked around for uses for its still-capable aircraft. Replacement of the TBF Canberras seemed an obvious answer. A proposal was tabled to employ 24 Valiants (three squadrons) to replace the 64 Canberras.

SACEUR reluctantly agreed to the proposal, but only on the basis that the Valiants could carry two weapons, thereby going some way to restoring the numerical loss of aircraft. At the same time, NATO recognised the improved all-weather capability offered by the Valiant.

On 1 January 1960 No. 207 Squadron at Marham was officially placed under the operational control of SACEUR, although training remained an RAF commitment. At the time the aircraft was still not cleared for dual carriage. As a NATO commitment the TBF would use US-supplied bombs, and there were three options: the Mk 5 (already in use as part of Project E), the small Mk 7 tactical weapon (as used by the Canberra) and the larger and more capable Mk 28 with variable-yield capability. In the event the latter was chosen, trials having already proven that dual carriage was possible.

No. 207 Squadron officially went on alert on 10 October 1960, probably with Mk 5 weapons. The TBF reached full strength the next year when No. 49 Squadron moved in from Wittering, chopping to SACEUR control on 1 July 1961. To complete the TBF No. 148 Squadron was assigned to NATO on 13 July. At this point it is believed that the dual-carriage Mk 28 came into use.

Peacetime TBF operations involved four aircraft being held on a 15-minute QRA, armed with live weapons. In time of

Photo-reconnaissance Valiants

Equipped with a battery of cameras in its belly, the PR version of the Valiant was an extremely useful tool for the V-force. No. 543 Squadron's main task was to map the approaches to the Soviet Union so that accurate target ingress/egress flightplans could be drawn up. It also flew the mapping/survey mission and was available for emergency reconnaissance following natural disasters, such as in the aftermath of a hurricane in British Honduras in 1961.

543 Valiants: above a B(PR)K.Mk 1 climbs out showing the 10 camera windows in the bomb bay doors. Below technicians pose with an F.52 camera and film magazines, while at bottom is a B(PR).Mk 1 in anti-flash markings.

During the 1950s the Valiant was a regular and welcome participant at the SBAC air show at Farnborough, which showcased the UK's aviation industry. As well as single-aircraft flying displays, the RAF also mounted set-piece demonstrations. No. 214 Squadron provided these four Valiants (plus a spare) for the impressive and noisy demonstration of a minimum-interval take-off scramble. The aircraft were parked alongside the runway, just as they would be on an Operational Readiness Platform at their dispersed bases during time of tension or a realistic war exercise.

tension the whole force would have stood by to attack targets in central Europe – two for each aircraft. In addition to NATO tasking the aircraft also retained a national role, for which they were assigned one or two targets (for attack with the national Red Beard weapon).

From April 1963 the Mk 43 bomb was supplied to supplant the Mk 28, also for dual carriage. This was a parachute-retarded lay-down weapon that allowed the Valiants to deliver weapons from low level. For their national role with Red Beards, however, the Valiants were still committed to vulnerable pop-up attacks from 12,000 ft (3658 m). The 24-aircraft TBF remained on alert until the dramatic withdrawal of the Valiant from any operations in January 1965. For the last seven weeks of their service careers, the TBF Valiants maintained a QRA, but did not fly.

The end of the Valiant in the MBF

With the creation of the TBF in July 1961 and the re-roling of No. 214 Squadron for tanker duties, Bomber Command's Medium Bomber Force retained three squadrons of Valiants in the bomber role: Nos 7, 90 and 138. By the early 1960s they were looking distinctly obsolete in the face of contemporary Soviet defences. Unlike the Vulcans and Victors they carried no electronic countermeasures, and were not fast enough to keep up with their V-force counterparts which might otherwise have offered some protection. They were vulnerable to both SAMs and fighters, and there were no worthwhile targets that were not well-defended.

A proposal to modify them for low-level operations was tabled, but the success of this was dependent on the supply of

US lay-down weapons such as the Mk 43. This would not be possible during the remaining life of the two squadrons, and the proposal was shelved. During 1962 the Valiant was therefore retired from the MBF. No. 90 Squadron converted to the tanker mission, while the other two squadrons disbanded. No. 138 Squadron, which had been the first to form on the Valiant, was also the first to go, on 1 April. No. 7 Squadron, also at Wittering, followed suit on 1 October, thereby ending the Valiant's career in its original role. As well as the tanker and TBF squadrons, No. 18 Squadron continued to fly the Valiant in the RCM role – but not for long – and No. 543 Squadron retained the type for reconnaissance duties.

Although perhaps best remembered for its weapons trials and other roles, the contribution of the Valiant to the RAF's MBF nuclear deterrent cannot be overestimated. It had performed its 'interim bomber' role impeccably, establishing the operational concepts that allowed the more capable Vulcan and Victor to step seamlessly into its shoes.

Low-level operations

The end of the Valiant's contribution to the MBF came before it could switch to low level, but the delivery of Mk 43s to the TBF allowed that force to adopt the new concept. As dictated by the advance in Soviet defences, all the V-force eventually made the change in the early 1960s, but again it was the Valiant that pioneered the new concept.

V-bombers were designed to attack from great height – up to 50,000 ft (15240 m) – where they were deemed to be immune to Soviet defences. Nevertheless, during 1956 trials were conducted to see if the Valiant was suitable for low-level flight. WZ383 was the aircraft chosen, and in March flew nine low-level sectors to accrue data on prolonged flight at altitudes down to 50 ft (15 m). These revealed that the Valiant could be flown safely at low level, although visibility was not ideal and crew comfort was poor. To reduce stick forces the pitch and roll artificial feel systems were altered, and other changes were made to make life more comfortable for the crew. Strain gauges and accelerometers were fitted to measure the effects on the airframe, from which Vickers deduced that a safe life at low level might be as little as 200 hours. However, in the late 1950s there was no requirement for such operations.

That changed in the 1960s. The advent of Soviet high-altitude surface-to-air missiles, graphically heralded by the

Vickers handled most of the maintenance, modification and repair work for the Valiants throughout their career. To allow engine run-ups to be conducted without unduly upsetting its neighbours at Wisley, the company built a pair of huge silencers.

Pegasus testbed

Between 1963 and 1964 the Valiant was used as a testbed for the Bristol Siddeley Pegasus engine. The brainchild of Sir Stanley Hooker, the BE.53 Pegasus was a ground-breaking turbofan that was used in the radical vertical take-off Hawker P.1127, which evolved into the Harrier. The engine was first run in August 1959 before it was installed in the prototype P.1127 for tethered hovering trials to begin on 21 October 1960. By September 1961 the P.1127 was transitioning between wing lift and engine lift. Despite this rapid progress, it was decided that a testbed aircraft could accelerate the engine's development. The Valiant was chosen.

WP199 was the aircraft in question, the first production aircraft. It had spent its career on a variety of trials. A 13,500-lb (60.07-kN) thrust Pegasus 3 was installed in a fairing with a large flattened intake, the whole pack being mounted in the bomb bay of the Valiant. Ground clearance was minimal. On 11 March 1963 WP199 took to the air again, flown by Tom Frost. During the third flight the Pegasus nacelle began to break up, damaging the engine and requiring extensive repair. Further problems dogged the programme, which ended on 17 September 1964 after 50 flights.

With its nozzles in the aft position, the Pegasus 3 test engine is seen in the bomb bay of WP199. Note the protective plate around the engine's hot rear nozzles (the front nozzles used cold air taken from the fan stage).

shooting down of Francis Gary Powers's U-2 on 1 May 1960, rendered the high-altitude mission dangerous to the point of obsolescence. The only recourse for an aircraft without ECM protection was to approach its target at low level, under the radar. While the approach was no problem, weapons delivery of the bombs then in service (such as Red Beard, Yellow Sun 1/2, Mk 5 and Mk 28) still required a pop-up to around 12,000 ft (3658 m) to allow the weapons to arm and the aircraft to escape the blast. When attacking high-value, well-defended targets the bombers would be shot down during the pop-up manoeuvre.

The answer was the lay-down bomb, which had parachutes to retard it to avoid it bouncing back and hitting the low-level bomber, and a time-delay fuze to give the aircraft adequate time to escape the blast. The first such weapons for the RAF were the US-supplied Mk 43s that were delivered to the TBF at Marham from April 1963, and the TBF Valiants were therefore the first V-bombers to switch to the low-level role.

The mission profile involved flying to the target area at medium/high level for maximum fuel efficiency, before dropping down to low level around 100 miles (160 km) from the target. This run-in was made as low as conditions and terrain

XD861 was a No. 214 Squadron BK.Mk 1, although it was loaned to No. 138 Squadron for their participation in the 1957 SAC bombing competition. This view shows the considerable size of the bomb bay, as well as the long bomb deflector plate behind it.

Valiant units in order of formation

No. 1321 Flight (trials)
Formed at Wittering on 3 August 1954 for weapons drop trials using Vickers aircraft. First aircraft assigned to the unit using WP201, arriving on 6 July 1955. 1321 Flight became 'C' Flight of No. 138 Squadron on 15 March 1956.

No. 138 Squadron (MBF)
This unit was reformed at Gaydon on 1 January 1955, receiving its first aircraft (WP206) on 8 February to begin Valiant training. 'A' Flight moved to Wittering on 6 July, conversion training complete. 'B' Flight followed in November. The squadron annexed 1321 Flight in March 1956, this unit becoming 'C' Flight. All eight of the squadron's aircraft took part in the Suez campaign. No. 138 Squadron remained in the Medium Bomber Force until it was disbanded on 1 April 1962.

No. 232 Operational Conversion Unit (OCU)
The Valiant training unit was formed at Gaydon on 21 February 1955. The first course comprised crew for No. 138 Squadron's 'A' Flight, which took their aircraft with them to Wittering. The OCU's first assigned aircraft was WZ362, which arrived at Gaydon on 22 July. No. 2 Valiant Conversion Course provided crews for No. 543 Squadron 'A' Flight, while No. 3 course provided the 'B' Flights for both No. 138 and No. 543. Valiant training continued at Gaydon until the withdrawal of the fleet on 26 January 1965. From November 1957 the unit also handled Victor conversion.

No. 543 Squadron (PR)
Having lain dormant since 1943, the photo-reconnaissance squadron was reformed on 1 April 1955 to operate the Valiant in the PR and survey roles. The first B(PR).Mk 1 (WP217) was delivered to Gaydon on 11 May 1955. 'A' Flight underwent training with No. 232 OCU before moving to its operational base at Wyton on 16 November. 'B' Flight joined it at Wyton on 16 December. Valiants were flown until the fleet withdrawal, No. 543 adopting Victors soon after.

No. 214 Squadron (MBF, tanker)
The squadron was reformed as a Main Force bomber squadron at Marham on 21 January 1956, acquiring its first aircraft (WZ379) on 15 March. Squadron aircraft participated in the Suez campaign. In May 1957 the first BK.Mk 1 tanker aircraft was delivered and the squadron's 'A' Flight was tasked with tanker evaluation duties. Although bombing was still practised, refuelling became an increasingly important part of the unit's work. It was officially re-roled as a tanker unit on 1 April 1962. It was disbanded on 28 February 1965 following the fleet grounding.

No. 207 Squadron (MBF, TBF)
Formed at Marham on 1 April 1956, No. 207 received its first aircraft (WP223) on 11 June. It undertook operations at Suez. On 1 January 1960 it was placed under SACEUR control, the first Valiant unit in the TBF. These duties lasted until the fleet withdrawal, after which No. 207 was disbanded on 5 May 1965.

No. 49 Squadron (trials, MBF, TBF)
Formed on 1 May 1956 at Wittering from No. 138 Squadron 'C' Flight, No. 49 was initially equipped with WP201 and WZ366 (WZ367 was later added) and it began training for atomic bomb tests. After Operation Buffalo (the air-dropped A-bomb) it received eight specially modified Valiants for the Grapple H-bomb trials, which occupied the unit throughout 1957 and 1958. After the Grapple trials were concluded the squadron became part of the Main Force. In June 1961 it transferred to Marham to become part of the Tactical Bomber Force under SACEUR control. It was disbanded there on 1 May 1965 after the fleet grounding.

No. 148 Squadron (MBF, TBF)
Formed on 1 July 1956 at Marham, No. 148's first aircraft was BK.Mk 1 XD814. The unit bombed Egyptian targets during the Suez crisis. After service in the MBF No. 148 was reassigned to the Tactical Bomber Force on 13 July 1961. It was disbanded at Marham on 28 April 1965.

No. 7 Squadron (MBF)
This unit formed on Valiants at Honington on 8 October 1956 as part of the Main Force. It was disbanded on 1 October 1962, although a flight stayed operational until the middle of the month to perform a flypast at Uganda's independence celebrations.

No. 90 Squadron (MBF, tanker)
Last of the bomber squadrons to form, No. 90 stood up on 1 January 1957 at Honington, receiving its first aircraft (XD862) in March. Following training by No. 214 Sqn, it was officially re-tasked as the second tanker squadron on 1 April 1962. It was disbanded on 16 April 1965.

No. 199 Squadron (RCM)
This squadron was reformed in July 1951 as a radio countermeasures unit at Hemswell operating Lincolns and Mosquitoes (soon replaced by Canberras). The first RCM-equipped Valiant (WP213) was taken on charge by 'C' Flight on 29 May 1957, operated from Honington. On 1 October, as the Valiant element at Honington built up, the Canberra and Lincoln operation at Hemswell became No. 1321 Flight, while the Valiants continued with the squadron numberplate until renumbered as No. 18 on 15 December 1958.

No. 18 Squadron (RCM)
Following on from the previous entry, No. 18 Squadron came into being at Honington on 16 December 1958 by the renumbering of No. 199 Squadron 'C' Flight. It operated its Valiants in the RCM role until disbandment on 31 March 1963.

WZ403 was the first to be painted, in early 1964.

The remainder of the V-force switched to low-level operations in the mid-1960s with the adaptation of the Blue Steel missile to low-level firings and the adoption of the British WE177 lay-down bomb. But, again, it was the Valiant that had proven the concept and that had 'taken the knocks' to ensure a smooth transition for the Vulcan and Victor.

The end of the Valiant

By the spring of 1964 the Valiant force had completed the transformation from its 1950s MBF posture to a new stance in the era following the re-equipment of the MBF with ECM-equipped Vulcans and Victors. As recounted above, the last MBF squadrons had been disbanded and the ECM specialists of No. 18 Squadron had also gone. Two squadrons (No. 90 at Honington and No. 214 at Marham) were fully operational in the tanker role and waiting for the Victor K.Mk 1/1A tanker, three squadrons (Nos 49, 148 and 207 at Marham) were operational in the low-level tactical role in SACEUR's TBF, and No. 543 Squadron at Wyton was flying PR missions in anticipation of receiving new equipment in the form of Victor SR.Mk 2s. No. 232 Operational Conversion Unit at Gaydon continued to train Valiant crews, alongside Victor aircrew.

These two BK.Mk 1s wear the markings of No. 90 Squadron at Honington, the aircraft below being seen during a visit to the US (note the RF-84F in the background). No. 90 gave up bombs for tanking equipment in 1962.

would allow. The aircraft aimed to arrive at the target undetected, giving the defences no time to bring their short-range weapons to bear. With the bomb gone the Valiant egressed at low level and high speed, staying low for some time until it could return to higher altitudes for the trip home. As befitted their new tasking the TBF Valiants adopted a grey/green disruptive camouflage on the top surfaces to make them less visible to enemy fighters that might be patrolling above.

For a few months the Valiant force settled into its new form and looked forward to a graceful retirement in the coming years, but – literally – cracks began to appear. As noted above, in 1957 the second prototype Valiant had been grounded following a cracking of the rear wing spar, but the failure was put down to the rigours of testing. However, during a repair to a Viscount airliner in 1964 cracks were discovered around a spar bolt fixing. As the Viscount used the same aluminium alloy for its spars as the Valiant, it was felt prudent to examine the bomber fleet. At this point Vickers was already aware of a potential fatigue problem, and had studied a resparring programme to prolong the Valiant's life. No. 543 Squadron had a foretatste of the problem in July, when one of its aircraft was found to have a rear spar crack. Further inspection revealed cracks in six of the squadron's remaining seven aircraft.

Matters came swiftly to a head on 6 August 1964. During a high-speed, high-altitude flight, a No. 232 OCU aircraft (WP217) with Flight Lieutenant J.W. Foreman in command experienced a loud bang and an associated shuddering of the airframe. The crew elected to return gingerly to Gaydon, but only the port flap would deploy. The crew nursed the aircraft back to a flapless landing. As it taxied in it was obvious that the aircraft was seriously damaged. One wingtip was higher than the other and there were numerous rivets missing. The upper wing skin was badly rippled. The lower rear spar boom in the starboard wing had broken completely, severing the flap controls. Thanks to the Valiant's multi-spar construction and

Valiants in the Tactical Bomber Force

As their days in the strategic Main Force came to an end, the Valiant was still felt to offer some capability for the tactical nuclear strike role, especially its accurate all-weather navigation ability. Despite something of a lukewarm reception, SACEUR accepted three squadrons of Valiants at Marham to replace Canberras in the Tactical Bomber Force. The aircraft were armed with two bombs each, and were assigned tactical targets behind the Iron Curtain, mostly in East Germany. At least four armed aircraft from each unit were maintained on a 15-minute QRA. The bombs were US-owned Mk 28s, and were held by US personnel. Due to the nature of its war role, the question of release of weapons by the US did not hamper the TBF as much as it had earlier affected the Main Force (which operated with rapid force dispersal as one of the cornerstones of its operational doctrine) when operating with US-supplied Project E weapons. For low-level operations the TBF received the lay-down Mk 43 from 1963, greatly enhancing the chances of survival.

their airmanship in bringing the Valiant gently back to base, the crew had survived what might have otherwise been a fatal inflight break-up.

Fleet inspection

The WP217 incident sent shock waves through the Valiant community. The aircraft had only flown for 55 per cent of its fatigue life. A rear spar inspection and fatigue index programme ensued which categorised the Valiant fleet. Initial results of the inspection were not good: only 12 aircraft were declared as Category 'A' – flyable to within 5 per cent of their fatigue life, while 19 were Category 'B' – flyable only in an emergency. Five more were Category 'C' and grounded immediately. Flying of the Cat 'A" and a few Cat 'B' aircraft continued, albeit with tight restrictions that made operational training

Above: Wearing the winged lion badge of No. 207 Squadron, WZ403 lands back at Marham. This squadron was the first to be assigned directly to NATO although, like the other TBF units, it did also have a national commitment for which it would have used the Red Beard bomb.

XD825 was on No. 49 Squadron's books and is seen here displaying the camouflage adopted with the low-level role.

Camouflaging the Valiant

Under Modification 3261 the Valiants of the TBF were camouflaged in accordance with a 5 September 1963 edict from the Air Ministry. In their anti-flash white schemes the Valiants were just too conspicuous against the ground when flying at low level. WZ403 of No. 207 Squadron was the first aircraft to be painted, re-appearing in the early summer of 1964. The remainder of the Marham Wing aircraft followed shortly after.

Valiants only flew in camouflage for a few months before the grounding. Above and below is No. 49 Sqn's XD825. At bottom is WP221, a B(PR).Mk 1 recently reallocated from No. 543 Squadron to No. 207 at Marham.

all but impossible. The TBF maintained its SACEUR alert and No. 543 Squadron maintained its PR commitment, albeit with just one PR-configured aircraft.

Initially, at least, it was thought possible to get as many as 40 Valiants from the then-total fleet of 61 back into the air with a relatively minor repair. Two Cat 'B' aircraft had been sent to St Athan for a detailed inspection, which essentially involved ripping the wings apart. This revealed a much worse problem: the front spars were even more cracked than the rear spars. In light of this the entire fleet was grounded on 9 December. Early opinions that the low-level operations were responsible for the cracks were dispelled when the tanker aircraft were found to be in just as bad a condition.

Retirement order

A complete re-sparring was the only answer, and although a trial installation was made to XD816, the cost was found to be prohibitive. Accordingly, on 27 January 1965 the Valiant fleet was retired and most aircraft were scrapped shortly after. Five were retained for trials work for a short while, including XD816. This aircraft, the only one to be re-sparred, flew on sporadically on fatigue-related trials. It made its last public appearance at the Abingdon air display in June 1968 to celebrate the RAF's 50th anniversary. It was eventually struck off charge in 1970 and broken up.

The loss of the Valiant fleet posed immediate but not serious problems for the national force. The Victor tanker conver-

Left: Only one Valiant was kept flying for any length of time beyond the fleet grounding. After the August incident which sparked the wing spar crisis, XD816 returned to Wisley for a trial re-sparring by Vickers and was therefore not affected by the fatigue problems experienced by the rest of the fleet. It is seen here in September 1964 at a time when the rest of the fleet was undergoing urgent inspection to assess their immediate future.

sion programme was accelerated so that the first aircraft entered service with No. 55 Squadron in May 1965. The same month No. 543 Squadron received its first Victor SR.Mk 2 aircraft to resume its reconnaissance commitment – and with a much better aircraft than the Valiants it had used for 10 years. However, the withdrawal of the TBF Valiants left a considerable gap in SACEUR's armoury. Initially it was planned to re-role Vulcan B.Mk 1s in the NATO role, but after much discussion the move was vetoed by the UK government. The TBF effectively disbanded with the official end of the Valiant in January 1965.

Most of the Valiants were scrapped with unseemly haste at their bases. Some argued that there was considerable life left in the bombers, especially in the TBF role, and that re-sparring would have been a viable option. It should be noted that the aircraft had already entered a phased retirement plan, and the TBF aircraft would shortly be rendered obsolete by the forthcoming BAC TSR.2 (itself cancelled on 1 April 1965). The decision to end the Valiant's career prematurely was not a difficult one, even if it did carry a heavy emotional burden.

Today only one Valiant remains, and given the speed with which the force was scrapped it is perhaps lucky that it survived at all. One of only five aircraft to escape the initial scrapping order, XD818 was saved at RAF Marham and placed on display at the main gate. A special effort had been made to save this machine as it was the aircraft that had dropped Britain's first H-bomb during Operation Grapple. By 1970 it was the only complete aircraft left. For many years this historic machine was displayed in the Bomber Command section of the Royal Air Force Museum at Hendon in London. In late 2005 it was moved to a new home in the RAF Museum's Midlands facility at RAF Cosford.

David Donald

The last Valiant

The re-sparred XD816 undertook a series of fatigue tests in 1967/68, at the end of which it was retired. Its last task was to fly during the RAF's 50th anniversary air show at Abingdon in June 1968. Here it is seen (above and left) during low-level passes over Wisley at around this time.

Below: Having been struck off charge on 1 March 1965, XD818 – the aircraft to drop the first Grapple weapon – was saved for use as Marham's gate-guard. It is now in the RAF's Cosford museum.

NASA's NB-52s
High and Mighty

Originally procured to act as carriers for the North American X-15 hypersonic research aircraft, NASA's NB-52s have been an enduring sight at Edwards AFB for more than five decades. The last of the original pair has recently retired after a career spanning involvement in many important research projects. Its replacement is .... another B-52.

Half a century ago, on 11 June 1955, Boeing RB-52B 52-0008 made its maiden production test flight from Boeing Field, Seattle, Washington. Its designation revealed it to be a dual-capable reconnaissance-bomber, but at the time no-one could have predicted that nearly five decades later, this same aircraft would still be flying and making significant contributions to aerospace science in a role completely removed from that for which it was designed.

After initially serving as an offensive/defensive systems test aircraft, 52-008 and sister ship, 52-003, (or 'Balls 8' and 'Balls 3' as they were commonly known), were selected to serve as carrier/launch aircraft, or 'motherships', for the North American Aviation (NAA) X-15 rocket-powered hypersonic research aircraft that was then under development. In their modified form, which essentially involved removal of all combat systems and installation of a wing-mounted pylon to carry the X-15, they were known as NB-52s, with the 'N' prefix denoting 'special test, permanent' in USAF nomenclature. This was most defi-

nitely a major career change – one that would be marked not by combat mission tallies on their fuselage sides, as carried by so many of their stablemates, but by X-15 mission markings. In the decades that followed, 'Balls 8' continued to amass an impressive number of other mission 'hieroglyphics' – a testament to the dozen or so aerospace programmes that it supported.

Motherships

The mothership role for carrying and launching rocket-powered aircraft was not a new one. The Bell X-1, which led the quest for supersonic flight immediately after World War 2, owed much of its success to the Boeing B-29 carrier aircraft. With a rocket motor burn time measured in seconds or minutes, a ground take-off was ruled out since most of the X-1's fuel would be consumed long before it could ever reach its optimum altitude required for supersonic flight. The solution was to carry it aloft for launching. The B-29, having a 20,000-lb (9072-kg) bomb load, was thus the ideal choice for carrying the 12,000-lb (5443-kg) X-1. It

NB-52A 003, piloted by Captain Jack Allavie and Squadron Leader Harry M. Archer (an RAF exchange officer), lifts off with X-15-1 (Flt 1-23-39) on 4 October 1961. On this first flight of the X-15 without the ventral fin, Robert Rushworth would evaluate its handling and stability characteristics. By this time 'Balls 3' had lost the original orange tail and nacelle markings, but had acquired a fuselage band.

is interesting to note that the Soviets, too, adapted the B-29 (albeit one abandoned by the USAAF after a combat mission in 1944) as the launch platform for their own rocketplane effort – the Samolet 346.

Modifications to the B-29 were made to carry the X-1 in the bomb bay, with three aircraft thus configured. Five versions of the X-1 were eventually supported, as well as the Lockheed X-7 unmanned rocket-powered test vehicle. It was

Right: NASA pilot Bill Dana pauses on Rogers Dry Lake to watch NB-52B 008 fly overhead following his HL-10 research flight on 20 May 1969. Both NB-52s were used during the lifting body programme, with 'Balls 8' taking the lion's share of the work.

There were several attempts to launch two X-15 missions on the same day. This one, on 4 November 1960, resulted in a successful flight of X-15-1 (1-16-29) for Robert Rushworth (his first), but an aborted flight (2-A-20) for Scott Crossfield. Pictured here are X-15-1 and -2 mounted on 'Balls 8' (left) and '3' (right).

from B-29 s/n 45-21800, appropriately named *Fertile Myrtle*, that Captain Chuck Yeager launched on his historic flight of 14 October 1947, piloting X-1 46-062 *Glamorous Glennis* through Mach 1.

As rocket-powered X-planes of the 1950s evolved in size, weight and performance, so too did the motherships required to carry them. The Boeing B-50, having a higher speed, altitude and payload (30,000 lb/13608 kg) capability than the B-29, was quick to fill in this requirement, with seven airframes eventually being modified to carry

The final duty for NB-52B 'Balls 8' was to drop the Pegasus booster fitted with the X-43 hypersonic research vehicle.

the X-7, Douglas D-558-2, Bell X-2, X-9 and later models of the X-1.

The last flight of the Bell X-2 on 27 September 1956 also brought to an end the piston-engined mothership era as far as manned X-planes was concerned. By the late 1950s the emergence of the revolutionary X-15 would bring rocket plane

Above: NB-52A 'Balls 3', with its vertical stabiliser removed, is shown with the X-15 No. 1 during ground vibration tests at Edwards AFB on 11 December 1958. Hangar mates include a B-57, F-102 and two F-101s of the Air Force Flight Test Center.

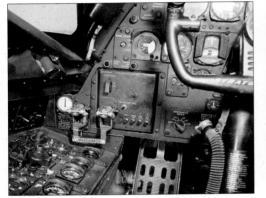

This is the cockpit of 'Balls 3' in 1959, shortly after the aircraft had been modified to NB-52A standard. The flight deck was standard B-52A except for the X-15 arming switch and emergency drop handle on the left console.

This is the Launch Panel Operator's station in 'Balls 8' in 1961. The levers on the right control the liquid oxygen top-off valve. Just to the left of them are controls for setting up the X-15's stable (INS) platform from the B-52's own navigation instruments.

research into the realm of hypersonic flight and to the fringes of space. A new mothership was needed – one capable not only of launching from higher speeds and altitudes, but one that could also carry the much heavier 31,000-lb (14062-kg) X-15 to a launch point some 500 miles (805 km) away.

The X-15 programme (1959-1968)

It was of course the X-15 that was the *raison d'être* for the NB-52 motherships. Between them

they carried the three X-15s aloft a total of 335 times, resulting in 199 launches.

The X-15 was the main component of a hypersonic research programme that was conducted by NASA (formerly the National Advisory Committee for Aeronautics, NACA) along with the Air Force, the Navy and North American Aviation (NAA). The aircraft flew from 1959 and 1968, and in the course of its flight test programme set the world's unofficial speed and altitude records. More important than records,

however, the X-15 investigated all aspects of piloted hypersonic flight, yielding data in over 700 research reports and returning benchmark hypersonic data on performance, stability and control, re-entry flight, structures, materials, shock interaction, hypersonic turbulent boundary layer, skin friction, reaction control, aerodynamic heating, and pilot performance and physiology. This data led to the acquisition of new piloted aerospace flight 'know-how' and contributed to the Mercury, Gemini and Apollo manned space programmes, as well as to the Space Shuttle.

X-15 design

North American designed and built three X-15s – 56-6670, 56-6671 and 56-6672 – commonly referred to as the X-15-1, -2 and -3, respectively. The basic aircraft was 50 ft 3 in (15.32 m) long, had a wing span of 22 ft 4 in (6.80 m) and a gross weight of 31,275 lb (14186 kg). The structure was fabricated primarily of Inconel X, a nickel-chromium alloy employed to withstand the effects of aerodynamic heating. The cockpit was insulated and the pilot wore a full pressure suit, seated in an ejection seat capable of supersonic escape.

The X-15 was powered by a throttleable Thiokol XLR-99 rocket engine producing 50,000 lb (222.5 kN) of thrust. At maximum thrust the standard 18,000-lb (8165-kg) propellant supply was exhausted in 85 seconds. Most of the X-15's internal volume contained fuel (anhydrous ammonia), oxidiser (liquid oxygen, LOX) and other gaseous/liquid tanks to drive its various systems.

The X-15 flew two types of mission profiles – horizontal flight through the atmosphere for maximum speed and heating studies, and high-altitude ballistic flight. Flight control in the atmosphere was provided by split rudder surfaces on the wedge-shaped vertical stabilisers, and by canted horizontal stabilisers. For flight in the thin upper atmosphere, a hydrogen peroxide reaction control system in the nose and wings provided pitch, yaw and roll control.

X-15 motherships

The X-15 was originally conceived to be carried in the bomb bay of the mammoth Convair B-36, in keeping with the then established practice of having the pilot enter his rocketplane from within the bomb bay prior to launch. But concerns were starting to mount that the B-36 could no longer meet the task. Its performance and handling qualities were being degraded as the X-15 and its systems grew in weight, and with the phase-out of the B-36 from the SAC inventory looming on the horizon, the loss of depot-level maintenance and spare parts (and spare aircraft) availability threatened its long-term supportability.

NAA, the designer and builder of the X-15, thus began to look at other alternatives, and after considering other aircraft such as the B-58 and KC-135, settled on the B-52. Not only did the B-52 offer superior performance over the B-36, but the spares, maintenance and training infrastructure that was then being established for the growing B-52 fleet promised years of continuous support – and as it turned out for over 50 years.

Mounting the X-15 within the bomb bay of the

Watched by an F-100F chase aircraft, 'Balls 3' carries the second X-15 on a flight which did not result in a launch (2-A-8). The NB-52A handled the early captive-carry flights and launches of the X-15. Note that it retained tail radar, although the gun turret had been removed.

Above: NB-52B 008 takes the X-15-3 up for launch. On this flight (3-21-32) Joe Walker took his X-15 to Mach 5.5 and climbed to 347,800 ft (106009 m), earning him astronaut status under NASA's 62-mile (100-km) rule. This was a high-altitude research mission carrying ultra-violet, infra-red and rarefied gas experiments.

Right: NB-52A 003 taxies out with X-15A-2 on the rocket-plane's first flight test fitted with external drop tanks. The purpose of this flight (2-43-75) was to evaluate the handling and separation characteristics of the empty tanks. Rushworth released them at Mach 2.25 over the Edwards AFB bombing range. The red and white markings on the tanks were to aid tracking.

B-52 was a non-starter due its close proximity to the ground, and the presence of fuselage-mounted landing gear. There was, however, ample space between the fuselage and the inboard engine nacelle, an area which NAA was already investigating to carry the Hound Dog air-to-surface missile. Placing the pylon at the 18 per cent span location provided a load-carrying capability of 50,000 lb (22680 kg) and thus a 60 per cent margin when carrying the X-15.

The wing pylon solution did, of course, have its own disadvantages. Pilot discomfort, while wearing a pressure suit, was one of them for this meant remaining enclosed in his cockpit, from pre-flight to launch following a 1.5-hour flight. In the case of a captive flight or an aborted mission this could be even longer. In the event of an emergency, the X-15 could be jettisoned if it posed a danger to the mothership, but this of course meant leaving the pilot to bail out on his own. The configuration did, however, permit for the pilot to eject from his X-15 while still mated to the B-52.

It was anticipated that the offset mounting and separation of the X-15 could pose some controllability challenges to the B-52 crew. This was just one of the areas investigated by wind tunnel testing at NACA Langley. On 19 June 1957 the B-52 was formally adopted as a carrier aircraft. Two aircraft were authorised for modification to ensure a smooth flight test programme.

The XB-52 and YB-52 prototypes were initially offered by the Air Force, but their non-standard configuration raised some concerns regarding their supportability. In August 1957 Strategic Air Command agreed to free-up a couple of its early production models, assigning B-52A 52-003 and RB-52B 52-008 to the X-15 programme in October 1957 and May 1958, respectively. At that time, both aircraft were already involved in test duties.

Avionics testbeds

'Balls 3' and 8 were part of the first production lot released under Letter Contract AF33(038)-21096 on 14 February 1951, which called for 13 B-52A-1-BO models. On 9 June 1952 the order was changed to cover three B-52A-1-BO (52-001 to 52-003), three RB-52B-5-BO (52-004 to 006) and seven RB-52B-10-BO (52-007 to 013).

'Balls 3' was delivered to the USAF on 30 September 1954, but retained by Boeing Seattle for Phase IV (performance) and A-3A defensive fire control systems tests. In November 1955, it was re-designated JB-52A ('J' for special test, temporary) and sent to the Wright Air Development Center, Air Research and Development Command (ARDC, became AF Systems Command in 1961) at Wright Patterson AFB, Ohio, the following January, returning to Seattle the next month.

On 29 November 1957, soon after being selected for the X-15 programme, it was delivered to Air Force Plant 42 in Palmdale, California, and stored until 4 February 1958 when it was moved to the NAA hangar at Plant 42 for modifications to carry the X-15. The modifications were so extensive that it was re-designated NB-52A. On 14 November it was flown to nearby Edwards AFB and assigned to the AF Flight Test Center (AFFTC).

It subsequently acquired a Dayglo orange paint scheme and nose art with the titles *The High and Mighty*, later amended to read *The High and Mighty One*.

'Balls 8' was delivered to the USAF on 22 March 1955 and it, too, was retained by Boeing Seattle, in this case to test the IBM MA-2 bomb-navigation system. The first MA-2 production prototype was delivered to Boeing in May 1955 and installed in 52-008. In September 1955, just three months after making its maiden flight, it was flown to Edwards AFB as a JRB-52B to conduct accelerated flight testing in order to qualify the system as quickly as possible as a replacement for the older K-3A and MA-6A equipment installed in production B-52s.

On 5 January 1959, 'Balls 8' made the short hop from Edwards AFB to the NAA facility at Plant 42 where it was to follow a similar modification route as 'Balls 3'. On 8 June it was returned, as an

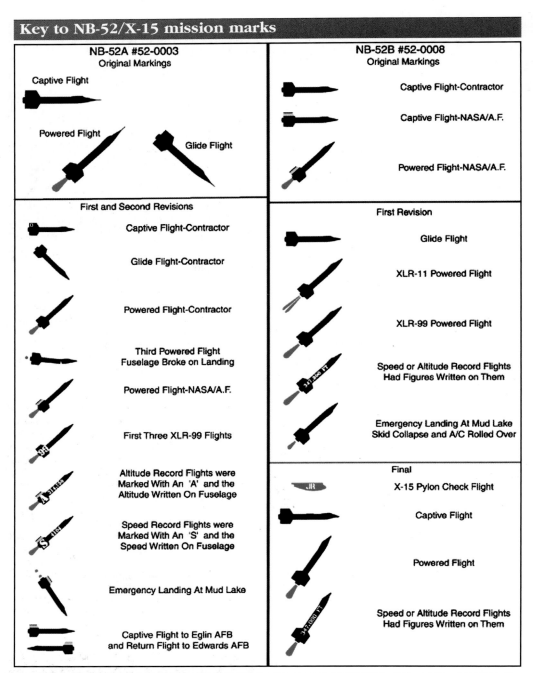

Key to NB-52/X-15 mission marks

NB-52A #52-0003
Original Markings

Captive Flight

Powered Flight

Glide Flight

First and Second Revisions

Captive Flight-Contractor

Glide Flight-Contractor

Powered Flight-Contractor

Third Powered Flight
Fuselage Broke on Landing

Powered Flight-NASA/A.F.

First Three XLR-99 Flights

Altitude Record Flights were
Marked With An 'A' and the
Altitude Written On Fuselage

Speed Record Flights were
Marked With An 'S' and the
Speed Written On Fuselage

Emergency Landing At Mud Lake

Captive Flight to Eglin AFB
and Return Flight to Edwards AFB

NB-52B #52-0008
Original Markings

Captive Flight-Contractor

Captive Flight-NASA/A.F.

Powered Flight-NASA/A.F.

First Revision

Glide Flight

XLR-11 Powered Flight

XLR-99 Powered Flight

Speed or Altitude Record Flights
Had Figures Written on Them

Emergency Landing At Mud Lake
Skid Collapse and A/C Rolled Over

Final

X-15 Pylon Check Flight

Captive Flight

Powered Flight

Speed or Altitude Record Flights
Had Figures Written on Them

NB-52B, to Edwards AFB, to join 'Balls 3'. And it too would soon acquire the AFFTC house colours of Dayglo orange, and for a brief period nose art and the titles *The Challenger*. The nose art depicted an eagle clutching an X-15 in its talons. Streaming from its beak was a banner with the titles 'UT VIRI VOLENT' (In Order That Men May Fly). Variations of this nose art continued until May 1963. By August 1964 it was replaced by a disc bearing an anthropomorphised B-52 preparing to throw an X-15 from its wingtip. Each NB-52 mothership was assigned its own ground crew, provided by the 6517th Flightline Maintenance Squadron (Bomber).

NB-52 modifications

The conversion from weapons system to mothership was a major affair, requiring first and foremost the removal of combat systems and the installation of a pylon and mission support equipment. The No. 3 fuel cell that occupied the space between the fuselage centre tank and the inboard engine pylon was removed to make room for the pylon, its attachment to the front and rear spars and structural reinforcements. The X-15 was secured to the pylon by means of two aft and one forward hook activated by a primary hydraulic (with a secondary pneumatic) release mechanism. The pylon housed numerous umbilical lines to the X-15.

The X-15 rendered the right inboard flap inoperative. The inboard flaps on both wings were thus bolted closed and their actuation mechanisms were disconnected. A U-shaped notch was cut into the right inboard flap to provide clearance for the X-15 vertical stabiliser. The forward body fuel tank (located just forward of the wing centre tank) was removed to provide access for the inspection and maintenance of various X-15 fluid and gas feed lines that ran through the wing and into the pylon. The mid-body fuel tank was also removed and 15 nitrogen and 9 helium storage cylinders were installed above the weapons bay for topping off the X-15's various systems and for cooling its stable (inertial navigation) platform.

Two stainless steel tanks containing a total of 1,500 US gal (5678 litres) of liquid oxygen (LOX) were installed in the weapons bay for topping off the X-15's tanks during the climb and cruise phases of the flight. The tanks could not be jettisoned from the weapons bay, but LOX could be vented or dumped through a line protruding from the forward left side. The standard B-52 landing gear was rated for 174 kt (322 km/h), but operating with zero flaps at heavier weights (while carrying an X-15) required the installation of high-speed wheels, tyres and brakes rated for 218 kt (403 km/h).

The need to monitor and photograph the X-15 from onboard the NB-52s necessitated a number of other additions. Television cameras in two streamlined fairings were added to the right side to permit the launch panel operator (LPO) to monitor the X-15. The forward camera faced aft towards the X-15 nose section. The aft one was equipped with a zoom lens and faced the rear section of the X-15, paying particular attention to the rocket chamber area. Four floodlights and three 16-mm motion picture cameras were also

X-15A-2 is dramatically captured as it falls away from NB-52B 008 on a check-out of the aircraft with full external tanks. This was Rushworth's 34th and last flight in the X-15. This flight (2-45-81) attained only Mach 1.7 but was used to evaluate patches of the MA-25S ablative coating and Maurer cameras.

Above: HL-10 (NASA 804) is shown mounted to NB-52B 008. Northrop had designed a special adapter to secure the HL-10 to the X-15 pylon. The X-15 mission markings on 'Balls 8' are noteworthy.

Right: The M2-F2 (NASA 803) is carried aloft by NB-52A 003 The High and Mighty One for a research flight on 2 September 1966.

Below: M2-F3 (NASA 803) falls away after being released from NB-52B 008 on 10 August 1971. Details of the adapter pylon are evident here.

installed – one in a side window at fuselage station 374, one in the hemispherical window at station 1217 and the other in the pylon facing downwards to record X-15 separation.

The LPO station was located on the aft upper flight deck in the space previously occupied by the electronic countermeasures operator. A hemispherical observation window was added above the forward TV camera fairing with two steel straps placed across it to provide safety in case of a blow-out. The B-52 flight deck was also provided with a master launch panel permitting the pilot to jettison the X-15 in an emergency. The B-52 breathing oxygen system was tapped to supply the X-15 pilot until launch.

All military systems, including the bomb-navigation system, defensive fire control system and tail turret were removed, as were provisions for carrying the reconnaissance pod in 52-008. Avionics changes included installation of an AN/APN-81 Doppler radar to provide ground speed and drift angle data to the X-15's stable platform, and an auxiliary UHF radio to improve communications with chase aircraft and ground stations.

Qualification testing

Soon after the NB-52A was delivered to Edwards, a series of ground vibration tests (GVT) were conducted on the pylon to verify the flutter characteristics of the mated pair with data already obtained from wind tunnel tests. This also built upon work already done by Boeing when integrating the Hound Dog missile to the B-52F. The GVT consisted of mounting a full-scale X-15 boilerplate model to the pylon and exciting the

assembly to various frequencies. This test was then repeated using the No. 1 X-15.

With the X-15 mounted just a short distance from the B-52's Nos 5 and 6 engines, acoustic fatigue on the X-15's empennage was a concern and to mitigate its effect it was decided to cut back the thrust of these engines to 50 per cent when flying with the X-15.

Operating with zero inboard flaps and restricted thrust settings were not part of the normal B-52 operational procedures. Neither was landing with a gross weight of some 300,000 lb (136080 kg) – this being the scenario of returning with a fully loaded X-15 following a mission abort. A flight test programme was thus conducted to investigate the NB-52's handling qualities and to develop new operating procedures for its crew.

USAF Captains Charles C. Bock, Jr, and John E. 'Jack' Allavie, along with NAA launch panel operator William 'Bill' Berkowitz, began a series of flight tests using the NB-52A. A number of changes were introduced to the flight manual that deviated from traditional B-52 handling procedures. For instance, when operating at gross weights of 300,000 lb a stabiliser trim nose setting of 2° was determined to be the optimum setting for take-offs.

Landing procedures were also revised. Instead of touching down on all four gears, which is how a B-52 normally landed, the NB-52 would land with its aft gears first. Little control in this landing configuration meant resulted in high vertical load factors of up to 1.8 *g* as the aircraft rotated to a level attitude and the front gear touched down. These load factors thus had to be taken into account when landing with a fully fuelled X-15.

Air brakes would then extend and the drag chute would deploy at 140 kt (259 km/h).

First flights with the X-15

The X-15 flight number consisted of the aircraft number (1, 2 or 3), the free flight number for that specific aircraft and the total number of flights that it had been carried aloft by either NB-52. If the flight was a captive or an aborted one, the free flight number was replaced by 'C' or 'A'.

Before NAA could deliver the X-15 to the government, it had to demonstrate each aircraft's airworthiness to flight above Mach 2 (flights above Mach 3 were in the government's domain). This phase ran from mid-1959 to mid-1960 and rested in the capable hands of NAA's pilot, Scott Crossfield (formerly of NACA). After five months of ground testing with X-15-1 mated to the NB-52A, the pair made its first flight, a captive one (No. 1-C-1), on 10 March 1959. Gross take-off weight was 258,000 lb (117029 kg) and after a ground roll of 6,200 ft (1890 m) lifted-off at 169 kt (313 km/h). This flight, which lasted 1 hour and 8 minutes, was not only used to test X-15 systems, but to validate the handling qualities of the NB-52 when carrying the X-15 as well.

On 8 June, after three aborted attempts, Crossfield conducted the first separation and glide test from the NB-52A, which was cruising at Mach 0.79 and 37,500 ft (11430 m). He landed at

The mounting arrangement of the M2-F2 on the NB-52 pylon is shown here. The leading edge of the NB-52's wing and left side of the pylon were painted with a flat black anti-glare finish.

The X-15 Follow-on Program, which was approved in March 1962, saw X-15s Nos 1 and 3 modified to fly numerous high-altitude and space research experiments, including mapping and reconnaissance cameras.

The X-15A-2

When X-15-2 was damaged during a landing accident on 9 November 1962 (seriously injuring McKay), it was decided to rebuild it as the high-performance X-15A-2. This was an attempt to increase the X-15's maximum potential and to eventually have it fly with a supersonic combustion ramjet to speeds of Mach 8. Its fuselage was increased by 29 in (0.74 m) to house a liquid hydrogen tank to power the ramjet, and two external propellant tanks were also added to increase the burn time of the XLR-99.

The fully fuelled X-15A-2 weighed 56,130 lb (25460 kg) – 1,000 lb more than what the NB-52 wing/pylon was stressed for. During 1966, the pylon and wing structure on the NB-52s were reinforced to carry a 65,000-lb (29484-kg) load to allow for the heaviest anticipated X-15A-2 flight, plus some reserve strength. But there was a slight performance penalty on the NB-52. Carriage of the X-15A-2 with tanks reduced its launch altitude by 1,500 ft (457 m), and restricted its launch speed to Mach 0.8.

On 25 June 1964, NB-52 003 carried the X-15A-2 up for its first free flight (2-32-55), less its external tanks. After additional envelope expansion flights had taken place, NB-52B 008 (having by now received its reinforced pylon) took up the X-15A-2 for its first flight (2-45-81) with full external tanks on 1 July 1966. NASA was by now preparing to apply an ablative coating over the X-15's Inconel X hot-structure to enable it to withstand the thermal loads experienced above Mach 6. The ablative coating was pink in colour and, after it had been applied to the aircraft, received a flat off-white wear finish.

On 3 October 1967, NB-52B 008 took off carrying the white X-15A-2 – its silver and orange external tank full of propellant. An hour later Pete Knight was released from 'Balls 8', ignited his XLR-99, jettisoned his tanks 67 seconds later and, after a burn time of 140 seconds, attained a speed

Rogers Dry Lake less than five minutes later. Separation from the NB-52A was flawless.

Development problems with the X-15's definitive propulsion system, the XLR-99, meant that X-15s Nos 1 and 2 would conduct the first powered tests using a pair of four-barrel XLR-11 rocket engines in its place. In preparation for this flight, NB-52A carried aloft X-15-2 with full propellant tanks for a captive flight (Flt 2-C-1) on 24 July 1959, culminating in a full powered flight on 17 September.

With two X-15s in flight status the test programme now began to accelerate. NB-52A 003 had carried X-15s a total of 14 times, and it was now the turn of NB-52B 008. Its first flight with the X-15 was with the No. 1 aircraft on 16 December, but this flight (1-A-6) was aborted due to pressurisation problems with the propulsion system. The first powered flight from 'Balls 8' finally occurred on 23 January 1960.

In the meantime, X-15-3 – which had arrived at Edwards on 29 June 1959 – was being fitted with the new and much more powerful XLR-99 in preparation for its first flight. During a ground run on 8 June 1960 the XLR-99 exploded, destroying the aircraft aft of the wing and hurling the forward fuselage, with Crossfield in it, 30 ft forward. Fortunately, he was not injured. This incident highlighted the dangers that faced the NB-52 crew on every X-15 mission.

Installation of the XLR-99 continued in X-15-2 and on 15 November 1960 it was taken up by the NB-52A, becoming the first X-15 to fly with this engine (Flt 2-10-21). The launch was at 46,000 ft (14021 m) and Mach 0.83, and even though Crossfield had set the XLR-99 throttle to

50 per cent, an altitude of 81,200 ft (24750 m) and Mach 2.97 was achieved. By December 1960, the contractor demonstration phase had come to an end and the keys to the government pilots were ceremoniously passed on.

Government tests begin

X-15-3 was rebuilt and made its first flight (Flt 3-1-2) with NB-52 003 and NASA pilot Neil Armstrong at the controls on 20 December 1961. (In 1969, Armstrong became the first man to set foot on the moon). All three X-15s were now on line and an ambitious flight test programme was well underway. Besides Armstrong the government pilot roster included Major Robert White (USAF), Joseph A. Walker (NASA), Commander Forrest S. Peterson (USN), John B. McKay (NASA), and Major Robert A. Rushworth (USAF). Over the next few years five more pilots would be added to replace those who had left the programme. The new pilots included Joe H. Engle (USAF), Milton O. Thompson (NASA), William J. Knight (USAF), William H. Dana (NASA) and Michael J. Adams (USAF). Walker was killed in 1966 when his F-104 collided with the XB-70.

By the end of 1961 the X-15 had achieved its Mach 6 design goal, and by the end of 1962 it was routinely flying above 300,000 ft (91440 m). On 17 July 1962, White flew X-15-3 to an altitude of 314,750 ft (95936 m), establishing an FAI world record that still stands today. The mothership for this flight (3-7-14) was NB-52A 003. A year later, on 22 August 1963, Walker would push this same aircraft to 354,200 ft (107960 m), establishing an unofficial record for winged vehicles – and once again 'Balls 3' would be part of this.

Above: The X-24B poses along side 'Balls 8' at Edwards AFB in 1974. Orange and white markings cover parts of its otherwise bare metal finish. Just visible below the canopy are the insignia of the AFFTC, NASA and Air Force Flight Dynamics Lab, sponsor of the X-24B. The X-15 and lifting body mission markings on 'Balls 8', which had by this time been consolidated, are in contrast to what were to appear 30 years later.

Left: The X-24B taxies out under the wing of NB-52B for its first flight on 1 August 1973.

The orange and white 3/8-scale F-15 remotely piloted research vehicle (RPRV) tested the Eagle's aerodynamics, and also performed spin research. It is seen above on its special adapter and at right being carried aloft by 'Balls 8' for a test flight on 25 October 1973, the year that the research programme began.

of 4,520 mph (7274 km/h) or Mach 6.7 – an unofficial record for winged craft that would stand until the return from orbit of the Space Shuttle *Columbia* in 1981. A post flight inspection showed that the ablator had worked well, but the structure around the ramjet had sustained significant thermal damage. The X-15A-2 would never fly again.

Up to this point in the programme, flight testing had been spoiled by few serious accidents and injuries. One accident would unfortunately mar the X-15's otherwise outstanding safety record. On 15 November 1967, Michael J. Adams was killed in X-15-3 when his craft disintegrated after entering a Mach 5 spin at 230,000 ft (70104 m). This flight (3-65-97) was carrying a number of scientific experiments.

During 1968 Bill Dana and Pete Knight took turns flying X-15-1. Plans to develop a delta wing X-15 never reached the hardware stage, and funding to continue the X-15 programme beyond 1968 was never authorised as the Apollo moon programme and the war in South East Asia continued to dominate NASA and USAF funding. On 24 October 1968, Dana completed the 199th and final X-15 flight after being carried aloft by 'Balls 3'. A number of attempts were made to conduct the 200th flight, but a variety of maintenance and weather problems conspired against it. On 20 December 1968, X-15-1 was de-mated from NB-52A 003 for the last time and, after nearly a decade of flight testing, the X-15 programme came to an end.

NB-52A 003 had carried aloft the X-15 174 times. Of these 93 resulted in drops, four were captive and 77 were mission aborts. NB-52B 008 carried the X-15 aloft a total of 161 times and of these six were captive and 49 aborted. The NB-52s became fixtures at the annual Edwards AFB open days, proudly displaying their increasing number of X-15 mission tallies with each passing year. Their high visibility orange markings had, unfortunately, largely disappeared by 1965. A total of 27 pilots flew the NB-52s in support of the X-15, the most prominent being Jack Allavie, Charles Bock and Major Fitzhugh 'Fitz' Fulton (USAF, of XB-70 fame). Between August 1961 and July 1963, Squadron Leader Harry M. Archer (RAF) was counted among them.

The end of the X-15 programme, however, did not mean the end of the NB-52 motherships. They would continue to soldier on supporting the lifting body programme. NB-52A 003 was delivered to the Military Aircraft Storage and Disposition Center (MASDC) at Davis-Monthan AFB on 15 October 1969, where it was placed in storage. In October 1974, it was formally dropped from the USAF inventory and transferred to the nearby Pima County Air Museum in September 1981. 'Balls 8' fared much better – the end of its flying career was still decades away.

Lifting bodies (1966-1975)

Just as the X-15 was exploring the high-speed, high-altitude regimes of flight, a new class of vehicles was emerging at NASA, designed to explore the other end of the aerospace flight spectrum – namely the ability of a wingless piloted vehicle to manoeuvre and safely return to a predetermined landing site following its re-entry from orbit. This is precisely what the Space Shuttle does today.

As the American space programme gathered pace, interest continued to mount regarding the development of a space vehicle which could combine the re-entry capability of spacecraft with the manoeuvring capability of an aerodynamic vehicle. One of the earliest attempts at such a vehicle was the USAF/Boeing X-20 Dyna-Soar (dynamic soaring), yet despite its cancellation in December 1963, research into aerodynamic re-entry shapes continued unabated. It is worth noting that, had the X-20 progressed into the hardware stage, the first B-52C (53-399) was earmarked for conversion to mothership for the drop test phase of the programme.

The lifting body programme combined research at the NASA Ames and Langley Research Centers with the flight test experience at Edwards. The body shapes of these vehicles were optimised for aerodynamic lift, but fins and control surfaces were also added to allowed the pilots to stabilise and control the vehicles throughout their landing profile. Three basic configurations of lifting bodies were flown in a test programme that lasted from 1963 to 1975. The test vehicles consisted of the M2-F1/F2/F3, HL-10, and X-24A/B. All except for the M2-F1 (which was towed and released by C-47), were launched from NB-52B 008, and a few by 003. Much like the X-15, data gathered from the lifting body programme contributed directly to the development of the Space Shuttle, X-33 and X-38.

Northrop M2-F2 and M2-F3

The success of the plywood M2-F1 towed glider led to the development of two heavier lifting bodies based on studies conducted at NASA's Ames and Langley research centres, the M2-F2 and the HL-10. Five companies entered bids to build these craft, and in June 1964 the contract was awarded to Northrop.

The M2-F2 was of aluminium construction, had a length of 22 ft 2 in (6.76 m), a span of 9 ft 7 in (2.92 m) and weighed 4,630 lb (2100 kg). Twin vertical fins provided yaw control and a full-span flap on the ventral fuselage controlled pitch. It was powered by an XLR-11 rocket engine of the same type that had powered the X-1 and initial X-15s. Northrop designed and built a special 22-ft (6.7-m) pylon adapter for the NB-52's existing X-15 pylon. On 23 March 1966 the M2-F2 (s/n

This 1 August 1977 view of a Firebee II DAST also clearly shows the wing notch cut into the NB-52B flap to accommodate the X-15's vertical tail. Note that the notch terminates at the wing rear spar. The flaps were rendered inoperative and bolted closed.

The HiMAT research vehicle is shown attached to the wing pylon of NB-52B 008 during a 30 December 1980 test flight. Notice that the adapter to the X-15 fairing, which resembles a Zodiac boat, also carried 125 US gal (473 litres) of JP-5 fuel to top up the HiMAT.

NASA 803) completed its first captive flight, followed by its first glide test on 12 July after being released at 45,000 ft (13716 m) from NB-52B 008 cruising at 450 mph (724 km/h). Milt Thompson, who had just come from the X-15 programme, was the pilot.

On 10 May 1967, during its 16th glide flight leading up to powered flight, a landing accident severely damaged the vehicle and seriously injured NASA pilot Bruce Peterson. It was film footage from this accident that was featured in the opening sequence of the 1970s TV series, *The Six Million Dollar Man*. NASA pilots and researchers

The F-111 Parachute Test Vehicle (PTV) was carried by the NB-52B in the former weapons bay. The LOX tanks from the X-15 era had by this time been removed.

realised the M2-F2 had lateral control problems, even though it had a stability augmentation control system. After it was rebuilt it was redesignated M2-F3, and a third vertical fin was installed to improve its handling characteristics.

After a series of captive and glide flights had been conducted, the first powered flight was made by Major Jerauld Gentry on 9 February 1971. The final flight was on 20 December 1972. The M2-F2/F3 was taken aloft 58 times, completing 43 flights – 16 as the -F2 and 27 as the -F3. The fastest speed achieved was Mach 1.6. NB-52A 003 took the M2-F2 aloft 11 times and participated in only three drop flights before being retired in 1969, with 'Balls 8' handling the remainder.

Northrop HL-10

The second of the NASA lifting bodies, the HL-10 (horizontal lander), was also built by Northrop. Unlike the M2-F2, the HL-10 had a flat bottom and rounded top, with three vertical fins and a flush canopy. For control it had a split rudder on the centre fin, blunt trailing-edge

elevons and upper surface and outer fin flaps for improved stability at supersonic speeds.

The HL-10 had an aluminium structure, was 22 ft 2 in (6.76 m) long, had a span of 15 ft 1 in (4.60 m) and weighed 9,000 lb (4082 kg) when fully fuelled. It, too, was powered by an XLR-11 rocket engine. As NASA 804 it made its first flight on 22 December 1966, with research pilot Bruce Peterson in the cockpit. Although an XLR-11 rocket engine was installed in the vehicle, the first 11 drop flights from the B-52 launch aircraft were powerless glide flights to assess handling qualities, stability and control. In the end, the HL-10 was judged to be the best handling of the three original heavy-weight lifting bodies

The HL-10 was taken aloft 53 times and made 37 flights, logging the highest altitude and fastest speed of the lifting body programme. On 18 February 1970 Air Force test pilot Peter Hoag piloted the HL-10 to Mach 1.86. Nine days later, NASA pilot Bill Dana flew the vehicle to 90,300 ft (27523 m). Once again, NB-52A 003 shouldered less of a burden than its sistership, carrying the HL-10 aloft 16 times and participating in only 11 drop flights.

Above: The NB-52B drops the Parachute Test Vehicle used to test parachutes for the Space Shuttle's recoverable booster rockets.

Martin X-24

The X-24 originated from the Air Force's earlier SV-5D/X-23 PRIME (Precision Recovery Including Maneuvering Entry) and SV-5P PILOT (Piloted Low-speed Tests) programmes. The basic X-24A shape, as the SV-5D, had successfully re-entered the atmosphere at 15,000 mph (24140 km/h) following launch by an Atlas missile. The Martin SV-5P became the X-24A in July 1967, with its prime goal being to determine if such a shape can fly down to a subsonic landing.

Built by Martin Aircraft, the X-24A was a bulbous vehicle shaped like a teardrop with three vertical fins at the rear for directional control. It had a length of 24 ft 6 in (7.47 m), a span of 13 ft 6 in (4.11m), weighed 6,300 lb (2858 kg) and was powered by a single XLR-11 rocket engine. Its gross weight with propellants was 11,450 lb (5194 kg). The X-24A (s/n 67-13551) was rolled out at Martin's Middle River, Maryland, plant on 11 July 1967.

The first unpowered glide flight was at Edwards AFB on 17 April 1969, with Air Force Major Jerauld Gentry at the controls. He later piloted the first powered flight on 19 March 1970. The X-24A was taken aloft 34 times and flown 28 times (31 and 25 of them respectively by 'Balls 8'). The fastest speed achieved was Mach 1.6 on 29 March 1971, reaching a maximum altitude of 71,400 ft (21763 m) on 27 October 1970.

The AF Flight Dynamics Lab at Wright-Patterson AFB had, in the meantime, developed a hypersonic vehicle, known as the FDL-8, which bore certain similarities to an earlier X-24 shape. NASA and the Air Force decided to 'glove' this FDL-8 shape over the basic X-24 at the conclusion of its test programme as an economic solution to evaluate this concept.

Martin conducted the modifications and, when it emerged in October 1972, it resembled a flat-iron with a double-delta planform that ended in a pointed nose. In its new guise it had a length of 37 ft 6 in (11.43 m) and spanned 19 ft (5.79 m). Its empty weight increased by 1,500 lb (680 kg) and its gross weight increased by 2,350 lb (1066 kg).

The first glide flight took place on 1 August 1973 with NASA pilot John Manke at the controls. The first powered flight followed a few months later on 15 November, again with Manke as the pilot. The final flight was on 26 November 1975, by which time the X-24B had gone up with NB-52B 008 a total of 49 times and had been released 36 times. The highest speed and altitude achieved in the programme were Mach 1.76 and 74,130 ft (22595 m), respectively.

In summary the X-24s were perhaps the most advanced of the lifting bodies tested at Edwards AFB and validated the application of the lifting body design to hypersonic transatmospheric vehicles. The X-24B further demonstrated the landing accuracies that could be achieved, and efforts to develop the Mach 6 X-24C never materialised. Thus ended the nine-year lifting body programme at Edwards and, once again, a central part of its success were the NB-52 motherships. The legacy of the X-24 did, however, re-appear years later as its shape helped influence that of the X-38 Crew Return Vehicle (CRV) technology demonstrator for the International Space Station.

With the completion of the lifting body programme, the Air Force considered retiring NB-52 008. NASA, however, had identified several potential programmes that had a requirement for an aircraft with an air launch capability. As a result, NASA and the Air Force entered a

Launching the Pegasus winged booster was an important programme for the NB-52B in the early 1990s. Here the aircraft takes off (above) and launches the Pegasus (right) in July 1991. The booster fell away from the mothership before its motor ignited.

loan agreement, signed in April 1976, placing 'Balls 8' on a loan status, whereby Dryden would maintain and fly the aircraft. It subsequently acquired the yellow 'NASA' band on the tail.

F-15 RPRV/SRV

Between October 1973 and 1982, the NB-52B was used to launch the F-15 RPRV/SRV, a remotely piloted sub-scale F-15 to investigate stability and control at high angles of attack. The F-15 RPRV was taken up 41 times resulting in 27 drop flights, and as the F-15 SRV it was taken up 32 times resulting in 26 drop flights.

In April of 1971, Assistant Secretary of the Air Force for Research and Development Grant Hanson sent a memorandum noting the comparatively small amount of research being conducted on stall and spin recovery. Early that year, the NASA FRC studied the feasibility of conducting flight research using a sub-scale fighter-type Remotely Piloted Research Vehicle (RPRV) in the stall-spin regime. In November 1971, flight research using a 3/8-scale F-15 RPRV was approved. It would measure aerodynamic derivatives of the aircraft throughout its angle-of-attack range and compare data with those from wind tunnels and full-scale flight test.

McDonnell Douglas designed and built three 3/8-scale unpowered F-15 RPRVs. They were 23 ft 6 in (7.16 m) long, had a 16-ft (4.88-m) wing span, weighed 2,500 lb (1134 kg) and were fabricated from aluminium, wood and fibreglass. The NASA FRC set up a dedicated RPRV control facility that featured a digital uplink capability, a ground computer, a television monitor and a telemetry system.

The first F-15 RPRV was launched on 12 October 1973 from NB-52B 008 cruising at 50,000 ft (15240 m). The initial flights were recovered in mid-air by helicopter, but later flights employed horizontal landings. The RPRVs thus allowed for a more rapid envelope expansion than that of piloted vehicles. The first 27 flights through the end of 1975 covered the angle-of-attack range of -20° and +53°, allowing researchers to test their theories. Data obtained thus gave a better understanding of the spin characteristics of the full-scale fighter. Angles of

attack were later expanded to the -70/+88° range.

There had been 36 flights of the 3/8-scale F-15s by the end of 1978 and 53 flights by mid-July 1981. By then it had been re-designated as the Spin Research Vehicle (SRV) to evaluate the effects of an elongated nose and nose strake on the airplane's stall/spin characteristics. Results of flight research with these modifications indicated that the addition of the nose strake increased the vehicle's resistance to departure from the intended flight path, especially entrance into a spin.

Large differential tail deflections, a tail chute, and a nose chute all proved effective as spin recovery techniques, although it was essential to release the nose chute once it had deflated in order to prevent an inadvertent re-entry into a spin. Overall, remote piloting with the 3/8-scale F-15 provided high-quality data about spin characteristics. Data gleaned thus allowed the Air Force to proceed with piloted spin trials in the actual F-15.

USAF strategic studies (1977)

In 1977 'Balls 8' conducted two flights for the AF Weapons Lab. Studies flown included 30 passes at the Idaho National Engineering Lab near Idaho Falls, Idaho, to assess the effects of engine and wingtip wake vortices on laser propagation. Missions were flown from Mountain Home AFB, Idaho.

SRB decelerator test (1977-1985)

Between 1977 and 1985, NB-52B 008 was involved in drop tests of the Space Shuttle Solid Rocket Booster (SRB) deceleration subsystem (DSS). Recovery and re-use of the SRBs once their propellant was exhausted was an essential part of Space Shuttle operating economics. A drop test programme was thus initiated to evaluate the DSS under conditions simulating actual recovery conditions.

The test programme involved a series of six air drops using full-scale, flight-type pilot, drogue and main parachutes. Data obtained were used to evaluate the design deployment performance and structural integrity of the parachutes. The parachute test vehicle (PTV) was approximately one-third the recovery weight of the SRB.

The first series of tests were conducted at the National Parachute Test Range at NAS El Centro, California, on 15 June, 4 August and 14 December 1977; and on 10 May, 26 July and 12 September

1978. The second series of tests involving eight drops, occurred between September 1983 and March 1985

One such test provided a memorable event. After a planned drop at the El Centro Range, pilot Fitz Fulton activated the release system but the 50,000-lb (22680-kg) PTV remained attached to the pylon. Activating the emergency release system did not work either. The launch panel operator, Ray Young, was however receiving a pylon 'hooks open' indication.

Several manoeuvres by Fulton produced negative results. The DTV was still attached to the pylon. The flight crew returned to Edwards and Fulton greased the landing. The problem was attributed to the hook release system building up enough friction over the years to prevent the hooks from opening to release their payload, but it allowed enough movement to activate the micro switches that sent a 'hooks open' indication to the launch panel. The situation was corrected prior to the next flight.

During early stages of the SRB/PTV test programme, the two rear hooks failed almost simultaneously during ground-towing of the NB-52 carrying the PTV after the test drop was cancelled due to unfavourable weather. Had the hook failed during take-off or flight, a catastrophic accident might have occurred. Careful examination revealed cracking due to flight test

An aerial view shows 'Balls 8' deploying an experimental Space Shuttle drag chute on 30 August 1990, one of eight such tests undertaken by the NB-52B.

stress cycling and corrosion. This little known event is discreetly recorded on the fuselage mission billboard as a small B-52 silhouette dangling a DTV, with the titles 'TFHB' (the f… hook broke)!

F-111 crew escape PTV (1978-1992)

Between 1978 and 1992, NB-52B 008 was involved in another parachute test programme, this one involving improvements to the F-111's crew escape capsule. With the X-15 era LOX tanks long removed, the parachute test vehicle (PTV) was carried in the former weapons bay. The tests were conducted in four series, with eight drops occurring between 1978-1979, 11 drops in 1982, 23 drops between 1987-1989, and 15 drops between 1991-1992.

Technology demonstrators

From the late 1970s to the early 1990s, most of the flight experiments conducted at NASA Dryden represented a sustained effort to improve the operating efficiency of aircraft and the manoeuvrability of military aircraft. The latter was fuelled by concerns that the Soviet air force possessed not only a quantitative advantage in air power, but that it was rapidly gaining a qualitative one as well. US military air planners, in cooperation with NASA, began to survey a number of promising technologies. They realised that combining them into a single technology demonstrator would be far more advantageous than to develop them separately. Two demonstrators that would emerge, the DAST and HiMAT, would also require the services of NB-52 008.

F-16 targeting studies (1979)

NB-52B 008 served as a large radar target for F-16 targeting studies

NB-52B 008 is seen configured for the SSC mission with J85s mounted on a special fixture in the bomb bay. It was celebrating its 40th anniversary of research flights in June 1995, accompanied by a NASA F-18.

Here the NB-52B carries the first Scaled Composites X-38 Crew Return Vehicle (V-131) on its first captive-carry flight on 30 July 1997.

DAST (1979-1983)

Between December 1977 and October 1983, the Dryden Flight Research Center, in concert with NASA Langley, conducted the DAST programme as a means to improve aircraft efficiency, and by extension fuel economy by using ground-controlled pilotless aircraft as 'wind tunnels in the sky'. DAST used surplus USAF Teledyne Ryan BQM-34E/F Firebee II supersonic target drones, specially modified to validate theoretical predictions under actual flight conditions. The DAST vehicle had a wingspan of 14 ft 4 in (4.37 m), a length of 28 ft 4 in (8.64 m) and had a gross weight of 2,200 lb (998 kg). It was powered by a Continental YJ69-T-406 turbojet.

During this flight test programme, the DAST vehicles were air-launched from NB-52 008, and later from a DC-130. A pilot in a ground station flew the DAST by remote control, putting it through a series of manoeuvres, including 5-*g* turns, and terminating each mission with a descent by parachute. The test programme tackled a complex problem, namely to determine the speed at which wing flutter would occur and to design and test a digital flutter suppression system. By controlling flutter electronically, wing structures could be designed with reduced stiffness and thus less weight. Flight research with this concept was extremely hazardous because an error in either the flutter prediction or control system implementation would result in wing structural failure and the loss of the vehicle. Flight demonstration using a sub-scale vehicle thus made sense, both from a safety and cost point of view.

NASA Langley modified a Firebee with an aeroelastic, supercritical research wing suitable for a Mach 0.98 cruise transport with a predicted flutter speed of Mach 0.95 at an altitude of 25,000 ft (7620 m). A total of four configurations were evaluated. Only 18 flights were achieved, eight of them captive. The NB-52B conducted 13 of the 18 missions, and 5 out of 10 launches (The remainder were conducted using a DC-130 Hercules). Two of these flights resulted in crashes – one on 12 June 1980, and the other in an alfalfa field on 1 June 1983. This latter event is recorded on the fuselage mission billboard as 'DAST /AIS' (alfalfa impact study). After the 1983 crash, the programme was abandoned. The low cost of the DAST programme did, however, give researchers the freedom to experiment boldly, and much of

the data collected found its way into other programmes.

HiMAT (1979-1983)

Between 1979 and 1983, NB-52B 008 carried aloft two HiMAT (Highly Maneuverable Aircraft Technology) vehicles on 40 missions, resulting in 26 launches. The programme was conducted jointly by NASA and the AF Flight Dynamics Laboratory, Wright-Patterson AFB, to develop high-performance fighter technologies that could be applied to future designs. The operational procedure called for launching from an NB-52 at 45,000 ft (13716 m), flying a set of manoeuvres controlled by a pilot in a ground cockpit, and ending with a horizontal landing on Rogers Dry Lake. Back-up control, in the event of lost ground control, was provided by a NASA TF-104G chase aircraft. Because the vehicles could be controlled from the ground, experimental technologies and high-risk manoeuvrability tests could be attempted without endangering a pilot.

HiMAT experiments provided information on integrated, computerised controls; design features such as aeroelastic tailoring, close-coupled canards and winglets; the application of new composite materials; a digital integrated propulsion control system; and the interaction of these then-new technologies with one another. The HiMAT featured rear-mounted swept supercritical wings with winglets, a digital fly-by-wire control system, aeroelastic tailoring, a composite structure, and close-coupled canards which gave it a turning radius twice as tight as that of then contemporary fighters; sustaining 8 *g* at near sonic speeds and at an altitude of 25,000 ft.

The vehicle was 23 ft 6 in (7.16 m) long and had a wingspan of 15 ft 4.75 m), roughly half the size of an F-16. It was powered by a 5,000-lb (22.25-kN) J85-GE-21 turbojet. At launch it weighed 4,030 lb (1828 kg). Its top speed was Mach 1.4. Approximately 30 percent of the materials used to construct it were composites, mainly fibre-glass and graphite-epoxy.

Right: Photographged in November 1996, the cockpit of 'Balls 8' is still vintage 1950s. Note the red pylon emergency release handle on the left console.

Below: A mix of current and 1950s equipment, this is how the Launch Panel Operator's position looked in 2004 during the X-43 programme.

Above: The X-38 V-132 is released on its first flight on 9 July 1999. The second X-38 vehicle was fitted with control surfaces to aid positioning prior to the parafoil being deployed.

Below: The X-38 V-131R is shown breaking away from NB-52B 008 on the programme's eighth and last free flight, on 13 December 2001. This 13-minute test flight was the longest and fastest.

In August 1975, Rockwell International won the competition to build two HiMAT aircraft. The maiden flight was on 27 July 1979, and the programme ended 25 flights later on 12 January 1983. The X-29 employed much of the technology developed from HiMAT research, including the successful use of the forward canard and the rear-mounted swept-wing constructed from lightweight composite materials

Pegasus (1989-1994)

During the early 1990s, NASA Dryden had become involved with a number of projects affiliated with the Access to Space initiative, by contributing to a novel launch endeavour by Orbital Sciences Corporation (OSC). OSC, under sponsorship by the Defense Advanced Research Projects Agency (DARPA), had defined an economical solution for placing small payloads in orbit in the wake of the *Challenger* disaster. The proposal called for using a winged booster, launched from a large transport aircraft, to place a satellite into orbit. OSC approached NASA Dryden about using the venerable NB-52 008 – already fitted with a pylon – as a first-stage launch platform. The winged booster was called Pegasus and was 49 ft (14.93 m) long, had a 22-ft (6.70-m) wingspan and was 50 in (1.27 m) in diameter.

As a result, Dryden arranged for three inert captive tests. In March 1990, NB-52 project pilot Gordon Fullerton conducted a proficiency flight to test the hook and safety pin system on the pylon adapter used to secure the Pegasus. On 5 April 1990 Fullerton initiated the launch sequence, released the Pegasus over the Western Test Range off the California coast and the first stage motor ignited. Twelve minutes later, the satellite was in orbit. Five more successful launches followed between July 1991 and August 1994. All launches were off the California coast, but the third one was from the Kennedy Space Center in Florida, marking the first time a commercial satellite (the OXP-1) had been sent into orbit by a launch vehicle dropped from an

aircraft. It is interesting to note that, in the early 1960s, a concept was proposed to launch a Bold Orion space vehicle from an X-15/B-52 combination. After the sixth flight, OSC acquired a Lockheed L-1011 Pegasus to act as a launch platform. A decade later, the Pegasus and NB-52 would team up again for the X-43 Hyper-X programme.

Space Shuttle chute tests (1990)

In 1990, NB-52B 008 was used to develop and test a drag chute to permit the Space Shuttle to land safely. Eight tests runs were conducted. It should be noted that, as far back as 1970, the aircraft had been used to simulate Space Shuttle landing approaches.

SCEES (1993-1997)

The Supersonic Cruise Emissions Environmental Studies test programme was conducted between 1993 and 1997. A special fixture was mounted in the weapons bay of NB-52B 008 which supported two J85 engines. The programme was funded by industry and test results were thus classified as proprietary to the customer. A total of 33 test flights was conducted.

X-38 CRV (1997-2001)

The X-38 was an advanced technology demonstrator for the Crew Return Vehicle. The CRV was intended as a space 'life raft' by which as many as seven astronauts inhabiting the International Space Station (ISS) could return to Earth. The X-38, which resembled the X-24, was 25 ft (7.62 m) long, 15 ft (4.57 m) wide and was an 80 per cent scale model of the CRV. Gross weight was 14,900 lb (6759 kg). Two prototypes, Vehicles 131 (later modified as the V-131R) and 132, were built by Scaled Composites of Mojave, California.

The CRV's planned descent from the ISS included re-entry from orbit, an unpowered glide and final descent by parafoil parachute. The purpose of the X-38 was to fly and evaluate the transition from lifting body to parafoil, assess the flight control system and record flight dynamics. In 1997, NB-52B 008 conducted a series of captive flights that lasted into 1998. On 12 March 1998 V-131 completed its first free flight but revealed problems with the parafoil. V-132, which was equipped with control surfaces, conducted its final free flight on 30 March 2000. It was released by 'Balls 8' at an altitude of 39,000 ft (11887 m), reaching a speed of over 500 mph before deploying its parachutes for a landing at Rogers Dry Lake.

Due to budgetary considerations, the X-38 programme was terminated after the final flight in December 2001. The V-131/131R and V-132 were carried aloft 21 times by 'Balls 8', resulting in eight free flights. A third vehicle, the V-201, was intended for actual space flight but was never built.

X-43A Hyper-X (2001-2004)

Three X-34A hypersonic research vehicles were carried aloft six times resulting in three launches. The X-43 Hyper-X was a technology demonstrator for a supersonic combustion ramjet. Unlike conventional turbine engines that compress air by means of a rotating compressor, the scramjet uses air compressed by shock waves created as an aircraft flies at hypersonic speed (greater than Mach 5). The X-43's fuel is hydrogen. The X-43 was mounted on the front end of a Pegasus booster.

The Pacific Ocean forms the backdrop as NB-52B 008 takes the Mach 7 X-43 No. 2 on a two-hour captive flight on 28 April 2001. The first launch attempt was made in June, but ended with the vehicle being command-destroyed by controllers.

Three unpiloted X-43A research vehicles were built. Each of the 12-ft (3.66-m) long, 5-ft (1.52-m) wide lifting body vehicles were designed to fly only once. The first and second vehicles were designed to fly at Mach 7 and the third at Mach 10. The first flight attempt in June 2001 failed when the booster rocket went out of control and the 'full stack' – the booster rocket and X-43A combination – was destroyed by ground controllers. The second attempt at Mach 7, in March 2004, was highly successful. The third and final flight was also the final flight of NB-52B 008. On 16 November 2004 the X-43 established a new world record by reaching Mach 9.6, or 7,000 mph (11265 km/h). Before the X-43 flights, the highest speed attained by a rocket-powered airplane, the X-15, was Mach 6.7. NB-52B 'Balls 8' thus has the distinction of launching these two extraordinary hypersonic vehicles – in effect serving as their first stage. The flight of the X-43A began with the stack being carried by a B-52B aircraft from NASA Dryden to a point 40,000 ft

NB-52B 008 takes the Pegasus booster fitted with the third X-43 on a captive flight on 27 September 2004, in preparation for its upcoming attempt at Mach 10. The vehicle achieved a remarkable Mach 9.6 on 16 November 2004, a fitting last flight for 'Balls 8'.

(12192 m) over the Pacific, 50 miles (80 km) west of the southern California coast. At that point, the stack was dropped, and the booster lifted the research vehicle to test altitude and speed.

At 95,000 ft (28956 m), the X-43A research vehicle separated from the booster and flew under its own power and pre-programmed control. The research vehicle was separated from the booster rocket by two small pistons. Shortly after separation, its scramjet engine operated for about 10 seconds, demonstrating forward thrust in flight and obtaining unique flight data for an airframe-integrated scramjet. When the scramjet engine test was complete, the vehicle went into a high-speed manoeuvring glide and collected nearly 10 minutes

of hypersonic aerodynamic data while flying to an impact point 450 miles (724 km) west of the Naval Air Warfare Center Weapons Division Sea Range off the California coast.

Last flight

The last flight of the X-43 also marked the last research flight for NB-52B 'Balls 8'. As the mothership for the first and latest hypersonic aircraft, this was a fitting end to a career spanning nearly 50 years. Despite its long calendar life, it was still a relatively young airframe considering that it had accumulated only 1,051 flights and logged 2,444 flight hours (the lowest time B-52). It conducted 405 drops (348 from the pylon).

NASA has already acquired a replacement for 'Balls 8', NB-52H 61-0025, painted in an overall gloss white scheme. Fore and aft camera fairings have now been added, as well as a payload pylon. It remains to be seen what plans NASA (or the Air Force) have in store for it.

Maintenance

Throughout their flying careers, 'Balls 3' and 8 made regular trips to Boeing Wichita and the Oklahoma City Air Logistics Area (Tinker AFB) to undergo depot level maintenance and modifications.

Air Force B-52Bs were retired in the 1960s and all the later models (C through G) were retired from the Air Force inventory in subsequent years, leaving only the B-52H models still flying. The question that always came up was: how is it that the NASA NB-52-008 was still flying and in good maintenance condition after all these years? The answers were provided by some of the people who were most closely associated with the aircraft. Crew Chief Dan Bain: "The maintenance team

NASA's new mothership, NB-52H 61-0025, poses alongside NB-52B 008 in October 2003. The NB-52H had already received a number of mods, such as a new pylon and camera fairings, but its role is as yet undefined. The new pylon (left) is installed using the former cruise missile pylon hardpoints.

looks at the aircraft with pride because of its historical significance. It's the last of its kind and we're all real proud to work on it. It's an honour because there's no other airplane like it in the world."

Special Assistant (operations directorate) Bill Albrecht: "DFRC has the capability to perform depot level maintenance on many B-52 parts and can often refurbish rejected items. Innovative tools and procedures have been created by Dryden maintenance personnel, which permit bench testing of many components. The most ambitious was a test stand which provides for ground test of a complete hydro pack, under load, using a GSE air start unit as a bleed air power source.

"The pipeline established several years ago, giving Dryden routine access to parts removed from stored B-52 aircraft at Davis Monthan AFB (AMARC), worked well. As a result, Dryden not only received parts required for immediate use on 008, but accumulated and stored a large number of spares for future support."

The aircraft also benefited from an effort to upgrade its systems to include configurations from later model B-52 aircraft. The result was a more modern aircraft and a greatly improved logistic situation. The aircraft could be considered a flying ambassador for the entire B-52 fleet since it has

parts and systems from six of the eight series of aircraft that were built. Some examples are: part of the gauges in the fuel quantity system are from B, C, D series; the ejection seats are D series; the drag chute system can accommodate G-series parachutes; and the landing gear and brakes are G/H series.

'Balls 8' retires

The 50 years of history since that launch of the X-15 rocket aircraft by NB-52 008 are reflected visually on the billboard-sized right side of the aircraft's fuselage. Fourteen major programmes have required the services of 008, including the Lifting Bodies, X-38 and X-43. One of the most unusual 'mission' markings is the title 'Dumpster Impact Study', painted over an 8-in (20-cm) square aluminium patch that repairs a hole put in the right lower fuselage by a runaway trash dumpster on a particularly windy day. It was fitting that just before its final flight, it would receive the titles 'Dryden Flight Research Center' alongside its mission billboard.

After a farewell retirement ceremony that included many of its former crew, NB-52B 008 was formally retired on 17 December 2004, just six months short of the 50th anniversary of its first flight. Its retirement home is the AFFTC Museum at Edwards AFB. Before doing so, NASA Dryden photographer Tony Landis and his colleagues painstakingly restored its famed artwork by hand, adding the famous 'Looney Toons' catchphrase..... "T-THA ... T-THA ... T-THAT'S ALL FOLKS!"

Terry Panopalis

'Balls 8' mission marks

This illustration depicts the mission marks on 'Balls 8' as they appeared at the end of the aircraft's career in November 2004. The early launches (X-15, lifting bodies etc) are summarised in the boxes at far right.

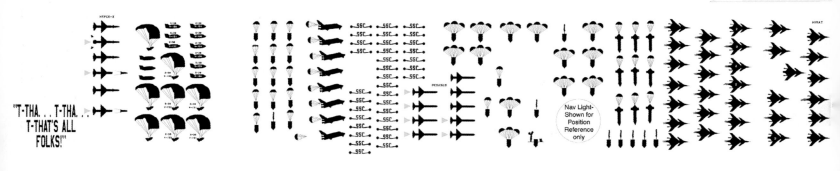

"T-THA... T-THA... T-THAT'S ALL FOLKS!"

Bomb Bay Doors- Shown for reference

Bomb Bay Doors- Shown for reference

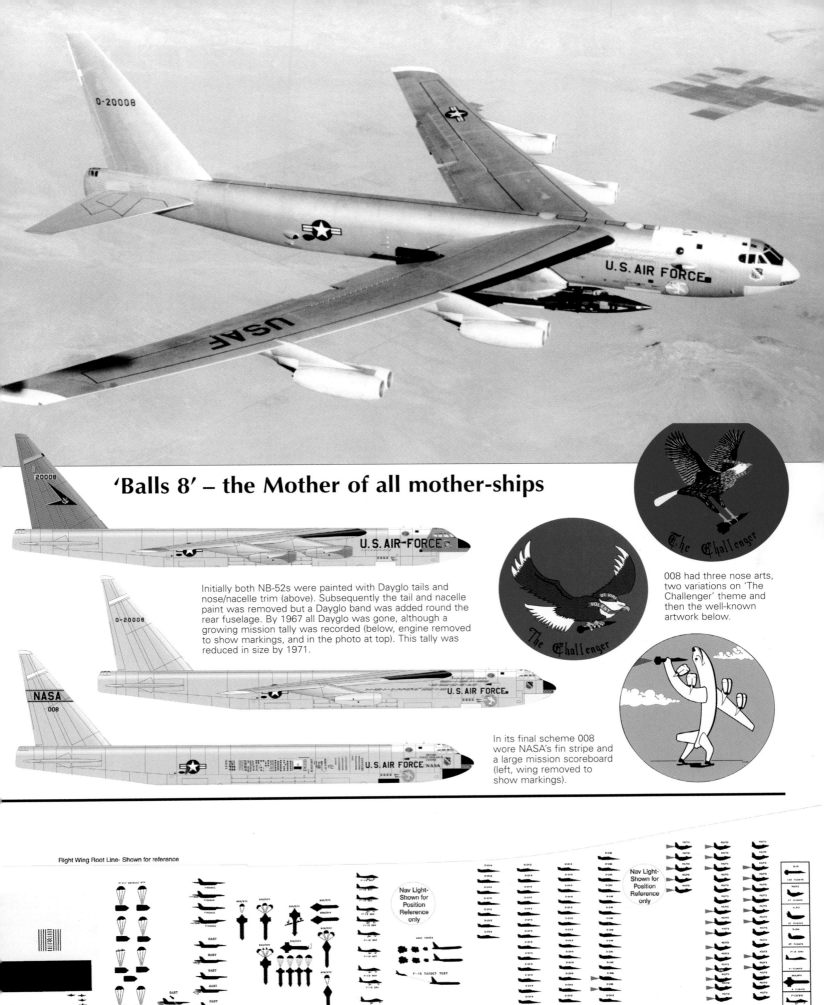

'Balls 8' – the Mother of all mother-ships

Initially both NB-52s were painted with Dayglo tails and nose/nacelle trim (above). Subsequently the tail and nacelle paint was removed but a Dayglo band was added round the rear fuselage. By 1967 all Dayglo was gone, although a growing mission tally was recorded (below, engine removed to show markings, and in the photo at top). This tally was reduced in size by 1971.

008 had three nose arts, two variations on 'The Challenger' theme and then the well-known artwork below.

In its final scheme 008 wore NASA's fin stripe and a large mission scoreboard (left, wing removed to show markings).

Right Wing Root Line- Shown for reference

Nav Light- Shown for Position Reference only

Nav Light- Shown for Position Reference only

Bomb Bay Doors- Shown for reference

Forward Main Gear Door- Shown for reference